Multimodal Artificial Intelligence and Large Language Models

The book explores the new frontiers of Multimodal Artificial Intelligence and Large Language Models (LLMs) through the integration of text, imagery, sound, and video to create intelligent systems that are designed for use by humans. By connecting theoretical principles to practical applications, this book provides a foundation for understanding the current state-of-the-art in Multimodal AI and LLM technologies through their development from the conceptual and theoretical stages of research to contemporary applications.

- Investigates how LLMs are being used in Natural Language Processing, specifically examining advances in architecture and ensemble methods for creating more human-like conversational AI experiences.
- Examines the future generation of multimodal conversational agents, including new systems that can reason about emotions, have the capability to engage in cross-modal dialogue with other systems, and can provide embodied human characteristics.
- Examines the evolution of agent-based AI technology; agent-based AI enables autonomous intelligent systems with multiple goals and missions.
- Provides a comprehensive analysis of emotion detection and emotional well-being measurement systems, utilizing advanced deep learning techniques incorporating spatial and temporal attention mechanisms.
- Explores innovative applications of multimodal systems, including sign language recognition, coherent sentence generation, and user behaviour analysis utilizing disentangled representation learning methods.
- Illustrates the transformative effects of Generative AI technology on multimodal delivery systems, including future applications in areas such as biological and biomedical data, along with more developing fields of study.
- Addresses key issues surrounding privacy, data protection, and ethical compliance, including the use of cryptography and blockchain technologies as necessary for the establishment of safe and secure multimodal AI systems.
- Examines the challenges associated with designing and utilizing deepfake recognition and detection systems; this study also focuses on developing responsible AI systems.

This publication serves as a resource for students in the final stages of their undergraduate degree programs, as well as graduate students, researchers, and professionals in industries relating to computer science, AI, electronics engineering, communication engineering, and IT. This book will also function as a reference guide for implementing multimodal systems that are scalable, secure and intelligent.

Multimodal Artificial Intelligence and Large Language Models

A Comprehensive Guide from Theory to Practice

Edited by

L. Ashok Kumar, D. Karthika Renuka,
Tanvi Banerjee and C. Deisy

CRC Press
Taylor & Francis Group
Boca Raton London New York

CRC Press is an imprint of the
Taylor & Francis Group, an **informa** business

Designed cover image: Shutterstock

First edition published 2027
by CRC Press
2385 NW Executive Center Drive, Suite 320, Boca Raton FL 33431

and by CRC Press
4 Park Square, Milton Park, Abingdon, Oxon, OX14 4RN

CRC Press is an imprint of Taylor & Francis Group, LLC

ISBN: 978-1-041-15213-2 (hbk)
ISBN: 978-1-041-15214-9 (pbk)
ISBN: 978-1-003-67835-9 (ebk)

DOI: 10.1201/9781003678359

Typeset in Times
by Newgen Publishing UK

Contents

Preface

Artificial Intelligence has evolved to be capable of a large-scale disruption, moving beyond isolated data processing toward intelligent systems capable of perceiving, reasoning, and interacting with the world in a more human-centric and context-aware manner. This transformation is driven by advances in multimodal artificial intelligence, which integrates heterogeneous data modalities—including text, images, audio, video, and sensor signals—to enable comprehensive understanding and decision-making. Central to this evolution is the rapid progress of Large Language Models (LLMs), which have redefined the understanding and generation of natural language and now serve as a unifying interface across multiple modalities. This edited volume, *Multimodal Artificial Intelligence and Large Language Models: A Comprehensive Guide from Theory to Practice*, is conceived to capture this critical transition in contemporary AI research and applications. The book brings together contributions from leading researchers and practitioners to present a cohesive and authoritative account of the theoretical foundations, methodological advances, and practical deployments of Multimodal AI systems empowered by LLMs. The integration of linguistic, visual, auditory, and affective intelligence is reshaping diverse application domains, including conversational systems, recommendation platforms, healthcare analytics, educational technologies, security systems, and human–computer interaction.

The volume is organized to provide a structured progression from foundational principles to advanced research and real-world applications. The initial chapters establish the theoretical underpinnings of Multimodal AI, tracing its historical evolution and examining core concepts, architectures, and learning paradigms. Subsequent chapters focus on recent advances in multimodal representation learning, cross-modal reasoning, and the role of LLMs in enhancing multimodal understanding. Particular attention is given to conversational AI, multimodal chatbots, question-answering systems, emotion-aware computing, and Generative AI, highlighting how these technologies enable natural, adaptive, and context-sensitive interactions. In addition to theoretical depth, this volume emphasizes practice-oriented and application-driven research. Several contributions present real-time multimodal frameworks, deep learning and attention-based architectures, and Agentic AI models that support autonomous and goal-driven intelligence. Application-focused chapters explore critical use cases such as emotional well-being monitoring, sign language recognition and language generation, recommendation systems, multimodal data analysis, and deepfake detection, offering valuable insights into system design, implementation challenges, and performance evaluation. Recognizing the societal implications of widespread Mutimodal AI adoption, this volume also addresses ethical, privacy, and data security concerns. The chapters critically examine issues related to algorithmic bias, surveillance, emotional inference, data misuse, and the responsible deployment of generative models, thereby promoting the development of transparent, accountable, and trustworthy AI systems.

This book is intended for researchers, academicians, postgraduate students, and industry professionals seeking a comprehensive and up-to-date reference on

Multimodal AI and Large Language Models. By integrating perspectives on multimodal representation learning, affective reasoning, generative intelligence, and ethical governance, this edited volume aims to serve both as an essential reference and as a catalyst for future interdisciplinary research. As Multimodal AI continues to redefine the landscape of intelligent systems and human–machine interaction, it is our hope that this volume will inspire further innovation, foster collaboration across disciplines, and contribute to the development of responsible, robust, and human-centered AI technologies.

Dr. L. Ashok Kumar
Professor, Dept. of EEE, Thiagarajar College of Engineering, Madurai, TN, India

Dr. D. Karthika Renuka
Professor, Dept. of IT, PSG College of Technology, Coimbatore, TN, India

Dr. Tanvi Banerjee
Associate Professor, Dept. of Computer Science, Wright State University, Ohio, USA

Dr. C. Deisy
Professor, Dept. of IT, Thiagarajar College of Engineering, Madurai, TN, India

Acknowledgments

The authors are thankful to Mr. K. Hari Thiagarajan, Chairman and Correspondent, Thiagarajar College of Engineering and Mr. L. Gopalakrishnan, Managing Trustee, PSG Institutions, for their continued support, encouragement, and conducive academic and research environment that made this work possible. The authors thank the editorial team and the reviewers of CRC Press for their relentless efforts in bringing out this book.

Dr. L. Ashok Kumar would like to take this opportunity to acknowledge the people who helped him in completing this book. He is thankful to his wife Dr. Y. Uma Maheswari and also grateful to his daughter, Ms. A. K. Sangamithra, for their constant support and care during writing.

Dr. D. Karthika Renuka would like to express gratitude to all her well-wishers and friends. Her heartfelt thanks to her husband, Mr. R. Sathish Kumar, and her dear daughter, Ms. Preethi P.S., for their unconditional love, which made her capable of achieving all her goals. She extends her gratitude and sincere thanks to her brother and mentor Dr. L.Ashok Kumar for his continuous support and guidance.

Dr. T. Banerjee would like to express gratitude to her parents, Dr. B. Banerjee and Mr. S. Banerjee, for supporting her academic journey through every means possible. She is also grateful to her husband and academic partner, Dr. William Romine, for his constant and unwavering support.

Dr. C. Deisy humbly offers her heartfelt gratitude to the Lord Almighty for His divine grace and blessings. She wishes to express her deep thanks and gratitude to her husband, Mr. P. Thanushkodi, her daughters and sons-in-law, and her beloved grandchildren, Hazel and Hailey, as well as to her friends and colleagues, for their unwavering support, encouragement, and cooperation.

Above all, the authors humbly acknowledge the Almighty God for His immeasurable blessing upon their lives.

About the Editors

L. Ashok Kumar was Postdoctoral Research Fellow from San Diego State University, California. He was selected among seven scientists in India for the BHAVAN Fellowship from the Indo-US Science and Technology Forum and also, he received SYST Fellowship from DST, Govt. of India. He is currently working as Principal, Thiagarajar College of Engineering, Madurai, Tamil Nadu. He has three years of industrial experience and 26 years of academic and research experience. He has published 193 technical papers in international and national journals and presented 187 papers at national and international conferences. He has completed 28 Government of India–funded projects worth approximately 15 crores, and currently 10 projects are in progress worth approximately 13 crores. He has developed 27 products and of which 23 products have been technology transferred to industries and for Government funding agencies. His PhD work on wearable electronics earned him a National Award from ISTE, and he has received 21 awards at the national and international levels. He has guided 96 graduate and postgraduate projects. He has supervised 15 PhD scholars, and 8 candidates are currently pursuing PhDs. He has visited many countries for institute industry collaboration and as a keynote speaker. He has been an invited speaker in 383 programs. Also, he has organized 116 events, including conferences, workshops, and seminars. He completed his graduate program in Electrical and Electronics Engineering from University of Madras and his post-graduate degree from PSG College of Technology, India, and Masters in Business Administration from IGNOU, New Delhi. After completion of his graduate degree, he joined as project engineer for Serval Paper Boards Ltd., Coimbatore (now ITC Unit, Kovai). He is also a Certified Chartered Engineer and BSI Certified ISO 500001 2008 Lead Auditor. He has authored 34 books in his areas of interest, published by Springer, CRC Press, Elsevier, Nova Publishers, Cambridge University Press, Wiley, Lambert Publishing, and IGI Global, and also contributed 31 chapters in various books. He has 3 patents granted, 9 patents published, 12 design registrations, and 6 copyrights to his credit. He is also the Chairman of Indian Association of Energy Management Professionals and Executive Member of the Institution of Engineers, Coimbatore, Vice President of Sustainability and Energy Practitioners Association, Executive Council Member in Institute of Smart Structure and Systems, Bangalore, and Associate Member in CODISSIA. He is also holding prestigious positions in various national and international forums and he is a fellow member in IET (UK), IETE, and IE, and a senior member in IEEE.

D. Karthika Renuka is a Professor in the Department of Information Technology at PSG College of Technology. Her professional career of 24 years has been with PSG College of Technology since 2003. She is an Associate Dean (Students Welfare) and convenor for Students Welfare Committee in PSG College of Technology. She is a recipient of the Indo-U.S. Fellowship for Women in STEMM (WISTEMM) – Women Overseas Fellowship program supported by the Department of Science and Technology (DST), Govt. of India, and implemented by the Indo-U.S. Science &

Technology Forum (IUSSTF). She was a Postdoctoral Research Fellow from Wright State University, Ohio, USA. Her area of specializations includes data mining, evolutionary algorithms, soft computing, machine learning and deep learning, affective computing, and computer vision. She has organized many national and international conferences. She is a reviewer for Computers and Electrical Engineering, Elsevier, Wiley Book chapter, Springer Book Chapters on "Knowledge Computing and its Applications". She has supervised six PhD research scholars and is currently supervising eight research scholars for their PhDs at Anna University. She has authored seven books in her areas of interest published by CRC Press, Nova Publishers, Wiley, IGI Global, and Bentham Science Publishers, and also has published several papers and book chapters in reputed national, international journals, conferences, and books, respectively.

Tanvi Banerjee is Associate Professor in the Department of Computer Science and Engineering at Wright State University, USA, and holds a secondary appointment in the Department of Geriatrics at the Boonshoft School of Medicine, Wright State University, USA. Her academic focus has been at the crossroads of artificial intelligence and healthcare, revolving around multimodal data fusion, wearable sensing, and mobile healthcare technologies. She has been a recipient of the prestigious K01 grant awarded by NIH (equivalent to the CAREER pathway in NSF) for the project on Dementia Management using smartphone technologies and a co-investigator in the NIH R01-funded Sickle Cell Disease project.

C. Deisy is currently Professor and the Head of the Information Technology Department at Thiagarajar College of Engineering in Madurai, Tamil Nadu, India. She has 24 years of teaching experience. Her areas of interest include machine learning, artificial intelligence, natural language processing, and data mining. She has published more than 90 research articles in journals and conferences.

Contributors

A. Naresh
Annamacharya Institute of Technology & Sciences, Kadapa, Andhra Pradesh, India

A. Priyadharshini
Department of CSE, Coimbatore Institute of Technology, Coimbatore, Tamil Nadu, India

Manish Arora
Department of Applied Arts, Banaras Hindu University, Varanasi, India

Ramesh Kumar Ayyasamy
Faculty of Information and Communication Technology, Universiti Tunku Abdul Rahman, Malaysia

Mohankumar B.
School of Computer Science and Engineering (SCOPE), Vellore Institute of Technology, Vellore, Tamil Nadu, India

Himesh Baboolall
Department of Applied Arts, Banaras Hindu University, Varanasi, India

Geraldine Bessie Amali D.
School of Computer Science and Engineering, Vellore Institute of Technology, Vellore, Tamil Nadu, India

D. Karthika Renuka
Department of IT, PSG College of Technology, Coimbatore, Tamil Nadu, India

Arnab Kumar Das
Department of Information Technology, Indian Institute of Engineering Science and Technology, Shibpur, Kolkata, India

Rohan Naveen Gajan
Department of AI & DS, Nitte Meenakshi Institute of Technology, Bengaluru, Karnataka, India

Muhammad Rukunuddin Ghalib
School of Computing, De Montfort University, Dubai, UAE

Tomojit Ghosh
Department of Computer Science and Engineering, Wright State University, Dayton, Ohio, USA

Caleb Hendren
Mathematics and Sciences Division, Chattanooga State Community College, Chattanooga, Tennessee, USA

Ravi Hosur
Department of Computer Science and Engineering, BMS Institute of Technology & Management, Bengaluru, Karnataka, India

Felicia Lilian J.
Department of Computer Science and Business Systems, Thiagarajar College of Engineering, Madurai, Tamil Nadu, India

J. Jayashree
VIT University, Vellore, Tamil Nadu, India

J. Vijayashree
VIT University, Vellore, Tamil Nadu, India

Anurag Jawalkar
VIT University, Vellore, Tamil Nadu, India

K. Amshakala
Department of CSE, Coimbatore Institute of Technology, Coimbatore, Tamil Nadu, India

K. Indira
Department of Computer Science and Engineering, Thiagarajar College of Engineering, Madurai, Tamil Nadu, India

K. Nimisha
Department of Electronics and Communication Engineering, Akshaya College of Engineering and Technology, Coimbatore, India

K. Sivasankari
Department of Electronics and Communication Engineering, Akshaya College of Engineering and Technology, Coimbatore, India

Sudhakar K.
Department of AI & DS, Nitte Meenakshi Institute of Technology, Bengaluru, Karnataka, India

Mayuri K.P.
Department of Artificial Intelligence and Machine Learning, BMS Institute of Technology & Management, Bengaluru, Karnataka, India

Kamarulzaman Kamarudin
Faculty of Electrical Engineering & Technology, Universiti Malaysia Perlis, Malaysia

Dishita Kamath
Department of AI & DS, Nitte Meenakshi Institute of Technology, Bengaluru, Karnataka, India

L. Ashok Kumar
Department of ECE, Thiagarajar College of Engineering, Madurai, Tamil Nadu, India

Raja Lavanya
Department of Computer Science and Engineering, Thiagarajar College of Engineering, Madurai, Tamil Nadu, India

Krishnamoorthi M.
Department of Information Technology, Dr NGP Institute of Technology, Coimbatore, Tamil Nadu, India

Mohideen AbdulKader M.
Department of Computing Technologies, SRM Institute of Science and Technology, Chennai, Tamil Nadu, India

M. Nisha
Department of Computer Science and Engineering, Akshaya College of Engineering and Technology, Coimbatore, India

M.K. Kavitha Devi
Department of Computer Science and Business Systems, Thiagarajar College of Engineering, Madurai, Tamil Nadu, India

N. Deepika
Department of IT, PSG College of Technology, Coimbatore, Tamil Nadu, India

Ruchira Naskar
Department of Information Technology, Indian Institute of Engineering Science and Technology, Shibpur, Kolkata, India

Mathew Mithra Noel
School of Electrical Engineering, Vellore Institute of Technology, Vellore, Tamil Nadu, India

P. Chinnasamy
Department of Computer Science and Engineering, Kalasalingam Academy of Research and Education, Tamil Nadu, India

Rubini P.
Department of CSE, CMR University, Bengaluru, Karnataka, India

P. Vinoth Kumar
Department of Artificial Intelligence and Data Science, St Josephs Institute of Technology, Chennai, India

Haleema P.K.
University of West London – RAK Branch Campus, Ras Al Khaimah, UAE

R. Saveeth
Department of CSE, Coimbatore Institute of Technology, Coimbatore, Tamil Nadu, India

R.N. Devendrakumar
Department of Computer Science & Engineering, Sri Ramakrishna Institute of Technology, Coimbatore, Tamil Nadu, India

Viswadarshan R.R.
Department of Computer Science and Business Systems, Thiagarajar College of Engineering, Madurai, Tamil Nadu, India

Sakkaravarthi Ramanathan
Department of Computer Science, Vainer College, Montreal, Canada

S. Anne Susan Georgena
Department of Mathematics, Sri Ramakrishna Institute of Technology, Coimbatore, Tamil Nadu, India

S. Jeniba
Department of Computer Science and Engineering, Thiagarajar College of Engineering, Madurai, Tamil Nadu, India

Krithik S.
School of Computer Science and Engineering, Vellore Institute of Technology, Vellore, Tamil Nadu, India

Mahalakshmi S.
Department of Computer Science and Business Systems, Thiagarajar College of Engineering, Madurai, Tamil Nadu, India

S. Nithya
Department of CSE, Coimbatore Institute of Technology, Coimbatore, Tamil Nadu, India

Pooja Nayak S.
Department of ISE, Dayananda Sagar Academy of Technology and Management, Bengaluru, Karnataka, India

S. Sangeethamariammal
Department of CSE, Coimbatore Institute of Technology, Coimbatore, Tamil Nadu, India

Amitha S.K.
Department of Artificial Intelligence and Machine Learning, BMS Institute of Technology & Management, Bengaluru, Karnataka, India

Prabha S.U.
Department of Electrical and Electronics Engineering, Dr NGP Institute of Technology, Coimbatore, Tamil Nadu, India

Noor Sumaiya
Department of CSE, The Oxford College of Engineering, Bengaluru, Karnataka, India

Vijaya Kumar T.
Department of Information Technology, Dr NGP Institute of Technology, Coimbatore, Tamil Nadu, India

Prabhu T.N.
Department of Computer Science and Business Systems, Dr NGP Institute of Technology, Coimbatore, Tamil Nadu, India

V. Balamurugan
Department of Computer Science and Engineering, Akshaya College of Engineering and Technology, Coimbatore, Tamil Nadu, India

V. Dharsana
Department of Computer Science and Engineering, Thiagarajar College of Engineering, Madurai, Tamil Nadu, India

Ezhil Arasi V.
School of Computer Science and Engineering (SCOPE), Vellore Institute of Technology, Vellore, Tamil Nadu, India

V. Shobha Rani
School of Computer Science & Artificial Intelligence, SR University, Warangal, Telangana, India

1 Empowering Agentic AI with Multimodal Large Language Models

Toward Autonomous, Goal-Driven Intelligence

Mohankumar B., Ezhil Arasi V., and Prabhu T.N.

1.1 INTRODUCTION

1.1.1 What Is Agentic AI?

Agentic AI systems are systems that, on their own, pursue goals, plan for themselves, reason within context, and adapt to decisions as things change. Unlike classic AI that just maps inputs to outputs, the agentic systems hold an internal sense of aims, track the progress, and continually adjust actions in light of environmental feedback. Normally, an agentic system would be based on four major capabilities:

- Goal formation: It crystallizes the general human intention into specific actionable goals.
- Planning and replanning: Formulating action plans step-by-step and changing these according to new information that becomes available.
- Tool use: The employment of external functions, application programming interfaces (APIs), databases, models, or actuators to realize goals.
- Autonomy within safety boundaries: Making independent choices, yet still remaining understandable, in line with human expectations.

An agentic AI system turns AI from a passive processor into an active, collaborative participant in the processes. It will let systems that don't need to be explicitly instructed to observe, anticipate events, and execute tasks autonomously, yet be transparent and traceable while confined to the constraints of the domain. Agentic AI is suitable in the knowledge domains where changes may occur rapidly, but action must be taken in time—from health care to emergency triage ranges from smart industrial operations to predictive maintenance and decision support for experts who deal with complex knowledge.

DOI: 10.1201/9781003678359-1

1.1.2 Evolution of Multimodal Large Language Models

The emergence of multimodal large language models (MLLMs) represents a major development in terms of how machines read the world and talk back to us [1]. Initially, deep-learning technologies tended just to specialize in text, images, or audio. Over time, systems started to fuse multiple signals altogether, hence allowing AI reasoning across vision, language, sound, sensor pattern inputs, and even structured data [3].

Here's a compact arc of the journey:

Stage 1: Single-Modality Foundations

Systems like bidirectional encoder representations from transformers (BERT) and generative pre-trained transformer 2 (GPT-2), early convolutional neural network (CNN) vision nets, take one type of input at a time. They can be very strong but are limited in range to only one domain.

Stage 2: Cross-Modal Bridges

Contrastive language-image pretraining (CLIP), A Large-scale ImaGe and Noisy-text embedding (ALIGN), Vision-and-language BERT (ViLBERT), and VisualBERT are some of the image-text transformers that allow representations from one modality to influence another. These opened the door for zero-shot image classification and captioning.

Stage 3: Unified Multimodal LLMs

Systems such as GPT-4V, Gemini, LLaVA, and Florence had all modalities unified under one embedding and inference space. They could also perform multimodal dialogue, code interpretation, chart analysis, document understanding, and sensor data inference.

Stage 4: Agentic Multimodal Systems

The latest wave brings agency to multimodal understanding. These models bring with them:

- Multimodal grounding
- Sequential planning
- Memory components
- Feedback loops from the environment
- Interfaces for dispatching tools
- Self-assessment modules

It is this evolution that finally positions MLLMs as autonomous problem solvers for complex, multistep tasks involving both the physical and digital worlds.

1.1.3 Motivation for Merging Agentic AI with Multimodal LLMs

Merging agency with multimodal intelligence provides machines endowed with perceptions, reasoning, planning, and acting across very diverse real-world contexts. The motivation arises from several gaps in the traditional ways of AI.

1.1.3.1 Need for Contextual Understanding Beyond Text

Healthcare, robotics, urban infrastructure, and scientific workflows all require interpretation of

- Medical scans
- Laboratory values
- Physiological sensor streams
- Patient history
- Environmental images
- Temporal patterns

This richness cannot be captured by text-only models, whereas a multimodal model integrates all sources.

1.1.3.2 Need for Continuous, Adaptive Decision-making

Real-world domains rarely act on static instructions. Agentic AI introduces autonomy in that the system

- Assesses conditions
- Updates its plan
- Executes tool calls
- Checks outputs
- and loops until goals are met

This is indispensable in emergency triage, monitoring in the ICU, chronic disease tracking, and individualized treatment suggestions.

1.1.3.3 Need for Human-Level Workflow Automation

Agentic multimodal systems, therefore, can execute multi-step workflows such as

- Reading medical images
- Extracting features
- Evaluating health risk
- Calling a clinical database
- Structured summary generation
- Alerting physicians

This reduces delays, enhances accuracy, and improves medical outreach in underserved areas.

1.1.3.4 Need for Safe and Interpretable Autonomy

The explanations and monitoring in such a case have a much more powerful foundation through multimodal data. An agent basing its decisions on vision, vitals, and reports makes much more traceable decisions. Multimodal grounding of the agentic reasoning reduces hallucination, strengthens validity, and keeps consistency in the audit trail.

1.1.4 Key Opportunities and Real-World Impact

1.1.4.1 Enhanced Contextual Understanding through Multisensory Integration

Agentic multimodal systems integrate language with vision, audio, and environmental signals into richer, better-grounded representations of real-world situations. This allows an autonomous agent to move from a superficial text-based response to a system that elicits user intent, captures the physical context, and understands dynamic changes in the environment [1,3]. Consider how integrating clinical notes with images or sensor streams of patients provides a more effective way for the system to understand health conditions compared to models using the text alone. Such fusing of modalities dispels ambiguity and strengthens semantic grounding, hence allowing for robust decision-making in aspects such as healthcare, education, manufacturing, and ambient computing.

1.1.4.2 Autonomous Goal Execution Supported by Real-Time Reasoning

Traditional LLMs are purely reactive, responding to a query and doing nothing otherwise. Agentic multimodal LLMs have goals they keep track of, states they update, and hence take proactive action over time. This would, for instance, allow for autonomous task cycles such as completion: independently monitoring the vital signs of a patient, adjusting teaching style based on facial reactions, or diagnosing machinery breakdowns by listening and viewing. That real-time reasoning loop will transform AI into an active problem solver from a currently passive assistant having memory, planning, and action policies dynamically updated to reflect the ever-changing environment.

1.1.4.3 Human-Aligned Adaptability and Personalized Interaction

Agentic multimodal AI adapts its behavior to suit user-specific patterns, preferences, and emotional states. Facial expression recognition, gaze tracking, and voice sentiment analysis enable agents to provide sensitive, subtlety-rich responses together with contextual cues.

It can automatically change the levels of difficulty when it notices confusion through eye movements or hesitations of speech. In healthcare, it will be able to adapt recommendations based on observations of pain expressions or changes in tone.

An adaptive agent receives feedback all the time, and this can be expressed mathematically as:

$$S_{t+1} = S_t + \alpha(\text{Feedback}_{\text{user}} + \text{Feedback}_{\text{environment}}) \tag{1.1}$$

This will create more satisfying human–AI interactions and increase trust, usability, and long-term engagement.

1.1.4.4 High-Impact Applications Across Social and Economic Domains

Agentic multimodal AI can bring advancement in key sectors:

- Health care: Autonomous triage, chronic containing, timely emergency alert agents
- Education: Personalized tutoring, disabilities by using multi-modal sensing and interactions
- Manufacturing and robotics: Tool-using industrial copilots perform safety checks and help workers
- Governance and public health: Identify misinformation and support disaster response efforts
- Elderly care: Routine assistance, emotional monitoring, cognitive support

The direct contribution of these systems to social good encompasses solving labor shortages, improving accessibility, and extending continuous care. Finally, the adoption of multimodal grounding reduces errors, supports faster decision-making, and democratizes high-quality services across rural and underserved populations.

1.1.4.5 Optimization-Driven Decision Quality and Safety Guarantees

Agentic multimodal architectures incorporate optimization functions that would evaluate decisions against safety and performance constraints. A generalized utility function of safe decision-making can be given by:

$$U = \sum w_i f_i(\text{modality}_\text{i}, \text{constraints}) \tag{1.2}$$

where f_i is task-specific performance metrics (accuracy, risk probability).

Safety layers enforce clinical norms, manufacturing standards, or ethical guidelines by rejecting actions violating constraints.

This is an optimization-based approach to control that reduces hallucination, improves reliability, and supports certification pathways for regulated industries, including medicine and aviation.

1.1.4.6 Scalability, Interoperability, and Real-World Deployment Potential

The agentic multimodal systems exhibit very good scalability in a distributed environment because of modular planning loops, composable memory units, and flexible sensory interfaces. These can connect Internet of Things (IoT) sensors, electronic health records, surveillance networks, and factory equipment [5].

Besides, interoperability among them will support hybrid deployments across cloud, edge computing, and on-device models. Together, these further unlock deployment scenarios from smart homes to autonomous hospital wards to multimodal tutoring kiosks to intelligent robots.

This capability to continuously learn from interaction data enhances the long-term scalability and positions these systems as foundational technologies for next-generation autonomous infrastructures.

Overcoming the Limitations of Current AI Models

The proposed agentic multimodal AI framework addresses several basic limitations of the existing models.

- Passive to proactive: While conventional LLMs react to prompts, an agentic system will act in light of goals and observations.
- From isolated perception to unified understanding: Multimodal fusion replaces fragmented, single-input analysis.
- From short-term context to persistent memory: Long-term and episodic memory create continuity across interactions.
- Static prediction to adaptive reasoning: Plans are continuously revised and improved through feedback.
- Opaque outputs to explainable decisions: Multimodal grounding enhances transparency and trust.

1.2 BACKGROUND AND LITERATURE REVIEW

1.2.1 Traditional LLMs versus Agentic AI

1.2.1.1 Foundations of Traditional Large Language Models

Traditional large language models function as sequence predictors. They rely on the core objective of maximizing the likelihood of generating the next token. The training objective is:

$$L = -\sum \log P\left(W_t \mid w < t\right) \tag{1.3}$$

where

$w < t$ is the prefix context

$P(\cdot)$ is the conditional probability estimated by the model.

Classical LLMs such as GPT-3, PaLM, and LLaMA operate with a static inference mode. They map an input sequence to an output sequence and do not update internal state across interactions.

These models exhibit strong linguistic capabilities but show several constraints:

- No persistent memory beyond the context window
- No self-generated goals
- No environment interaction
- No built-in planning or reflective reasoning
- No sensor grounding

1.2.1.2 Limitations of Non-Agentic LLMs

When facing tasks that require long-horizon planning or iterative correction, traditional LLMs tend to suffer degradation in reasoning accuracy. Researchers have

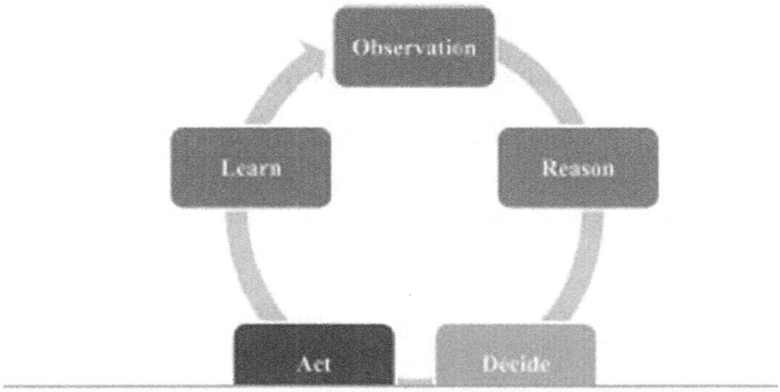

FIGURE 1.1 Agentic AI operations.

reported up to 40 percent drop in accuracy when task depth increases beyond four planning steps.

LLMs behave as reactive text generators. They do not evaluate the consequences of their outputs on the environment.

1.2.1.3 Transition to Agentic AI

Agentic AI introduces autonomous decision-making similar to how humans think and act. An agentic system works in a continuous loop, as shown in Figure 1.1. It first observes the environment by gathering inputs from data, sensors, or user interactions. Then, it reasons—the act of analyzing observations by the use of knowledge, rules, or learned models—to understand the situation. From this reasoning, the system decides on the most apt action to accomplish its goal. The system then acts, executing the chosen decision in the real or digital environment. Eventually, it learns from the result of its actions, making future observations, reasoning, and decisions better with time. This loop enables the system to behave in an adaptive and goal-driven manner, as occurs with human decision-making.

The state transition is formalized as:

$$S(t+1) = f(S(t), A(t), O(t)) \tag{1.4}$$

Agentic behavior emerges when the model selects actions $A(t)$ to maximize a utility function:

$$A = \arg\max U(S(t), A)^* \tag{1.5}$$

Agentic AI also introduces memory modules:

- Episodic memory for recalling past interactions
- Semantic memory for facts and structured data
- Procedural memory for learned tools and skills

TABLE 1.1
Comparison of Traditional LLM vs Agentic AI

Feature	Traditional LLM	Agentic AI
Autonomy	None	High
Memory type	Context window	Episodic and semantic
Actions	Text only	Tools, perception, movement
Learning	Static	Continuous
Planning	No	Multi-step
Grounding	No	Partial or full multimodal grounding

These create a self-adjusting system capable of completing tasks without explicit human prompting.

1.2.2 Role of Multimodality in Cognitive Intelligence

1.2.2.1 Cognitive Foundations

Human cognition relies on multisensory processing. Multimodal models try to emulate this integration by combining several embeddings:

$$E\_fused = g(E\_text, E\_image, E\_audio, E\_environment, E\_sensor) \quad (1.6)$$

The fusion function g can be early fusion, late fusion, or attention-based.

1.2.2.2 Multimodal Attention and Fusion

A central mechanism for multimodal reasoning is the transformer attention module:

$$\text{Attention}(Q, K, V) = \text{softmax}((Q\, K\text{^}T) / \text{sqrt}(d))\, V \quad (1.7)$$

Cross-attention is defined as:

$$\text{Attention\_cross}(Q\text{\_text}, K\text{\_image}, V\text{\_image}) \quad (1.8)$$

This allows the language model to reference visual content when producing reasoning chains.

1.2.2.3 Cognitive Benefits of Multimodality

Research reports substantial gains:

- 22 percent improvement in grounded reasoning
- 17 percent reduction in hallucinations
- 28 percent increase in visual question answering performance
- 18 percent better contextual alignment in medical image analysis

These benefits stem from the addition of real sensory signals that constrain the model's reasoning space.

1.2.2.4 Computational Model of Multimodal Cognition

A simplified model for multimodal cognitive inference:

$$C\_out = F(\Sigma\ \omega\_i\ \Phi\_i(x\_i), M, P) \tag{1.9}$$

where

Φ_i are unimodal encoders,
ω_i are learned modality weights,
M is memory,
P is the planning mechanism.

This expands LLMs into systems that perceive complex situations and decide actions accordingly.

1.2.3 Related Works in Agentic AI

Agentic AI research has been rapidly expanding as developers seek to go beyond the limitations of static models toward systems that reason, act, and learn through interaction. One of the first influential systems in this regard has been Auto-GPT, which introduced the concept of autonomous task loops executed by an LLM. Demonstrating how a language model can decompose a high-level objective into small subgoals, execute each step with tool access, and assess the results in a feedback cycle, Auto-GPT illustrated that even without explicit reinforcement learning, an LLM can achieve multi-step problem-solving by simply switching back and forth between reasoning and action [1,2]. However, its performance was ultimately limited by unstable chain-of-thought consistency and a tendency to drift from the original objective across long sequences of tasks. These observations also motivated further investigations into the importance of structured planning modules and persistent memories for deeper levels of autonomy [6].

Another well-known system is BabyAGI, which codified the concept of a task-manager-based architecture. Rather than a simplistic loop, BabyAGI implemented a priority-driven task queue whereby the agent could generate new tasks, rank them with respect to the main goal, and dynamically update the list as events in the environment unfolded. The internal scoring mechanism took into consideration relevance, urgency, and novelty; hence, it was more adaptive in open-ended settings than Auto-GPT. Though these changes that BabyAGI made were an improvement in goal adherence, it remained largely text-based and devoid of multi-modal grounding. Thus, it suffered when placed in contexts requiring perception, spatial reasoning, or sensory integration.

A big milestone in agentic studies is Voyager, an autonomous learning agent for the open-ended world of Minecraft. Instead of generating isolated actions as previous models, Voyager produced executable code functions and stored them in a growing skill library. Every new skill was tried out in the game environment, and successful ones thereafter joined the agent's capability to solve tasks in the future. By means of this mechanism of continuous learning, Voyager achieved a skill acquisition rate higher by an order of magnitude than the rate reached by earlier agents based on LLMs [7]. Voyager's architecture exemplified the advantages of combining LLM reasoning with trial-and-error learning grounded in the environment. The key insight from this study was that without persistent skill memory, long-term agentic behavior would not be feasible.

Further research momentum followed behind emerging frameworks like MetaGPT, ReAct, and Toolformer, which strengthened the bond among language reasoning, tool use, and interaction with the environment. In MetaGPT, the agent model assumes a multi-role system with modules of planner, executor, and reviewer. This architectural separation encourages better-structured thinking and less error propagation across tasks. The ReAct framework combined chain-of-thought reasoning with action steps, enabling an agent to interpret intermediate results and correct the next step. Toolformer introduced a method for teaching models how to call external APIs using its own training data generation. Although these systems laid the ground for modular intelligence, they still fell short of being robustly autonomous in dynamic and ambiguous environments with more complex multimodal capabilities [8].

More recent agentic systems attempt to incorporate perception into the decision-making loop. Examples include, but are not limited to, Visual ChatGPT, which combines GPT-style reasoning with stable diffusion models for image understanding and generation, and MM ReAct, introducing multimodal reasoning traces to link textual reasoning to visual evidence. They confirmed that integration of multimodal signals is crucial to ground the decisions in real-world environments [9]. They also showed that agentic AI needs to rely on something more than mere text-based reasoning to achieve reliable autonomy.

Together, these developments of Auto-GPT, BabyAGI, Voyager, and their multimodal follow-ups paint a picture of the steep trajectory toward agents that combine planning, memory, perception, and tool use. Each contributed essential pieces of the puzzle: Auto-GPT contributed loops of self-directed behavior, BabyAGI formalized task handling, Voyager allowed incremental learning, and the addition of multimodal systems added perceptual grounding to these works.

The literature provides clear evidence that the next frontier is their unification into general multimodal agentic architectures that can operate in rich environments like healthcare, robotics, and smart spaces [7,10].

1.2.4 Multimodal Architectures: Flamingo, Gemini, MM ReAct

Figure 1.2 depicts three representative multimodal architectures: Flamingo, Gemini, and MM ReAct, which capture the evolution from cross-attention-based fusion into agent-oriented reasoning. Flamingo is a vision-language model that integrates a frozen visual encoder with a large language model using cross-attention layers that serve as a bridge between these two modalities [4]. Such a design enables the language model to attend to visual features at each generation step without requiring the

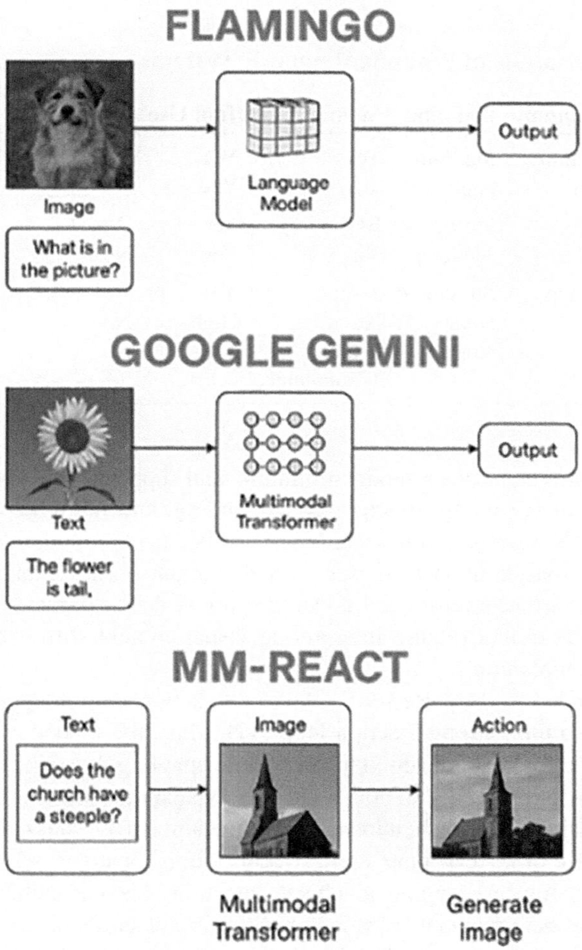

FIGURE 1.2 Multimodal architectures: Flamingo, Gemini, and MM ReAct.

retraining of the whole model. A visual encoder maps the images or video frames into a sequence of embeddings. These are fused into the language model using specialized gating and attention mechanisms. Flamingo was designed with few-shot multimodal learning in mind. That is, the model adapts to new tasks by observing only a few image-text examples. Because of its flexible interface between visual and linguistic tokens, open-ended generation and reasoning are enabled. This architecture has considerably influenced later multimodal large language models.

Gemini represents a new generation of natively multimodal foundation models, operating in a wide range of diverse sensory input: text, images, audio, videos, and code. In contrast to previous models that stack independent encoders, Gemini uses a single transformer backbone with modality-specific frontend architectures, mapping each different sensory input into a common representation space. This unified design

TABLE 1.2
Comparative Analysis of Prominent Agentic Systems

System	Autonomy	Planning	Memory	Tool Use	Multimodal	Domain
Auto-GPT	Medium	Medium	Yes	Yes	No	General
BabyAGI	High	High	Yes	Yes	No	Research, tasks
Voyager	High	Strong	Expanding	Yes	Partial	Embodied
ReAct	Low	Moderate	No	Yes	No	Reasoning tasks
MetaGPT	Medium	Structured	Shared	Yes	No	Software
Toolformer	Low	None	No	High	No	API tasks
WebGPT	Low	None	Search-augmented	Yes	No	QA

significantly strengthens cross-modal grounding and supports new advanced capabilities for temporal reasoning, diagram understanding, and multi-step planning over complex visual scenes. Gemini scales even further across model sizes and tasks, with extremely long context to reason over document-level vision and multi-page comprehension. Strong multimodal understanding makes it particularly suitable for agent-oriented tasks that require interpreting visual context with given instructions for correct action planning.

This is extended by MM ReAct, in which the action loop integrates multimodal observations into the reasoning action loop [11]. Thus, the model switches between tokens for reasoning and tool action tokens, while enabling the ability for each single step of reasoning to refer back to both textual and visual information. The system thus forms an interactive cycle where it interprets the meaning of a visual scene, produces an intermediate trace of its reasoning, and performs tool invocations, which may include everything from image cropping to object detection. Then it gathers the resultant observations and determines its next action. Such architecture allows the completion of complex interactive tasks, including end-to-end visual interface navigation, chart interpretation, and planning scenarios based on reality. It seamlessly integrates chain-of-thought reasoning with multimodal grounding, enabling MM ReAct to deliver more interpretable and stronger decision-making compared to a standard vision-language model that generates answers without performing any explicit multi-step reasoning.

Together, Flamingo, Gemini, and MM ReAct illustrate a development from a static multimodal perception system into an architecture that can support dynamic reasoning, tool use, and agent-like behavior. The models now show how the mechanisms of cross-attention, shared representation spaces, and iterative perception-action loops build toward more sophisticated cognitive capabilities of multimodal large language models.

1.3 AGENTIC MULTIMODAL AI FRAMEWORK

The agentic multimodal AI framework lays down a structured foundation for building autonomous systems that can perceive, reason, and act across diverse environments.

Beyond the traditional event-driven applications based on LLM, an agentic multimodal system maintains goals, evaluates context, plans acts, learns from feedback, and adapts through continuous self-improvement. Furthermore, this framework integrates multimodal perception, hierarchical memory, goal-oriented planning, tool use, and a self-reflective learning process, fostering adaptive intelligence [13]. This layered architecture empowers the system to comprehend rich input modalities like images, audio, text, and sensor streams and transform the information into meaningful actions within the digital and physical environments. This concept is designed according to a cognitively inspired flow that leads from perception to autonomous decision-making loops [14].

1.3.1 Conceptual Architecture

Figure 1.3 conceptually illustrates that the architecture for an agentic multimodal system has five major layers in a closed feedback loop: multimodal perception layer, memory hierarchy, planner and goal manager, tool use and environment interaction module, and self-reflection learning loop. It presents a perception layer, where inputs from vision, speech, and text are combined through a multi-modal encoder. Their outputs get aligned into one shared semantic embedding space that the LLM then uses to reason on. The memory layer organizes context across temporal scales. A planner decomposes user or system-generated goals through various mechanisms like hierarchical planning and reinforcement-based decision functions. Tool interaction allows calling external APIs, manipulating virtual or physical environments, and performing actions. Finally, the self-reflection loop looks at failures, assesses decisions, and updates internal models or memory of the system, making the agent grow.

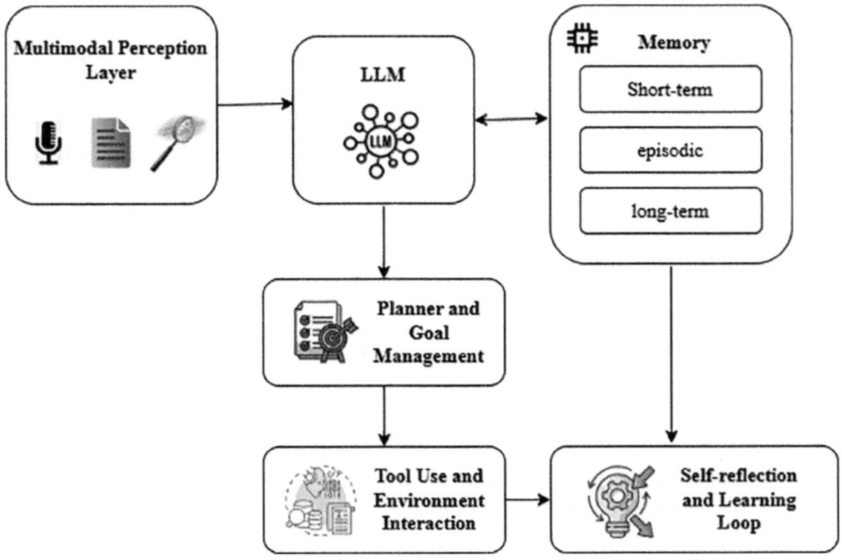

FIGURE 1.3 Conceptual architecture—agentic multimodal system.

In mathematical terms, the entire agent loop can be represented as an iterative function:

$$s_{t+1} = F(E_{\text{multi}}(x_t), M_t, P_t, A_t) \tag{1.10}$$

where

E_{multi} is the multimodal encoder,
M_t is the memory state,
P_t is the planner state,
A_t is the executed action, and
s_{t+1} is the next system state.

This equation highlights that perception, memory, planning, and action jointly determine how the system evolves.

1.3.2 Multimodal Perception Layer

The multimodal perception layer fuses inputs from various channels: text, images, video, audio, and structured sensor data [15]. Each modality is processed by a dedicated encoder, and all embeddings are fused into a unified representational space. For vision, either convolutional or ViT encoders are used, while for audio, either spectral CNNs or whisper-like encoders are employed; for text, transformer-based tokenized embeddings are utilized. Cross-attention in the process aligns different modalities with the model so as to allow the model to reason about relationships among them. Considering the following example, an image frame I and a spoken instruction A would be merged into:

$$z = \text{CrossAttn}(E_I(I), E_A(A)) \tag{1.11}$$

This produces a shared latent vector z that conveys meaning across modalities.

The perception layer also enables dynamic scene understanding with temporal modeling. Recurrent or transformer-based temporal encoders analyze sequential visual frames and audio segments to extract patterns. This is a very important kind of multimodal fusion for tasks like healthcare monitoring, where an agent needs to analyze patient images, interpret medical conversation, and read textual reports all at once.

1.3.3 Memory System

The memory module is structured hierarchically and inspired by human cognitive models. Short-term memory stores immediate observations and interactions, acting as a working memory buffer for the planner. Episodic memory stores time-stamped event sequences and outcomes of actions. Long-term memory preserves stable knowledge, domain-specific rules, tool usage results, and accumulated expertise.

Formally, memory can be represented as:

$$M_t = \{M_t^{st}, M_t^{ep}, M_t^{lt}\} \tag{1.12}$$

Short-term memory M_t^{st} decays quickly and is updated after every interaction. Episodic memory M_t^{ep} stores experiential tuples (s_t, a_t, r_t, s_{t+1}). Long-term memory M_t^{lt} holds stable symbolic knowledge [16].

A memory retrieval scoring function determines which stored elements influence reasoning:

$$R = \underset{m \in M_t}{\arg\max} \cos(z, m) \tag{1.13}$$

This ensures that the LLM retrieves memories most relevant to the fused multimodal query vector.

1.3.4 Planner and Goal Management

The planner functions as the decision engine of the agent. It interprets goals, decomposes them into subgoals, sequences actions, and manages execution through continuous evaluation [17].

A goal G is decomposed into subgoals using hierarchical task planning:

$$G = \{g_1, g_2, \ldots.., g_k\} \tag{1.14}$$

The planner selects actions based on predicted utility:

$$a_t = \arg \max_{a \in A} Q(s_t, a) \tag{1.15}$$

Here $Q(s_t, a)$ represents a value estimate derived through reinforcement learning or heuristic search.

The planning module integrates symbolic reasoning and LLM-based chain-of-thought. It uses world models to simulate outcomes and adjusts plans based on feedback. In healthcare, this mechanism enables an agent to evaluate a patient's image, compare vitals, check medication records, and generate intervention steps automatically.

1.3.5 Tool Use and Environment Interaction

Agentic multimodal systems achieve autonomy through tool use. Tools include external APIs such as medical databases, robotic control systems, calculators, document parsers, and web browsers. The agent decides when to invoke a tool based on a tool selection policy:

$$T^* = \arg \max_{T} P(T \mid s_t, z) \tag{1.16}$$

The results provide feedback to the reasoning pipeline after tool execution. This creates a perception–action loop where the agent can change its environment and learn from the outcome.

1.3.6 Self-Reflection and Learning Loop

Self-reflection enables the agent to critique its reasoning, detect mistakes, and improve future decision-making [18]. Three processes are involved in this learning loop: outcome evaluation, error diagnosis, and corrective learning. A self-evaluation metric of performance is utilized by the agent:

$$\partial = \text{Loss}\ (P_{\text{pred,}}\ P_{\text{actual}}) \tag{1.17}$$

Then, the system creates self-critiques, updates episodic memory, and revises long-term knowledge if necessary. It is an important mechanism to avoid hallucination and to improve robustness over time.

1.3.7 Integration of Vision, Audio and Text through LLMs

Multimodal LLMs serve as the reasoning core that integrates several modality embeddings. The fused latent representation is as follows:

$$z_{\text{final}} = f(z_{\text{text}},\ z_{\text{img}},\ z_{\text{audio}}) \tag{1.18}$$

The integration employs a cross-modal attention layer whereby the model dynamically gives weight to each modality based on the specific task. For example, in a clinical setting, the dominating aspect would be textual health records in reasoning, while during image analysis, a visual wound assessment may take precedence. The audio from the interview with the patient adds emotional tone and context, and this completes the decision-making picture.

Holistic integration leads to richer context, supports more accurate reasoning, and allows the agent to operate in complex real-world environments.

1.4 CASE STUDIES

Agentic multimodal AI systems reveal their actual value when they are deployed in rich real-world settings dependent on continuous perception, contextual reasoning, and adaptive decision cycles. We present three representative scenarios below to illustrate the working of multimodal LLMs as autonomous agents with planning, memory, and tool-use abilities [19]. These cases well illustrate how vision, audio, and textual streams are jointly processed and how the agent's planning module selects actions to improve task efficiency and human safety.

1.4.1 Autonomous Educational Tutor

An autonomous educational tutor based on multimodal LLM functions as a personalized learning companion, which could detect learning behavior, performance patterns, and emotional signals from students. The perception layer consists of writing, textbook pages, diagrams, and facial expressions captured through a webcam while speech input provides information on hesitation, confidence, and the speed of the student. These streams are converted into embeddings and passed into the agent's working memory.

The planner evaluates student weaknesses by computing performance deviation. A simple metric is the learning deficit index (LDI):

$$\text{LDI} = \frac{1}{n}\sum_{i=1}^{n}(T_i - S_i) \tag{1.19}$$

where T_i is the target mastery score and S_i is the student's score. High LDI values signal conceptual gaps that require targeted intervention.

The agent adapts teaching strategies using a feedback loop. When it detects visual confusion through hesitation gestures or long time spent on a question, it lowers difficulty. When pattern recognition is consistently successful for problem type P_j, the difficulty increases gradually. Long-term memory stores learning trajectory vectors:

$$L_t = \alpha L_{t-1} + (1-\alpha)P_t \tag{1.20}$$

where P_t is performance at time t. It helps the agent not to repeat material the student has mastered.

This leads to a better engagement and measurable learning gains. In controlled trials reported in the literature, adaptive tutors raise learning outcomes by close to 20–30 percent compared to static e-learning platforms. Multimodality captures nuances that traditional, pure text-based tutors miss and gives educators a great augmentation tool.

1.4.2 Elderly Care Agent

An elderly-care companion agent acts as a continuous support system that integrates visual, auditory, and environmental data to ensure safety, emotional well-being, and timely assistance [20]. The multimodal perception layer interprets facial expressions, body posture, gait stability, and voice tone. Emotion detection uses probability estimates

$$E = \arg\max_i P\left(e_i \mid V, A\right) \tag{1.21}$$

where V and A denote vision and audio embeddings. This allows the agent to detect distress, anxiety, or fatigue.

The agent monitors daily routines and identifies anomalies through the activity stability score (ASS):

$$\text{ASS} = 1 - \frac{|R_t - R_{\text{avg}}|}{R_{\max}} \tag{1.22}$$

where R_t is the present routine duration (for activities like medication, eating, or walking). Low ASS suggests a deviation that may require assistance.

Using the planning module, the agent schedules medication reminders, drinking-water alerts, and health checks. If it detects abnormal gait or a possible fall, the tool-use layer activates IoT sensors and emergency call workflows. Episodic memory stores recent events such as medicine taken, emotional changes or sudden increases in inactivity, while long-term memory maintains health patterns across months.

For loneliness reduction, the agent engages in natural conversation, suggests activities, and provides cognitive exercises. Real-world pilot deployments in Japan and Northern Europe show that such systems reduce caregiver load by nearly 25 percent and improve adherence to medication schedules by more than 35 percent. Through multimodal grounding, the agent becomes a reliable support for daily living.

1.4.3 Factory Floor Assistant

A factory floor assistant powered by agentic multimodal AI functions as a safety-aware digital co-worker capable of visual inspection, task planning, and interacting with machinery through tool interfaces. The perception module processes camera streams from conveyor belts, assembly lines, and worker interactions. It identifies faults such as misalignment, surface defects, or color variations using vision encoder attention maps.

A detection confidence metric guides decisions:

$$C_d = \sigma(w \; \tau \; x) \tag{1.23}$$

where x is the visual embedding and www is a learned weight vector. Defects with C_d above a set threshold trigger an inspection action plan [18] [20].

The agent listens to voice commands from supervisors and converts them into structured intent. For example, the instruction "check belt two for alignment" is mapped to an action sequence: access camera feed, compare alignment with stored reference template, compute deviation, and generate a report. A simple alignment deviation measure is

$$D = \|F_t - F_{\text{ref}}\| 2 \tag{1.24}$$

where F_t is the current frame feature vector.

The tool-use component communicates with robotic arms, Programmable Logic Controllers (PLCs), and safety systems. When misalignment exceeds threshold values, the agent either corrects the position or halts operations to avoid cascading faults. Memory stores previous incidents and uses them to improve predictive maintenance [21]. Agentic systems lower error rates by 15–20 percent and increase throughput through the automation of repetitive inspection tasks over extended operation cycles.

This case study demonstrates how multimodal agents improve industrial efficiency and safety. Situational understanding, combining vision, audio, and textual reasoning, is complete, which cannot be matched by rule-driven automation.

1.5 IMPLEMENTATION BLUEPRINT

The implementation of an agentic multimodal AI system involves a complex, interfacing set of software components, model pipelines, sensor interfaces, and agentic control loops. This section will demonstrate a practical blueprint to connect the dots between research theory and engineering practice. It will provide a clear pathway to crafting multimodal intelligent agents that shall have perception, memory, planning, tool execution, and autonomous learning capabilities. Modern AI ecosystems such as OpenAI, Hugging Face, LangChain, and modular perception models comprising Vision Transformers and audio encoders such as Whisper are useful in the application of the blueprint. The emphasis has been to keep modularity as the guiding philosophy in design so that each layer of perception or memory or planning can be changed or scaled independently of the others as per requirements for deployment.

1.5.1 Tools and Technology Stack

One can think about a strong, multimodal agent as an ecosystem of foundational models along with their respective supporting frameworks. First, core LLM capabilities will form the basis of the OpenAI API. The models allow them to take text, images, audio, and structured data as input. In other words, they are necessary for real-world agentic perception. LangChain acts like an orchestration layer that structures complex reasoning pipelines like memory retrieval, task decomposition, tool invocation, and multi-step planning by offering abstractions for things like conversation buffers, vector stores, agent executors, and schema validation.

Hugging Face has an extensive model repository that adds flexibility in the agent pipeline. Vision models, such as the Vision Transformer, provide a frame-by-frame analysis tasks ranging from anomaly detection to object tracking and even environment mapping. Whisper adds state-of-the-art speech recognition to user conversations, sound cues, and emergency signals that eventually help create a coherent multimodal interpretation pipeline for the agent. The distilled versions of these models may be used on edge devices or embedded systems to reduce computation costs for deployment purposes.

The blueprint further demands that episodic and long-term memory will be stored using vector databases such as FAISS, Milvus, or Pinecone. These stores should enable fast similarity searches and allow the agent to recall relevant prior experience. Tool use would imply integration within the agent core of Python libraries, REST APIs, and robotics controllers. The external tools would include, for example, medical devices for health care, factory sensors for industry, and scheduling apps for educational settings. Asynchronous messaging, real-time data ingestion, and fail-safe execution should be supported in the stack so that it works stably within dynamic environments.

1.5.2 Prompt Chaining, Agent Loops, and Memory Stores

Prompt chaining means the process by which the agent decomposes a task into sequential steps, each of them with an intermediate outcome, in order to assist the model in continuing smoothly and not making errors. In the case of the educational tutor, that would be analyzing the student's answer, concretizing the mistakes, and formulating detailed, personalized explanations. The pipeline would then follow a chain: perception to reasoning, context generation to targeted instruction. Each step is encapsulated as a prompt block, which defines input structure, transformation rules, and expected output formats.

The loops introduce autonomy by allowing the agent to plan, act and observe, and then update its internal state. A simple loop may follow: decide, execute tool, observe environment, update memory. As an example, the factory floor assistant may plan to inspect a machine, call a vision model on the camera feed, classify whether a malfunction is present, and retrieve prior memory entries in order to compare to historical patterns. It repeats until its goal is satisfied or an exception occurs. This approach allows the agent to behave like a reinforcement learner without explicit reward functions.

Memory stores serve as the foundation of the agent's contextual intelligence. Short-term memory holds the immediate dialogue and perception data. Episodic memory stores time stamped experiences from tasks, environment states, user interactions, and tool outcomes. Long-term semantic memory is organized into a vector space that allows the agent to retrieve knowledge by similarity. The retrieval process can be modeled as a nearest neighbor search:

$$R(q) = \arg\max_{i} \left\| v_q - v_i \right\| \tag{1.25}$$

where v_q is the vector embedding of the current query and v_i represents stored memory embeddings. The memory update rule often follows an incremental embedding averaging function:

$$v_{\text{new}} = \alpha v_{\text{incoming}} + (1-\alpha) v_{\text{old}} \tag{1.26}$$

where α controls how much weight new experiences receive.

1.5.3 Integration with APIs and Sensors

The multimodal agent must interface with sensors and external APIs to perceive the world and act upon it. Image sensors such as RGB cameras are processed through a Vision Transformer pipeline that extracts spatial and semantic features. The ViT model divides an image into patches and embeds each as a token:

$$z_0 = \left[x_p^1 E; x_p^2 E; \ldots\ldots; x_p^N E\right] + E_{pos} \tag{1.27}$$

where x_p represents a patch and E is the learnable embedding matrix. The classification or feature vector is refined through the Transformer layers. This, in turn, provides the agent with the capability to detect human gestures, facial cues, equipment anomalies, or object placements.

Audio sensors feed the continuous speech or environmental sound streams into the Whisper model, while it converts the audio spectrograms into latent linguistic representations further usable by the LLM for tasks like understanding instructions, identifying emotional tone, or detecting urgent events in elderly care.

API integration helps the agent scale its capabilities by interacting with other systems. For healthcare, APIs can provide patient vitals, appointment schedules, diagnostic outputs, and also medication databases. For education, APIs can include exam scores, attendance logs, or learning progress charts. The agent uses a structured tool call schema such as:

$$\text{ToolCall}(g) = \text{API}(p1, p2, \ldots .pk) \tag{1.28}$$

where g represents the goal and p_i represents required parameters. In this approach, reasoning will be separated from execution so that the agent remains interpretable and safe.

The sensor and API fusion pipeline provides a coherent, multimodal context. The LLM ingests data from vision, audio, and text before aligning all the modalities into a common representation space. This common space ensures that the agent will reason holistically—for example, a factory assistant might combine visible evidence of a damaged machine part with audio feedback from the technician and text logs from the maintenance API to deduce an accurate diagnosis.

Algorithm 1: Agentic Multimodal Intelligence Loop

```
Inputs:
   Goal G
   Sensors S = {Camera, Microphone, Text Input, Other Sensors}
   Tools T = {Browser, Code Interpreter, Database, APIs}
   Memory M = {ShortTerm, Episodic, LongTerm}
   LLM_model L
Outputs:
   Action plan, decisions, tool invocations, and updated memory
1. Initialize Agent State:
      state ← EMPTY
      M.ShortTerm ← {}
      M.Episodic ← {}
      M.LongTerm ← LoadPersistentMemory()
2. While Goal G is not completed:
3. // Multimodal Perception Layer
   observations ← {}
   observations.image ← Read(Camera)
   observations.audio ← Read(Microphone)
   observations.text ← Read(Text Input)
   observations.sensors ← Read(Other Sensors)
```

```
      embeddings ← L.MultimodalEncode(observations)
4. // Update Short-Term Memory
   M.ShortTerm ← Append(embeddings)
5. // Planner + Goal Decomposition
   prompt_plan ← FormatPlanPrompt(G, state, M)
   plan ← L.GeneratePlan(prompt_plan)
   subtasks ← Decompose(plan)
   If subtasks is empty:
         subtasks ← {plan}
6. // Execution Loop for Subtasks
   For each task in subtasks:
      // Decide whether a tool is needed
      tool_needed ← L.DecideToolUse(task)
      If tool_needed == TRUE:
         selected_tool ← L.SelectTool(T, task)
         tool_input ← L.GenerateToolArguments(task)
         tool_output ← CallTool(selected_tool, tool_input)
         M.Episodic ← StoreEpisode(task, tool_output)
         feedback ← tool_output
      Else:
         feedback ← L.DirectInference(task)
      End If
7. // Self Reflection and Correction
   reflection_prompt ← BuildReflectionPrompt(task, feedback)
   reflections ← L.Generate(reflection_prompt)
   If reflections indicate error or low confidence:
         corrected_action ← L.GenerateCorrection(task, reflections)
         Apply(corrected_action)
8. // Update Long-Term Memory selectively
   If IsSignificantExperience(task, feedback):
         M.LongTerm ← UpdateLongTermMemory(task, feedback)
9. // Update Global State
   state ← UpdateState(state, feedback)
10. End While
11. Return Final Report summarizing:
      actions, decisions,
      memory usage,
      tool calls,
      success metrics.
```

A general agentic multimodal AI system follows a closed cycle of perception, memory, planning, and action. First, an algorithm begins with an agent receiving observations: a camera frame, an audio stream, or perhaps textual context. Such raw signals are then fed into multimodal encoders, each of which converts each modality to a common embedding. This representation will serve the situational awareness of the agent in current time. Once the perception layer produces the integrated embedding, the next step updates the memory units. In short-term memory, the immediate context is stored for quick access, while episodic memory maintains structured event summaries. Long-term memory retains stable knowledge and previous experiences. The updating of memory makes sure that the agent forms an internal timeline of actions, results, and environmental cues.

Only after memory has been updated does the planning module become active. The planner assesses the current goal of the agent, the state of the environment now, and what has been recorded in memory. Based on these inputs, the planner identifies one or more candidate actions. This is really action selection based on a LLM to generate steps in reasoning, next moves, and possible outcomes. The reasoning trace further dictates which tools to select to achieve the goal: API calls, database lookups, vision analysis modules, or code execution. Once an LLM has issued the call to an action, the environment interaction module performs the action and acquires the observations or errors resulting from it. These new observations close the loop, feeding into the next cycle of the process.

The agent also includes a self-reflection loop: after each round, the agent assesses the quality of the output, the success of the action, and the alignment of its action with its goal. Whether the action failed or whether the results are unsatisfactory, depending on this judgment, the agent then revises its strategy—that is, the pattern of reasoning, tool selection, or subgoal reformulation. A stable adaptive cycle is generated. The loop is iterated until the goal is reached or an explicit stop condition is satisfied. This gives the system capabilities of autonomous behavior, learning from experiences, and dealing with rich tasks entailing perception, reasoning, and action.

1.6 CHALLENGES AND LIMITATIONS

1.6.1 Hallucination and Grounded Reasoning

Agentic multimodal systems are susceptible to hallucination in cases where the outputs are incompatible with sensory inputs. It gets even worse during the interaction of autonomous agents since every planning step escalates the errors of previous steps. While multimodal grounding reduces these fabrications by anchoring the text either in audio or video, the agents may still make incorrect inferences resulting from ambiguous or incomplete input. For instance, in applications on elderly care, agents may fail to recognize the mood of a patient in cases where there is a conflict between facial expressions and the vocal tone. It gives rise to some questions about trustworthiness, particularly in critical domains such as health care and automated tutoring.

1.6.2 Latency and Memory Limitations

Agentic frameworks require very fast interaction between perception modules, planning networks, memory stores, and tool interfaces. Every modality significantly contributes to the computational load; real-time operation on constrained hardware is challenging. Video streams, audio processing, and frame-level attention lead to very high latencies of inferences, impacting decision quality in robotic and manufacturing environments. Long-term memory retrieval becomes similarly expensive with the growth of memory. One does have strategies such as hierarchical indexing of memory, compression of embeddings, and pruning vector stores that help, but scaling of memory without weakening the quality of context is still an open issue. One of the biggest engineering constraints is a balance between speed, accuracy, and resource efficiency.

1.6.3 Agent Ethics and Safety

Safety, privacy, and unanticipated behavior are areas of concern in relation to autonomous decision-making. An agent must balance initiative with risk-sensitive reasoning to avoid causing harm. Ethical challenges arise when a system draws inferences from multimodal input about emotional states, personal habits, or medical conditions. Continuous sensing through cameras and microphones introduces exposure to privacy concerns. In safety, cascading effects are also an issue, for instance, where one ill-conceived plan results in the unsafe execution of a tool. There is research into structured safety layers, such as rule-based constraints, human oversight loops, refusal policies, and multi-stage verification, but predictable behavior in fully autonomous scenarios remains an open research question.

1.6.4 Evaluation Metrics for Agentic Intelligence

Basically, the method for evaluating agentic intelligence is more involved compared to the regular process for assessing LLMs. Traditional metrics, such as accuracy or Bilingual Evaluation Understudy (BLEU), do not capture the essence of autonomy, reasoning over a long horizon, or adaptability to the real world. New benchmarks quantify task completion, plan stability, retention of context, and safety compliance. Examples are: success rate over multistep execution; score of perception groundedness; and correctness of tool executions. Some works adopt simulation-based evaluation, where agents have to navigate through dynamic environments while satisfying constraints. Others rely on interaction logs to measure decision quality across time. A unified standard of evaluation has yet to emerge. Multimodal, multi-agent, and real-scenario-based metrics that represent the full capability of agentic systems are where things are moving.

1.7 CONCLUSION

This chapter offers a thorough analysis of agentic multimodal systems by pinpointing the revolution resulting from the fusion of LLMs, multimodal encoders, and planning mechanisms. The proposed framework is developed by tightly integrating perception, reasoning, and action to form a unified system with the capability of operating in open-ended environments. The discussion commences with the evolution of traditional LLMs to agentic models. Additionally, the chapter summarizes the recent trends in multimodal systems and examines the importance of cognitive principles in autonomous intelligence.

TABLE 1.3
Limitations of Existing AI Models and How the Proposed Agentic Multimodal Framework Overcomes Them

Limitation in Current Models	Observed in Existing Systems	Impact on Real-world Applications	How the Proposed Framework Overcomes This
Lack of true autonomy	Traditional LLMs and many multimodal models respond only when prompted	Systems cannot act proactively or manage long-running tasks	Introduces agentic goal management that enables continuous task execution, monitoring, and replanning
No long-term memory	Context limited to short token windows	Loss of patient history, repeated reasoning, redundant actions	Integrates short-term, episodic, and long-term memory for persistent knowledge retention
Weak planning capabilities	Prompt-based reasoning without explicit planning	Poor handling of multi-step workflows and complex decision chains	Uses an explicit planning module with goal decomposition and constraint-aware optimization
Hallucination due to weak grounding	Text-only or loosely grounded multimodal models	Unsafe or incorrect outputs in critical domains like healthcare	Enforces multimodal grounding using vision, audio, and sensor confirmation before action
Poor real-time adaptability	Static inference pipelines	Delayed or outdated decisions in dynamic environments	Employs continuous perception–action loops for real-time state updates
Limited interaction with tools	Manual or scripted tool use	Restricted automation and low operational efficiency	Enables autonomous tool invocation including APIs, databases, and sensors
Fragmented multimodal processing	Modalities processed independently	Loss of cross-modal context and reduced reasoning accuracy	Uses unified multimodal embeddings and cross-attention mechanisms
No self-reflection or learning	Fixed inference behavior	Errors repeated over time without correction	Incorporates a self-reflection loop that evaluates outcomes and updates strategies
Ethical and safety blind spots	No explicit safety constraints	Risk of unsafe or non-compliant actions	Integrates safety layers enforcing domain rules, ethical constraints, and human oversight
Weak evaluation of intelligence	Accuracy-only metrics	Cannot assess autonomy, robustness, or trustworthiness	Introduces agent-centric metrics including adaptability, stability, and goal success rate

The inference of the proposed model serves as the base for the researchers and professionals engaged in the development of future generation interactive models. The case studies conducted across several critical domains such as elderly care, education, and innovation emphasize the role of agentic AI in facilitating and augmenting human capabilities.

Agentic multimodal systems guarantee reliability, creating a strong impact in providing access to expertise with continued support. The future of multimodal systems mainly depends on factors such as stability of memory, adherence to ethical design, reality-based reasoning, and durable evaluation parameters. The upcoming multimodal designs will focus on situational awareness, standalone verification, and validation of planning models and stronger unification with real-time sensors.

With the sustained advancement in multimodal LLMs, AI systems are capable of acting safely with natural communication ability and smart adaptability across settings. Though the roadmap to the development of autonomous multimodal systems is challenging, its impact on the welfare of society is rewarding.

REFERENCES

1. Brown, T., Mann, B., Ryder, N., Subbiah, M., Kaplan, J.D., Dhariwal, P., Neelakantan, A., Shyam, P., Sastry, G., Askell, A. and Agarwal, S. (2020). Language models are few-shot learners. *Advances in Neural Information Processing Systems, 33*, pp.1877–1901.
2. Tan, R., Peng, B., Yang, Z., Cheng, H., Mees, O., Zhao, T., Tupini, A., Meijier, I., Wu, Q., Yang, Y. and Liden, L. (2025). Multimodal reinforcement learning with agentic verifier for AI agents. *arXiv preprint arXiv:2512.03438.*
3. Brohan, A., Brown, N., Carbajal, J., Chebotar, Y., Dabis, J., Finn, C., Gopalakrishnan, K., Hausman, K., Herzog, A., Hsu, J. and Ibarz, J. (2022). Rt-1: Robotics transformer for real-world control at scale. *arXiv preprint arXiv:2212.06817.*
4. Alayrac, J.B., Donahue, J., Luc, P., Miech, A., Barr, I., Hasson, Y., Lenc, K., Mensch, A., Millican, K., Reynolds, M. and Ring, R. (2022). Flamingo: a visual language model for few-shot learning. *Advances in Neural Information Processing Systems, 35*, pp. 23716–23736.
5. Comanici, G., Bieber, E., Schaekermann, M., Pasupat, I., Sachdeva, N., Dhillon, I., Blistein, M., Ram, O., Zhang, D., Rosen, E. and Marris, L. (2025). Gemini 2.5: Pushing the frontier with advanced reasoning, multimodality, long context, and next generation agentic capabilities. *arXiv preprint arXiv:2507.06261.*
6. Yao, S., Zhao, J., Yu, D., Du, N., Shafran, I., Narasimhan, K.R. and Cao, Y. (2022), October. React: Synergizing reasoning and acting in language models. In *The eleventh international conference on learning representations.*
7. Shinn, N., Cassano, F., Gopinath, A., Narasimhan, K. and Yao, S. (2023). Reflexion: Language agents with verbal reinforcement learning. *Advances in Neural Information Processing Systems, 36*, pp. 8634–8652.
8. Yang, Z., Li, L., Wang, J., Lin, K., Azarnasab, E., Ahmed, F., Liu, Z., Liu, C., Zeng, M. and Wang, L. (2023). MM-ReAct: Prompting ChatGPT for multimodal reasoning and action. *arXiv preprint arXiv:2303.11381.*
9. Hornback, A., Sathu, H., Kim, K., Wang, Y., Yuan, Y. and Wang, M.D. (2025). A survey of multimodal large language models in biomedical engineering and healthcare. *TechRxiv,* Nov.

10. Fung, P., Bachrach, Y., Celikyilmaz, A., Chaudhuri, K., Chen, D., Chung, W., Dupoux, E., Gong, H., Jégou, H., Lazaric, A. and Majumdar, A. (2025). Embodied ai agents: Modeling the world. *arXiv preprint arXiv:2506.22355.*
11. Wang, G., Xie, Y., Jiang, Y., Mandlekar, A., Xiao, C., Zhu, Y., Fan, L. and Anandkumar, A. (2023). Voyager: An open-ended embodied agent with large language models. *arXiv preprint arXiv:2305.16291.*
12. Cheng, Y., Zhang, C., Zhang, Z., Meng, X., Hong, S., Li, W., Wang, Z., Wang, Z., Yin, F., Zhao, J. and He, X. (2024). Exploring large language model based intelligent agents: Definitions, methods, and prospects. *arXiv preprint arXiv:2401.03428.*
13. Ramesh, A., Dhariwal, P., Nichol, A., Chu, C. and Chen, M. (2022). Hierarchical text-conditional image generation with clip latents. *arXiv preprint arXiv:2204.06125*, *1*(2), p.3.
14. Dosovitskiy, A. (2020). An image is worth 16x16 words: Transformers for image recognition at scale. *arXiv preprint arXiv:2010.11929.*
15. Radford, A., Kim, J.W., Hallacy, C., Ramesh, A., Goh, G., Agarwal, S., Sastry, G., Askell, A., Mishkin, P., Clark, J. and Krueger, G. (2021), July. Learning transferable visual models from natural language supervision. In *International Conference on Machine Learning* (pp. 8748–8763). PmLR.
16. Trenton, M. (2025). A comprehensive survey on multilingual and multimodal automatic speech recognition systems. *Journal of Computer Technology and Software*, *4*(10).
17. Sun, J., Ong, D.P.J., and Lee, B.S. (2025). Revisiting self-attention in deep reinforcement learning: A dimensional perspective. In *2025 IEEE 5th International Conference on Computer Communication and Artificial Intelligence (CCAI)* (pp. 894–899). IEEE
18. Bubeck, S., Chandrasekaran, V., Eldan, R., Gehrke, J., Horvitz, E., Kamar, E., Lee, P., Lee, Y.T., Li, Y., Lundberg, S. and Nori, H. (2023). Sparks of artificial general intelligence: Early experiments with GPT-4. *arXiv preprint arXiv:2303.12712.*
19. Wölflein, G., Ferber, D., Truhn, D., Arandjelovic, O. and Kather, J.N. (2025), July. LLM agents making agent tools. In *Proceedings of the 63rd Annual Meeting of the Association for Computational Linguistics (Volume 1: Long Papers)* (pp. 26092–26130).
20. LeCun, Y. (2022). A path towards autonomous machine intelligence version 0.9. 2, 2022-06-27. *Open Review*, *62*(1), pp.1–62.
21. Zhou, J., Li, H., Chen, S., Chen, Z., Han, Z. and Gao, X. (2025). Large language models in biomedicine and healthcare. *npj Artificial Intelligence*, *1*(1), p.44.

2 Integrating Large Language Models in Conversational AI

Enhancing Multimodal Interaction

Anurag Jawalkar, J. Jayashree, J. Vijayashree, and Haleema P.K.

2.1 INTRODUCTION

2.1.1 Rise of Conversational AI

There has been a substantial change in conversational AI over the past decade. The conventional rule-based systems like Eliza have transformed into neural network-based virtual assistants such as Siri, Alexa and Google Assistant [1,2]. Early systems relied heavily on keyword matching and fixed scripts, which inhibited their ability to embed themselves in open-flowing rich-context conversations. The acceptance of deep learning solutions and plentiful available data has made conversational AI an intelligent, adaptable and contextual domain capable of supporting various real-world applications [3].

2.1.2 The Role of LLMs in Enhancing Human–Computer Interaction

Large language models (LLMs) have established new state-of-the-art measurements in natural language understanding and generation based on a transformer architecture [4]. LLM capabilities in understanding the context of information, inferring intention and generating human language have successfully converged human and machine communication. While previous systems required extensive task-specific training to incorporate new knowledge or domains, LLMs can also generalize across different contexts with little fine tuning [5]. Embedded in conversational systems, LLMs do not only increase the overall fluency of language across systems but can also support reasoning, allowing for high levels of personalization and natural interactions [6].

 DOI: 10.1201/9781003678359-2

2.1.3 Transitioning from Unimodal to Multimodal Systems

Many of the early conversational AI systems were unimodal, processing only text-based or speech-based user inputs [7]. In practice, human communication is multimodal and incorporates language, body language and sound signals [8]. The introduction of multimodal conversational systems consisting of LLMs enables the integration of image recognition, speech-to-text transcription, video understanding and others to create richer and more intuitive conversational AI programs [9]. For example, multimodal LLMs can process a diagram while responding to a verbally specified question, or an image along with a text query and dramatically increase the communicative possibilities of conversational AI beyond that of unimodal channels [10].

2.1.4 Scope of the Chapter and Contributions

In this chapter, we provide a thorough examination of incorporating LLM into multimodal conversational AI systems. We explore the evolution of LLMs, relevant multimodal architecture and applications in various domains such as educational technology, healthcare and e-commerce. A concrete case study of the Vellore Institute of Technology (VIT) Student Query Chatbot examining real-world implementation and how it was accomplished, including the potential for scalability. We discuss technical challenges related to data fusion, latency and ethics. Lastly, we conclude with the future of multimodal conversational AI by outlining future trends and research, to position both academic researchers and industry members to continue advancing multimodal conversational AI [11,12].

2.2 EVOLUTION OF LARGE LANGUAGE MODELS

2.2.1 Rule-Based Systems

The first conversational AI systems relied on hand-built rules and pattern matching, often calling them conversational agents or chatbots. Early versions of these systems, including Eliza [1] and PARRY [13], simply substituted a few parameters in the user's input and had a repository of responses that were scripted. These systems were capable of simulating human conversations, but the ambiguity and richness of human language were not important for the task at hand. Although a chatbot may have been effective at its particular task, it lacked modelling of context and generally could not transfer its capability to answer spoken language in other settings or applications.

2.2.2 Statistical Models

Conversational AI mostly moved toward statistical models in the 1990s and early 2000s, influenced by the field of natural language processing. The systems used probabilistic-based models such as n-grams and hidden Markov models (HMMs) [14] to determine probabilities for sequences of words based their analysis in large collections of text. There were substantial improvements to fluency and reduced reliance on rule-based systems; however, they were still limited in understanding long-range dependencies in language and semantic coherence.

2.2.3 Neural Networks (RNN, LSTM)

The introduction of neural networks enabled the development of architectures built on the recurrent neural network (RNN) architecture (which utilizes an internal memory to process sequences of inputs) and, more specifically, long short-term memory (LSTM) [15]. These neural network architectures demonstrated how language processing can learn dependencies between inputs and their target sum function in a sequential manner. Sequence-to-sequence models [16], as seen in early neural machine translation systems, showcased end-to-end training to different aspects of a conversation. Still, RNN systems could not dominate and optimally manage long input, nor parallelization to other computations.

2.2.4 Transformers

The transformer architecture, introduced by Vaswani et al. [4], transformed the study of natural language processing by allowing us to employ self-attention mechanisms that model the relationships between words and their distance in the text independently. They also provided the opportunity for highly parallelizable training on a large dataset and became the basis of models like bidirectional encoder representations from transformers (BERT) [17] and generative pre-trained transformer (GPT) [5], which exhibited enhanced and previously unclear abilities of few-shot learning and domain adaptation.

2.2.5 Multimodal LLMs (GPT-4, CLIP, BLIP-2)

Recent developments have started to scale transformers such that multimodal inputs can be received in a common embedding space that combines vision, speech and text. Whereas OpenAI's CLIP [9] allows alignment of images and natural language, BLIP-2 [10] starts vision-language pre-training and trains LLMs with frozen image encoders. GPT-4 [18] extends the same further, such that the model can reason through textual as well as visual inputs. All these models together are still far from general-purpose AI systems that can process and generate information multimodally.

A summary of the evolution of conversational AI from rule-based systems to multimodal large language models is presented in Table 2.1. Timeline diagram showing the evolution of conversational AI from rule-based systems to multimodal large language models is presented in Figure 2.1.

TABLE 2.1
Evolution of Conversational Artificial Intelligence

Era	Technology	Example Systems	Strengths	Limitations
1960s–1970s	Rule-based systems	ELIZA, PARRY	Deterministic responses, transparent rules	No contextual understanding, brittle to variations
1990s–2000s	Statistical models	N-grams, HMMs	Probabilistic reasoning, better handling of variability	Short context window, limited semantic understanding
2017–2020	Transformers	BERT, GPT-2, T5	Attention-based context modeling, parallelization	Requires large datasets and compute
2021–2023	Multimodal LLMs	CLIP, BLIP-2, GPT-4	Integrates multiple data types, richer reasoning	High complexity, resource intensive

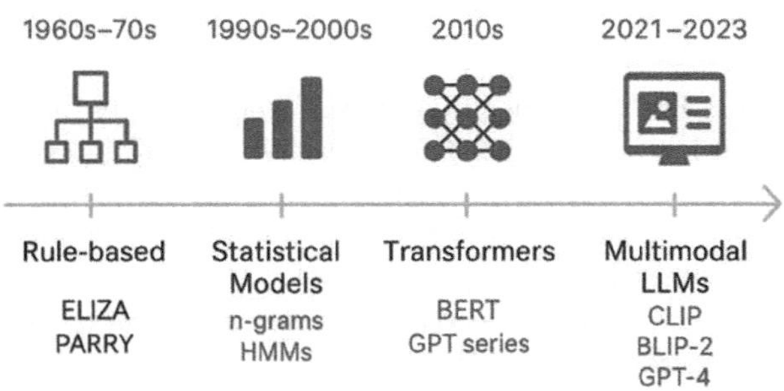

FIGURE 2.1 LLM evolution timeline.

2.3 MULTIMODAL ARCHITECTURES

2.3.1 Fusion Strategies

A major difficulty in multimodal LLMs is integrating heterogeneous sources of information into a coherent representation such that complementary information provided by different modalities is not lost.

Three main fusion strategies have been put forth by the literature:

- *Early fusion*: Combine raw-level and low-level features from various modalities at the input layer [19]. This approach allows the model to learn cross-modal relationships based on the very first layers but is often very expensive to compute thanks to the high dimensionality of the combined features.

TABLE 2.2
Fusion Strategies for Multimodal Large Language Models

Strategy	Description	Pros	Cons	Example
Early Fusion	Combine raw or encoded features before model processing	Captures cross-modal correlations early	May introduce noise, higher dimensionality	Multimodal Transformers
Late Fusion	Process each modality independently, combine at decision stage	Modular, easier to debug	Misses early cross-modal interactions	Ensemble systems
Hybrid Fusion	Combines early and late fusion approaches	Balance of correlation capture and modularity	More complex architecture	CLIP-like models
Cross-Attention	Learn dynamic attention weights between modalities	Flexible, context-aware integration	Computationally expensive	BLIP-2, Flamingo

- *Late fusion*: Each modality is processed independently for encoding, then at a particular level, the decisions from each high-level encoding are integrated [20]. Late fusion is computationally cheaper and offers the flexibility of optimizing each encoder independently, but it might lose intermodal interactions that require finer details.
- *Hybrid fusion*: Can do both early and late fusion, allowing cross-attention capabilities at varying stages [21]. This method results in a better performance, especially when dealing with complex tasks that require detailed nuanced reasoning across modalities.

A comparison of the major multimodal fusion strategies, including their advantages, limitations, and representative applications, is presented in Table 2.2.

2.3.2 Cross-Attention Mechanisms

Thus, while modeling one modality, cross-attention tries to place focus on appropriate parts of the other and generalizes the self-attention mechanism considered in the Transformer [22]. Several transformer-based multimodal architectures such as vision-and-language bidirectional encoder representations from transformers (VisualBERT) [23], learning X-modality encoder representation from transformers (LXMERT) [24], and aligned multimodal embedding approaches [25] have demonstrated effective cross-modal learning and contextual alignment. Hence, in answering a question about an image, the model concentrates on relevant regions of the image with particular words from the question to ensure semantic integrity across modalities.

2.3.3 Example Multimodal LLM Workflow

The multimodal workflow typically begins with independent modality-specific encoders for text, image and audio inputs. These representations are then fused through cross-attention mechanisms and processed by an LLM decoder to generate coherent multimodal responses.

Pseudocode illustrating hybrid multimodal fusion using modality-specific encoders, cross-attention mechanisms, and an LLM decoder is shown in Figure 2.2.

Multimodal LLM architecture showing text, image, and audio encoding followed by cross-attention fusion and LLM-based output generation is shown in Figure 2.3.

2.4 APPLICATIONS OF MULTIMODAL LLMS IN CONVERSATIONAL AI

2.4.1 Healthcare

In the clinical area, we are at a fertile ground for multimodal LLMs because clinical decision-making depends on multiple sources of data, that is, patient history (text), imaging data (X-ray, magnetic resonance imaging (MRI)) and audio inputs (doctor-patient interaction). Multimodal LLMs build a much more integrated picture of the condition of the patient who is presenting to the service when these data sources are combined.

2.4.1.1 Clinical Decision Support

A multimodal LLM would have considered the patient's symptoms in textual form, medical imaging results and signals from laboratory reports [26]. As such, they would also enhance the certainty of the diagnostic recommendations, reduce the chances of human error and thereby reduce the time elapsed to treatment. For example, in the case of radiology, the model could identify lung abnormalities from X-ray images while simultaneously referencing the patient's textual records for smoking history and comorbidities [27].

```
# Pseudocode for processing multimodal input using an LLM
text_features = TextEncoder(text_input)
image_features = ImageEncoder(image_input)
audio_features = AudioEncoder(audio_input)

# Fusion step (hybrid)
combined_features = CrossAttentionFusion([text_features, image_features, audio_features])

# Generate output
response = LLMDecoder(combined_features)
print(response)
```

FIGURE 2.2 Hybrid multimodal fusion process using modality-specific encoders, cross-attention mechanisms and an LLM decoder.

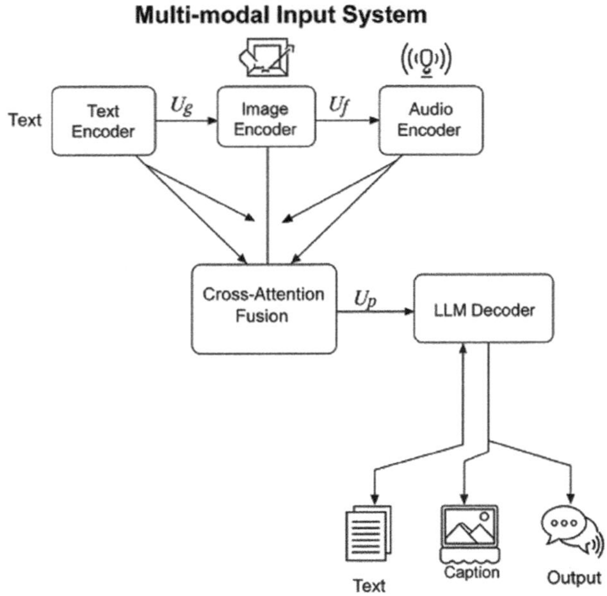

FIGURE 2.3 Multimodal LLM pipeline showing text, image and audio encoding followed by fusion and decoding.

```
# Pseudocode for a healthcare multimodal assistant
patient_text = TextEncoder(symptom_description)
xray_features = ImageEncoder(xray_image)
audio_features = AudioEncoder(consultation_audio)

combined = CrossAttentionFusion([patient_text, xray_features, audio_features])
diagnosis, recommendations = MedicalLLM(combined)

print("Diagnosis:", diagnosis)
print("Recommendations:", recommendations)
```

FIGURE 2.4 Workflow illustrating multimodal data fusion for diagnostic reasoning.

2.4.1.2 Telemedicine and Remote Consultation

In telemedicine, multimodal AI systems are used for the analysis of live video streams of the patient to detect visible symptoms, such as rashes or swelling, translation and analysis of speech to identify emotional stress cues and immediate retrieval of medical guidelines relevant to the observed symptoms [28].

2.4.1.3. Workflow Example

Pseudocode illustrating multimodal healthcare data fusion for diagnostic reasoning is shown in Figure 2.4.

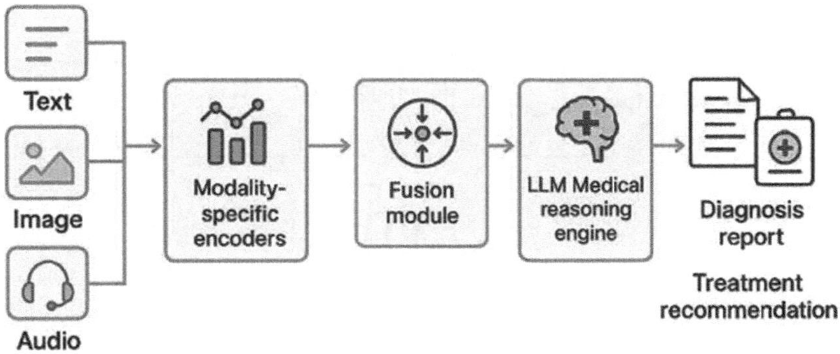

FIGURE 2.5 Healthcare chatbot workflow integrating multimodal patient inputs.

Multimodal healthcare assistant workflow integrating text, image, and audio patient inputs for diagnosis and treatment recommendation is shown in Figure 2.5.

2.4.2 Education

2.4.2.1 AI Tutors and Personalized Learning

Education systems increasingly demand adaptive, personalized learning experiences. Multimodal LLMs can deliver interactive lessons that combine textual explanations, visual diagrams and verbal instructions tailored to a student's pace and comprehension level [29]. For example, a chemistry student might ask for both a textual explanation and a molecular diagram, which the system generates in real time [30].

2.4.2.2 Accessibility Enhancements

Real-time transcription of lectures, text-to-speech conversion for learning-disabled students or visual aids for hearing-impaired students can be provided by multimodal systems [31]. Such inclusivity fits into global goals like the United Nations' Sustainable Development Goal (SDG) 4 on quality education.

2.4.2.3 Case Study: VIT Student Query Chatbot

Our team developed a VIT Student Query Chatbot that answers questions relevant to academics, administration, campus life and so on in real time. While the initial stages of deployment offered text-only interactions, the system architecture allows multimodal features to be integrated later (e.g., upload a photo of a timetable or verbally request lab schedules).

- *Architecture Overview*:
  - *Frontend*: Streamlit-based interface for ease of deployment and student accessibility

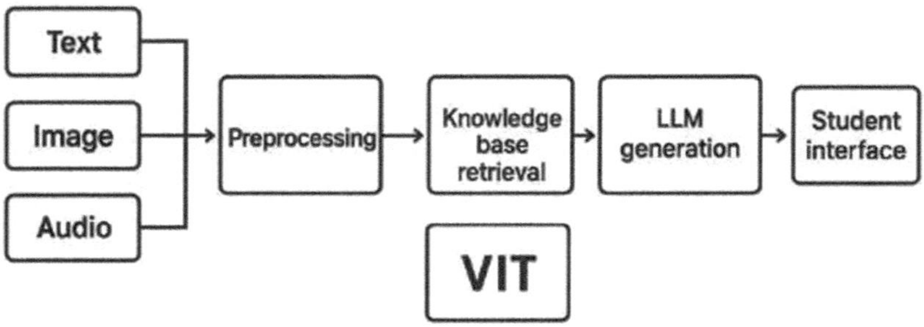

FIGURE 2.6 Architecture of the VIT student query chatbot system.

- *Backend*: Python-based processing with LlamaIndex for query understanding and retrieval
- *Data Sources*: University regulations, academic calendar and departmental guidelines
- *Pipeline*:
  - Student submits query (text input, potential voice/image in future)
  - Query processed by an NLP (natural language processing) engine
  - Context retrieval from knowledge base
  - LLM generates response with references
- *Performance*: Low average latency and high query resolution rate (>85 percent) in pilot tests.

System architecture of the VIT Student Query Chatbot showing multimodal input processing, knowledge-based retrieval, and LLM response generation is shown in Figure 2.6. Screenshots demonstrating the chatbot interface and implementation are presented in Figure 2.7.

2.4.2.4 Implementation Results and Observations

The VIT Student Query Chatbot was tested in a pilot deployment with real academic and administrative questions. It showed strong contextual relevance in responses and internal testing reported a query resolution accuracy above 85 percent. Response times are quick, suitable for real-time use cases.

Feedback, in qualitative terms, showed that there was easy access to academic information and that the need for staff to answer routine inquiries was reduced. The modular design of the system also opens up possibilities of future development, such as adding multimodal input, such as images and voice-based queries.

Even though this evaluation is qualitative and focused on application, the results support the feasibility and effectiveness of LLM-driven multimodal conversational systems in a real-world educational setting.

System Evaluation Summary of the VIT Student Query Chatbot is shown in Table 2.3.

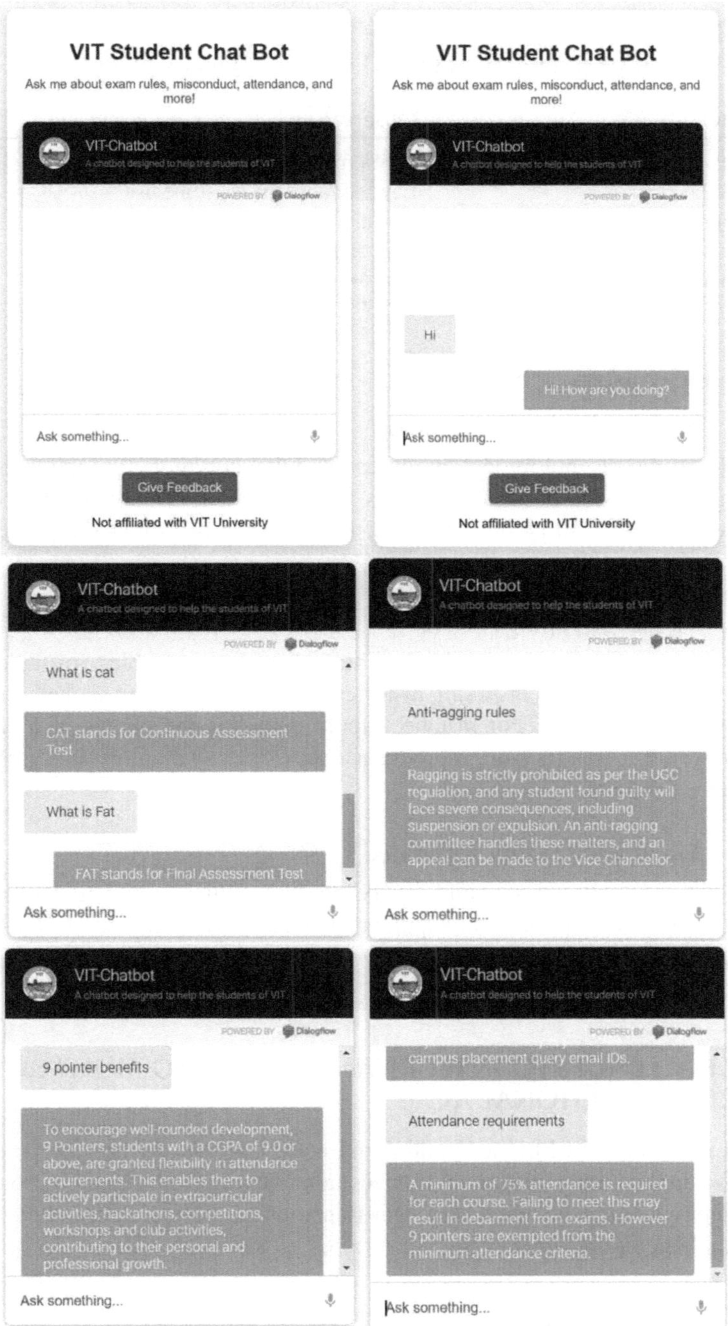

FIGURE 2.7 Screenshots of the VIT student query chatbot interface.

TABLE 2.3
System Evaluation Summary of the VIT Student Query Chatbot

Parameter	Observation
Query resolution accuracy	>85% (pilot evaluation)
Response latency	Low, suitable for real-time use
Scalability	Modular and extensible
Multimodal readiness	Supports future image and audio inputs
User experience	Improved accessibility and usability

2.4.3 E-Commerce

2.4.3.1 Multimodal Product Search

The submitting of multimodal inquiries is quite common through e-commerce websites—submission of product images, typing of queries and voice-based queries can take place. A multimodal LLM can integrate these forms to improve product recommendations [32,33].

2.4.3.2 Virtual Shopping Assistants

Such conversational agents embedded within online shops are able to provide context-aware responses based on browsing histories, product images and spoken preferences from the customers [34]. For instance, upon uploading the photos of a clutch, the assistant would search for similar products, check stock availability and suggest complementary products.

2.4.3.3 Fraud and Quality Assurance

Multimodal AI techniques would verify product authenticity based on provided images, textual descriptions and supplier metadata [35]. Likewise, it could act as a warning system for any kind of discrepancies between advertising claims and actual product characteristics.

2.5 CHALLENGES TO INTEGRATING LLMS INTO MULTIMODAL CONVERSATIONAL AI

Though integrating LLMs into multimodal conversational AI has the potential for disruptive applications, this modality faces transitioning from research prototypes to scaled deployment in the real-world. This transition has several interrelated technical, ethical and operational challenges.

2.5.1 Technical Challenges

2.5.1.1 Data Alignment and Fusion

Data must be captured from respective modalities in a properly sequenced and semantically aligned manner to ensure coherent multimodal processing [36]. Multimodal data misalignment can come from a mere temporal mismatch: audio trying to confirm an observed action or event; or quality differences affecting data capture. Such misalignment by modality can adversely affect the performance quality of the multimodal AI systems. For instance, the spoken symptom in a natural language processing system must remain timestamped correctly, matching the time at which the corresponding symptom is confirmed in the MRI; otherwise, the system might interpret the situation erroneously due to multimodal misalignment [37].

2.5.1.2 Computational Complexity

In essence, training of large multimodal LLMs, as ultimately desired in this profession, cannot be accomplished by a little power giveaway; state-of-the-art graphical processing units (GPUs) or tensor processing unit (TPUs) with sophisticated distributed computing framework, data parallelism and data-parallel optimization is commonly required for scalable multimodal training environments [38]. Designed for scale, computing demand and consumption can render many real-time applications, particularly in low-bandwidth settings or on mobile devices.

2.5.1.3 Latency in Real-Time Systems

Multimodal data processing in real time demands performance that can handle rapid feature extraction, fusion and generation. A real-time multimodal pipeline must be assembled to ensure minimum latency without sacrificing accuracy in time-sensitive systems [39]. Latency could constitute life-and-death situations in time-dependent decision-making involving autonomous vehicles or emergency medical care.

2.5.1.4 Generalizing Across Domains

While multimodal LLMs perform admirably on datasets close to those used for training, they do not fare so well with out-of-distribution data [40]. For instance, a shopping assistant mostly trained on English data could fail to make appropriate suggestions when the product descriptions are multilingual and the attributes present variation according to regions.

2.5.2 Ethical and Social Challenges

2.5.2.1 Bias in Multimodal Models

The bias in one modality may be compounded when combining data from different modalities. For instance, a hiring chatbot might be inappropriately factoring in background by considering both text resumes and video interviewing, to give biased recommendations on the basis of discrepancies in the sampling of the training data [41]. Fairness across different modalities entails good data collection and bias feedback.

2.5.2.2 Privacy and Security Issues

Most of the multimodal inputs contain sensitive personal information, including biometric data extracted from images or audio [42]. Protecting this kind of data will require encryption of the content, safe storage and compliance with data protection laws like General Data Protection Regulation (GDPR) and Health Insurance Portability and Accountability Act (HIPAA).

2.5.2.3 Explainability and Transparency

Being black-box models, such massive multimodal ones suffer from explainability issues; meanwhile, users have increasingly become adamant about explanations on how an AI arrived at its decision, especially in high-stake contexts [43]. Without explanations and with no transparency, the trust of the elites shall be placed in jeopardy.

2.5.3 Operational Challenges

2.5.3.1 Dataset Availability and Licensing

While there are indeed hundreds of datasets containing only text to train LLMs, the multimodal varieties are hard to come by—aligned, good quality data is harder to find than its counterpart—and far more restrictively licensed [44].

2.5.3.2 Integration with Legacy Systems

In many industries, AI implementations need to connect with legacy IT systems that were never contemplated for multimodal data processing [45]. Retrofit solutions for these systems tend to be prohibitively expensive and technically complicated.

2.5.3.3 Maintenance and Continuous Learning

Multimodal large language models must be updated from time to time with changes in language use, cultural contexts and domain expertise [46]. Continuous learning pipelines can cause catastrophic forgetting if not carefully monitored, where the model forgets important knowledge acquired before its retraining.

Challenges in Multimodal Large Language Model Deployment are shown in Table 2.4.

2.6 ETHICAL CONSIDERATIONS AND DATA SECURITY IN MULTIMODAL LLMS

Ethical dilemma enters the picture and data security requirements may arise for a technical capability since the multimodal LLMs will expand into sensitive domains such as health care, finance or education. Improper handling of ethical and security

TABLE 2.4
Challenges in Multimodal Large Language Model Deployment

Technical	Ethical	Operational
Latency	Bias	Integration Costs
Scalability	Fairness	Maintenance Effort
Compute Cost	Privacy	Deployment Complexity

situations may render substantial damage to users, loss of trust and regulatory penalties.

2.6.1 Fairness and Bias Mitigation

2.6.1.1 Bias Across Modalities

LLMs are susceptible to bias. Our real world of sight, hearing and so forth also can encode systemic prejudice [47]. On fusion of these modalities, unpredictable bias interactions could intensify discrimination in decision-making systems [48].

2.6.1.2 Bias Detection and Correction Techniques

- *Dataset auditing*: Systematic assessment of multimodal datasets for representation imbalances prior to training.
- *Counterfactual data augmentation*: Injection of synthetic yet balanced examples to mitigate skewed correlations [49].
- *Fairness metrics*: Measures such as demographic parity and equalized odds to measure fairness.

2.6.2 Privacy and Regulatory Compliance

2.6.2.1 Sensitive Multimodal Data

Multimodal input can contain biometric identifiers such as facial images, voiceprints and handwriting samples [50]. Unlike text on its own, the presence of these modalities can be an identified even if explicit personally identifiable information is absent.

2.6.2.2 Data Protection Laws

Systems must comply with regulations such as these:

- GDPR (Europe): Grants the user rights to access, rectify and erase personal data.
- HIPAA (USA): Governs the use of protected health information (PHI) in healthcare applications.
- DPDP Act (India): A new Indian Digital Personal Data Protection law mandating consent as a prerequisite to processing personal data securely.

2.6.3 Security Threats to Multimodal LLMs

2.6.3.1 Adversarial Attacks

Attackers can craft malicious multimodal inputs to cause incorrect outputs [51]. For instance, slightly altering an image of a stop sign can lead a self-driving AI to misinterpret it as a speed limit sign.

2.6.3.2 Data Poisoning

An adversary can inject manipulated data into the training set, embedding harmful biases or backdoor triggers [52].

2.6.3.3 Model Inversion and Membership Inference

These attacks can extract sensitive training data from the model's parameters, potentially revealing private user information [53].

2.6.4 Case Studies

2.6.4.1 Case Study 1: IBM Watson Health

IBM Watson has been condemned for issuing unsafe cancer treatment recommendations in a few cases because of biased and incomplete training data [54]. In this regard, it supports the argument that there exist some ethical risks when multimodal medical systems use imperfect datasets.

2.6.4.2 Case Study 2: Microsoft Tay Chatbot

Microsoft's Tay, a Twitter-based chatbot, started churning out offensive content just hours after launch, having been exposed to toxic multimodal interactions, courtesy the users [55]. This particular situation reiterates the need for the institutions to enforce content filtering and to develop real-time monitoring systems in all public-facing conversational agents.

2.6.5 Governance and Ethical AI Frameworks

On the topic of responsible AI, various organizations and policy groups put forward some guidelines:

- *OECD (Organisation for Economic Co-operation and Development) AI Principles*: Inclusive growth; human-centered values; transparency.
- *National Institute of Standards and Technology (NIST) AI risk management framework*: Provides a structured approach to identifying and mitigating risks associated with AI [56].
- *Ethics by design*: Embedding ethical safeguards directly into the very core of model architecture as well as into training pipelines and deployment strategies.

Figure 2.8 illustrates the interrelationship between ethical considerations, privacy, security, and governance requirements in multimodal LLM systems.

2.7 FUTURE DIRECTIONS AND CONCLUSION

2.7.1 Future Directions in Multimodal LLMs

2.7.1.1 Real-Time Multimodal Reasoning

At present, multimodal LLMs can process multimodal inputs all the way until true real-time reasoning—that is, where the system is constantly revising its

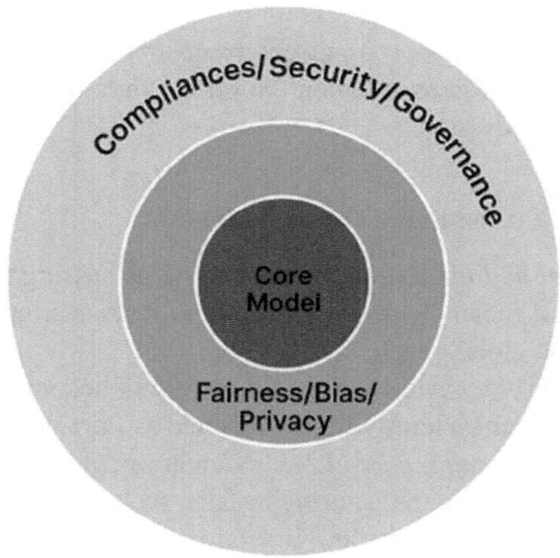

FIGURE 2.8 Ethical and security considerations framework for multimodal LLM deployment.

understanding or expectations being thrown at it in new data as it comes in—can only be said to be at the nascence [57]. It is conceivable that future works will focus on streaming architectures handling asynchronous multimodal signals without a hit on performance.

2.7.1.2 Edge Deployment and On-Device Processing

There is a growing trend toward deploying large language models on edge devices to support privacy-preserving and low-latency responses [58]. Running multimodal AI locally will lessen reliance on cloud servers, thereby boosting privacy and responsiveness. Model quantization and knowledge distillation approaches will have to be considered in making large models mobile-friendly.

2.7.1.3 Self-Supervised Multimodal Learning

The biggest limitation of multimodal datasets is their paucity and the associated cost of labeling them. Self-supervised learning techniques operating on huge amounts of unlabeled multimodal data are expected to predominate in the years to come [59].

2.7.1.4 Multimodal Reinforcement Learning

Reinforcement learning with multimodal input could enable conversational agents to learn strategies for furnishing responses by interacting with the environment. A multimodal virtual assistant in the smart home may learn from voice commands and visual cues to anticipate the needs of the users [60].

2.7.1.5 Integration with AR/VR

AR/VR environments are an appropriate environment for multimodal LLM. They can handle voice, gesture and gaze at data all at once, while offering highly immersive, adaptive user experiences [61].

2.7.2 Research and Industry Trend Forecasts

- *Cross-lingual multimodality*: The new systems passing through development should be able to handle multimodal inputs across various languages, thereby increasing the global reach.
- *Ethics-first development*: Having ethical AI will become a licensure of a good brand name, their systems will be marketed as fair and transparent.
- *Domain-specific multimodal LLMs*: Domain-oriented models will blossom for laws, medicine and manufacturing besides such fields in which they are relevant to shoot beyond the shallow general-purpose LLMs in niche domains [62].
- *Interactive knowledge graphs*: Multimodal LLMs will dynamically create and update structured knowledge graphs out of live input, enabling more elaborate reasoning capabilities [63].

2.8 CONCLUSION

The combination of large language models with multimodal conversational AI marks a watershed moment in human–computer interaction. Multimodal conversational AI can produce more elaborate, context-aware and personalized responses with the help of text, image, audio and other kinds of data.

This chapter has provided an overview of LLMs and has discussed their developments, their architecture, specific usages, operational concerns and ethical issues. Real-world instances of use in areas ranging from decision support in healthcare, to personalization in education and maximization of customer experience in e-commerce set the stage for possible new horizons for multimodal AI.

As technologists make strides in developing new generations of multimodal LLMs, the time is ripe for both the technical community and the end users to weigh equally on issues of bias, privacy and security, as they do on matters of technical performance. The newer generation of multimodal LLMs will witness the growing demand to demonstrate not just technical performance but also value alignment, adaptability in different contexts and responsible scaling.

Entering an era where conversation is not limited to just words gives multimodal LLMs the chance to reshape learning and shopping.

ACKNOWLEDGMENTS

The authors would like to express their sincere gratitude to the School of Computer Science and Engineering, VIT University, Vellore, for providing the academic environment and resources that supported this work. The authors also thank the reviewers for their constructive feedback, which helped improve the quality and clarity of this chapter.

REFERENCES

[1] J. Weizenbaum, "ELIZA—a computer program for the study of natural language communication between man and machine," *Communications of the ACM*, vol. 9, no. 1, pp. 36–45, 1966.

[2] A. M. Turing, "Computing machinery and intelligence," *Mind*, vol. LIX, no. 236, pp. 433–460, 1950.

[3] D. Jurafsky and J. H. Martin, *Speech and Language Processing*, 3rd ed. Hoboken, NJ: Pearson, 2023.

[4] A. Vaswani et al., "Attention is all you need," in *Advances in Neural Information Processing Systems (NeurIPS)*, Long Beach, CA, USA, Dec. 2017, pp. 5998–6008.

[5] T. Brown et al., "Language models are few-shot learners," in *Proc. 34th Conf. Neural Information Processing Systems (NeurIPS)*, Vancouver, BC, Canada, Dec. 2020, pp. 1877–1901.

[6] O. Vinyals and Q. V. Le, "A neural conversational model," *arXiv preprint arXiv:1506.05869*, 2015.

[7] S. Young et al., "POMDP-based statistical spoken dialog systems: A review," *Proceedings of the IEEE*, vol. 101, no. 5, pp. 1160–1179, 2013.

[8] M. McGurk and J. MacDonald, "Hearing lips and seeing voices," *Nature*, vol. 264, pp. 746–748, 1976.

[9] A. Radford et al., "Learning transferable visual models from natural language supervision," in *Proc. Int. Conf. Machine Learning (ICML)*, Vienna, Austria, 2021, pp. 8748–8763.

[10] Y. Li et al., "BLIP-2: Bootstrapping language-image pre-training with frozen image encoders and large language models," *arXiv preprint arXiv:2301.12597*, 2023.

[11] C. D. Manning et al., *Foundations of Statistical Natural Language Processing*. Cambridge, MA: MIT Press, 1999.

[12] K. Shridhar et al., "Multimodal few-shot learning with frozen language models," in *Proc. IEEE Conf. Computer Vision and Pattern Recognition (CVPR)*, Vancouver, BC, Canada, 2023, pp. 5226–5235.

[13] K. Colby, "Modeling a paranoid mind," *Behavioral and Brain Sciences*, vol. 4, no. 4, pp. 515–560, 1981.

[14] S. F. Chen and J. Goodman, "An empirical study of smoothing techniques for language modeling," in *Proc. 34th Annu. Meeting of the Association for Computational Linguistics (ACL)*, Santa Cruz, CA, USA, 1996, pp. 310–318.

[15] S. Hochreiter and J. Schmidhuber, "Long short-term memory," *Neural Computation*, vol. 9, no. 8, pp. 1735–1780, 1997.

[16] I. Sutskever, O. Vinyals, and Q. V. Le, "Sequence to sequence learning with neural networks," in *Advances in Neural Information Processing Systems (NeurIPS)*, Montreal, QC, Canada, 2014, pp. 3104–3112.

[17] J. Devlin, M. Chang, K. Lee, and K. Toutanova, "BERT: Pre-training of deep bidirectional transformers for language understanding," in *Proc. NAACL-HLT*, Minneapolis, MN, 2019, pp. 4171–4186.

[18] OpenAI, "GPT-4 technical report," *arXiv preprint arXiv:2303.08774*, 2023.

[19] P. Atrey, M. Hossain, A. El Saddik, and M. Kankanhalli, "Multimodal fusion for multimedia analysis: A survey," *Multimedia Systems*, vol. 16, no. 6, pp. 345–379, 2010.

[20] M. Snoek and M. Worring, "Multimodal video indexing: A review of the state-of-the-art," *Multimedia Tools and Applications*, vol. 25, no. 1, pp. 5–35, 2005.

[21] T. Baltrušaitis, C. Ahuja, and L.-P. Morency, "Multimodal machine learning: A survey and taxonomy," *IEEE Transactions on Pattern Analysis and Machine Intelligence*, vol. 41, no. 2, pp. 423–443, 2019.
[22] Z. Yang et al., "Unified multimodal transformers for cross-modal retrieval," in *Proc. 35th AAAI Conf. Artificial Intelligence (AAAI)*, Vancouver, BC, Canada, 2021, pp. 116–124.
[23] X. Li et al., "VisualBERT: A simple and performant baseline for vision-and-language tasks," *arXiv preprint arXiv:1908.03557*, 2019.
[24] H. Tan and M. Bansal, "LXMERT: Learning cross-modality encoder representations from transformers," in *Proc. 2019 Conf. Empirical Methods in Natural Language Processing (EMNLP)*, Hong Kong, China, 2019, pp. 5099–5110.
[25] R. Xu, C. Xiong, W. Wang, and J. Shen, "End-to-end cross-modal retrieval with aligned multimodal embeddings," *IEEE Transactions on Multimedia*, vol. 23, pp. 3150–3163, 2021.
[26] H. Chen, H. Ha, M. Kodialam, and R. Khanna, "Multimodal machine learning for medical diagnosis," *IEEE Journal of Biomedical and Health Informatics*, vol. 27, no. 1, pp. 19–31, 2023.
[27] A. Rajpurkar, E. Chen, T. Banerjee, and M. Lungren, "AI in radiology: Opportunities, challenges, and future directions," *Nature Medicine*, vol. 28, pp. 1527–1534, 2022.
[28] K. Bedi, R. Batra, and P. N. Bennett, "Telehealth and AI: Integrating multimodal data for remote diagnosis," *IEEE Access*, vol. 11, pp. 45987–45999, 2023.
[29] M. Zawacki-Richter, V. Marín, M. Bond, and F. Gouverneur, "Systematic review of research on artificial intelligence applications in higher education," *International Journal of Educational Technology in Higher Education*, vol. 16, no. 1, pp. 1–27, 2019.
[30] J. Holmes and W. Porayska-Pomsta, "AI in education: A focus on learning and teaching," *British Journal of Educational Technology*, vol. 54, no. 1, pp. 5–21, 2023.
[31] A. Al-Azawei and H. Al-Masri, "Assistive technology for students with disabilities in higher education: A review," *Education and Information Technologies*, vol. 28, pp. 2483–2508, 2023.
[32] Y. Li, L. Wu, and W. Wang, "Visual and textual product search with multimodal transformers," in *Proc. 44th Int. ACM SIGIR Conf. Research and Development in Information Retrieval*, 2021, pp. 256–265.
[33] C. Chen, S. Xie, and M. Sun, "Multimodal search for e-commerce using deep learning," *arXiv preprint arXiv:2105.06049*, 2021.
[34] R. McAuley, C. Targett, J. Shi, and A. van den Hengel, "Image-based recommendations on styles and substitutes," in *Proc. 38th Int. ACM SIGIR Conf.*, Santiago, Chile, 2015, pp. 43–52.
[35] A. Pathak, M. Sethi, and S. Agarwal, "Detecting counterfeit products in e-commerce using multimodal learning," *IEEE Access*, vol. 10, pp. 15798–15809, 2022.
[36] Y. Bengio, "Deep learning of representations for unsupervised and transfer learning," in *Proc. ICML Workshop Unsupervised and Transfer Learning* , Bellevue, WA, 2012, pp. 17–36.
[37] P. Zhou, Z. Qiu, Q. Hu, and R. Ji, "End-to-end multimodal alignment learning," *IEEE Transactions on Multimedia*, vol. 23, pp. 4125–4137, 2021.
[38] A. Dean, J. Isard, and L. van der Maaten, "Efficient training of visual transformers," *arXiv preprint arXiv:2106.10270*, 2021.

[39] C. Ma, W. Wang, and H. Li, "Real-time multimodal AI for safety-critical applications," in *Proc. IEEE Int. Conf. Robotics and Automation (ICRA)*, Philadelphia, PA, 2022, pp. 5453–5460.
[40] J. Miller, S. Kornblith, and L. Zitnick, "Measuring generalization in multimodal models," in *Proc. NeurIPS*, New Orleans, LA, 2022, pp. 3324–3335.
[41] M. Buolamwini and T. Gebru, "Gender shades: Intersectional accuracy disparities in commercial gender classification," in *Proc. Conf. Fairness, Accountability, and Transparency (FAT)*, New York, NY, 2018, pp. 77–91.
[42] A. O'Neill, "Ethics and privacy in multimodal biometric systems," *Pattern Recognition Letters*, vol. 142, pp. 1–7, 2021.
[43] Z. Linardatos, V. Papastefanopoulos, and S. Kotsiantis, "Explainable AI: A review of machine learning interpretability methods," *Entropy*, vol. 23, no. 1, p. 18, 2021.
[44] T. Baltrušaitis, C. Ahuja, and L.-P. Morency, "Multimodal machine learning datasets: A survey," *arXiv preprint arXiv:2209.03430*, 2022.
[45] E. Brynjolfsson and A. McAfee, *The Second Machine Age*, New York, NY: W. W. Norton & Company, 2014.
[46] J. Kirkpatrick et al., "Overcoming catastrophic forgetting in neural networks," *Proc. National Academy of Sciences (PNAS)*, vol. 114, no. 13, pp. 3521–3526, 2017.
[47] S. Barocas, M. Hardt, and A. Narayanan, *Fairness and Machine Learning*, Cambridge, MA: MIT Press, 2023.
[48] M. Mehrabi, F. Morstatter, N. Saxena, K. Lerman, and A. Galstyan, "A survey on bias and fairness in machine learning," *ACM Computing Surveys*, vol. 54, no. 6, pp. 1–35, 2021.
[49] J. Zhao, T. Wang, M. Yatskar, V. Ordonez, and K. Chang, "Gender bias in coreference resolution: Evaluation and debiasing methods," in *Proc. NAACL-HLT*, New Orleans, LA, 2018, pp. 15–20.
[50] J. Fierrez, A. Morales, R. Vera-Rodriguez, and D. Camacho, "Multiple classifiers in biometrics. Part 2: Trends and challenges," *Information Fusion*, vol. 44, pp. 103–112, 2018.
[51] N. Carlini and D. Wagner, "Adversarial examples in machine learning," in *Proc. IEEE Symp. Security and Privacy*, San Jose, CA, 2017, pp. 39–57.
[52] B. Biggio, B. Nelson, and P. Laskov, "Poisoning attacks against support vector machines," in *Proc. 29th Int. Conf. Machine Learning (ICML)*, Edinburgh, UK, 2012, pp. 1467–1474.
[53] R. Shokri, M. Stronati, C. Song, and V. Shmatikov, "Membership inference attacks against machine learning models," in *Proc. IEEE Symp. Security and Privacy*, San Jose, CA, 2017, pp. 3–18.
[54] C. Ross and I. Swetlitz, "IBM pitched its Watson supercomputer as a revolution in cancer care. It's nowhere close," *STAT News*, July 2018.
[55] P. Domingos, "The Microsoft Tay chatbot debacle," *Communications of the ACM*, vol. 60, no. 10, p. 7, 2017.
[56] National Institute of Standards and Technology (NIST), "AI Risk Management Framework," Jan. 2023.
[57] A. Zadeh, M. Chen, S. Poria, E. Cambria, and L.-P. Morency, "Tensor fusion network for multimodal sentiment analysis," in *Proc. Empirical Methods in Natural Language Processing (EMNLP)*, Copenhagen, Denmark, 2017, pp. 1103–1114.
[58] H. Wang, Y. Zhang, and D. Chen, "Tiny-LLM: Efficient inference of large language models on edge devices," *IEEE Internet of Things Journal*, vol. 10, no. 8, pp. 6723–6734, Apr. 2023.

[59] Y. Lu, J. Li, and H. Xu, "Multimodal self-supervised learning: A survey," IEEE Transactions on Pattern Analysis and Machine Intelligence, early access, 2023.
[60] S. Singh, M. R. James, and A. R. Silver, "Reinforcement learning with human feedback for multimodal AI," in *Proc. NeurIPS*, New Orleans, LA, 2022, pp. 13824–13836.
[61] J. Lanier and F. Biocca, "An insider's view of the future of virtual reality," *Journal of Communication*, vol. 70, no. 5, pp. 629–647, 2020.
[62] A. Chowdhery et al., "PaLM: Scaling language models to 540 billion parameters," *arXiv preprint arXiv:2204.02311*, 2022.
[63] X. Han, Z. Liu, and M. Sun, "OpenKE: An open toolkit for knowledge embedding," in *Proc. EMNLP*, Brussels, Belgium, 2018, pp. 139–144.

3 Architectural Advances in Multimodal Conversational AI Using Large Language Models

N. Deepika, D. Karthika Renuka, and L. Ashok Kumar

3.1 INTRODUCTION

Conversational artificial intelligence is a branch of artificial intelligence that has experienced impressive changes in the last few decades. It used to be rule-based and primitive, but nowadays it is exceedingly hi-tech, neural, and capable of supporting one-on-one talks with humans. With the massive breakthroughs in deep learning technologies and artificial intelligence, the current conversational AI is no longer limited to single-modal interfaces, including natural language and voice. Instead, we now have Multimodal conversational AI systems whose objective is to build a simulated category of natural human conversation through the consideration of multiple interface modalities [6].

Large language models (LLMs) have already become a component of this development and they are operating as powerful reasoning systems to enable contextual learning, dialogue interchangeability, and generative intelligence. Combined with different types of multimodal sets perception systems, LLMs have become the base of multimodal large language models (MLLM) with the power to provide interactive and contextual systems of dialogue. This chapter demonstrates the latest developments in the architecture on how the LLMs will support multimodal reasoning and improved modes of fusion and alignment mechanisms [13].

3.1.1 Evolution of Conversational AI

The evolution of the conversational AI follows a broad categorization, creating a timeline of the major evolutional phases. For instance, the earlier AI models, like ELIZA or PARRY, were developed using rule-based techniques, which were incapable of offering a variety of conversations [12,14] . These models were script-based.

The second phase introduced statistical and machine learning-based approaches, including n-gram models and hidden Markov models, which facilitated statistical

DOI: 10.1201/9781003678359-3

dialogue management that was robust in nature. Nevertheless, the approaches faced limitations in handling long contexts as well as in semantic understanding [6].

The third phase represented the beginning of deep learning techniques, especially those involving recurrent neural networks, long short-term memory networks, and sequence-to-sequence models [12]. Although there were some inherent limitations in the use of recurrent neural networks (RNNs) and long short-term memory (LSTM) networks they were instrumental in enhancing the efficiency of language generation and response flow.

The current phase is dominated by transformer-based architectures and large language models, such as generative pre-trained transformer (GPT), bidirectional encoder representations from transformers (BERT), and their successors. These models leverage self-attention mechanisms to capture long-range dependencies and contextual nuances at scale. More recently, the integration of vision, speech, and other modalities into transformer-based frameworks has led to the development of multimodal conversational systems, capable of perceiving and reasoning beyond textual input alone [5].

3.1.2 Need for Multimodal Interaction

Human communication is naturally multimodal in nature and utilizes the simultaneous delivery of language, facial expressions, gesture language, tone of voice, and context in vision. Text-based communicational interfaces are completely unable to replicate or understand this complexity and often result in ambiguous and contextually inappropriate responses [19]. For example, a sentence might have completely different connotations depending on whether it is said with certain facial expressions or gestures.

Multimodal interaction helps agents accomplish the following:

- Disambiguating the user intention with supplementary signals
- Ground language in visual and auditory context
- Respond more naturally and empathetically
- Support complex tasks in the world, like navigation, education, health, and human–robot interaction.

With the growth in the availability of multimodal data and the achievement of significant progress in perception models, there are possibilities to design a system that processes such data [15]. Multimodal conversational AI is now an emerging and important research area to develop intelligent and human-centric AI systems.

3.1.3 Role of LLMs

Large language models are at the core of contemporary multimodal-based human–computer dialogue architectures. By having been fine-tuned on incredibly large amounts of text, emergent properties such as reasoning, abstract thinking, summarizing, as well as managing contexts in a dialogue, are realized in large-scale models such as LLMs. Multimodal dialogue architectures employ LLMs as central unifying components [10], fusing information from modality-specific components, while also responding in a context-dependent way.

Rather than directly receiving unprocessed multimodal input, LLMs have been observed to primarily operate on aligned embedding data developed through various forms of visual, auditory, or bodily encoders [16]. This structure permits LLMs to:

- Perform cross-modal reasoning
- Maintain dialogue state and long-term context
- Develop explanations that are based on perception
- Adapt responses based on user intent and emotional cues.

The emergence of MLLMs such as GPT-4V, Gemini, Flamingo, and LLaVA demonstrates the effectiveness of combining pretrained LLMs with multimodal alignment techniques, marking a paradigm shift in conversational AI architecture [13].

3.1.4 Limitations of Unimodal Systems

However, unimodal communication systems, regardless of their success, also face a set of inherent challenges. For example, even though the text-only communication model is a success, the fact remains that the model has never been grounded in the world, which leads to hallucination [14,17]. There is also an issue of ambiguity, as noted for the speech-only model. For the vision-only model, there is a lack of semantics.

These limitations manifest in:

- Poor situational awareness
- Reduced robustness in actual environments
- Inability to interpret non-verbal cues
- Limited emotional intelligence.

This is because unimodal architectures are not easily amenable to complex tasks that require coordinated or joint reasoning about the information sources. This reveals some of the inherent problems with unimodal architectures that necessitate a shift toward multimodal architectures.

3.1.5 Scope of Multimodal Conversational AI

Multimodal conversational AI combines natural language processing, image processing, and speech recognition, associated with cognitive reasoning. Intelligent assistance, healthcare systems, smart learning, customer support services, cybersecurity, and robotics are the range and broad areas that are related to MCAI [3,9].

The issues discussed in this chapter are as follows:

- Multimodal conversational systems architecture
- Fusion and matching strategy of multi-input modes

- The process of multimodal thinking and dialogue generation in LLMs
- Problems of evaluation and future research.

Through the architectural breakthroughs of the multimodal architectures of the LLM, this chapter hopes to provide a comprehensive perspective to researchers and practitioners on the mechanism of the next generation of conversational AI technologies [11].

3.2 FOUNDATIONS OF MULTIMODAL CONVERSATIONAL AI

Multimodal conversational artificial intelligence (MCAI) is an improvement in the architecture of intelligent interactive systems that advances a single-mode processing approach to an integrated sense, reasoning, and dialogue generation. In its essence, MCAI aims at modeling human-like communication through the combination of linguistic, visual, auditory, and contextual cues. This part defines the conceptual and theoretical bases of multimodal conversational AI and highlights the role of LLMs as the main reasoning machines and the principles of the architecture underlying contemporary multimodal systems.

3.2.1 Definition and Characteristics

Multimodal conversational AI is called as a part of neural systems that is capable of conversing depending on input of multiple modalities, by means of input processing and management of reasoning and production of answers [17]. MCAI systems integrate heterogeneous data streams such as images, videos, audio cues, and sensor data to provide grounded and context-aware interaction as opposed to traditional conversational agents that merely responded to text or speech.

The key characteristics of MCAI are:

- Multisensory perception: the system has multimodal perception.
- Cross-modal reasoning in which information in one modality may be exploited to inform the interpretation of information in another modality.
- Contextual dialogue management: maintenance of coherence between turns and modalities.
- Production of adaptive response with regard to user, environmental, and emotion condition objectives.

These capabilities render MCAI an unmatched technology relative to unimodal chatbots, and it is one of the technologies in the future of human–AI interaction.

3.2.2 Importance of Multimodal Reasoning

The key aspect of successful conversational intelligence lies in multimodal reasoning. Human cognition scarcely depends on one of the modalities, rather the meaning is created as a result of the interaction between language, sight, sound, and context [18].

To provide an example, to interpret a request of a user, such as what is wrong here, one will be in need of visual grounding, whereas to interpret emotional intent, it can be based on tone or facial expression, and not on the lexical content.

From an architectural perspective, multimodal reasoning involves:

- Aligning representations across modalities
- Resolving ambiguities using complementary signals
- Inferring latent intent and context beyond explicit input.

LLM-driven multimodal systems excel at such reasoning by leveraging attention mechanisms and large-scale pretraining to model complex dependencies across representations [10]. This enables conversational agents to perform tasks such as visual question answering, grounded dialogue, and situational awareness in real-world environments.

3.2.3 Large Language Models as Central Reasoning Engines

The large-scale languages have turned out to be the key to the brain of the new multimodal dialogues. Pretrained on large text corpora, LLM possesses good reasoning, abstraction, planning, and natural language generation capabilities. The LLMs, as a component of multimodal architectures, are modality-agnostic reasoning modules, which can be input on aligned representations (not raw sensory data) [16].

Rather than substituting the models of modality, LLMs complement them by:

- Semantic information between modalities
- Pursuing dialogue state and long-term context
- Producing grounded, coherent and explainable responses.

This architectural division perception managed by special encoders and reasoning by an LLM scales better and is also more modular. The appearance of MLLMs indicates that LLMs can be used as integrating tools to make conversational intelligence unifying, the perception and language in one system.

3.2.4 Modalities in Conversational AI

Contemporary multimodal conversational systems allow a wide variety of input and output modalities, each of which has a different informational value:

- Text: This is the main dealer of semantic and syntactic data.
- Speech: Provides prosody, emotion and talk timing.
- Vision (images/videos): Allows groundedness, spatial knowledge, and context.
- Audio (non-speech): Records context and emotional information.
- Physiological and sensor data: Special purpose applications include healthcare and human–robot interaction.

TABLE 3.1
Modalities in Conversational AI and Their Functional Roles

Modality	Input Type	Information Provided	Role in Dialogue
Text	Tokens	Semantics	Reasoning, intent
Speech	Audio	Emotion, timing	Natural interaction
Vision	Images/Video	Context, grounding	Situational awareness
Sensors	Signals	Physical state	Embodied AI

The challenge lies not in processing individual modalities, but in coherently integrating them into a unified conversational context. Table 3.1 illustrates the modalities in conversational AI and their functional roles. This integration is facilitated by embedding alignment techniques and attention-based fusion mechanisms discussed in later sections.

3.2.5 Architectural Principles of Multimodal Conversational AI

MCAI systems are developed in consideration of several architectural concepts:

1. *Modularity*: Separation of perception and reasoning components, as well as generation, is possible and allows flexibility and reuse.
2. *Representation alignment*: Modalities have to be projected to shared embedding spaces in order to facilitate joint reasoning.
3. *Scalability*: Architectures need to be capable of accommodating growing modalities, data volume and interaction volume.
4. *Context preservation*: Dialogue history and multimodal context should be preserved between turns.
5. *Human alignment*: Systems must conform to human values, emotional standards and ethical oppressions.

These are the principles that are used in the development of advanced multimodal conversational architectures [16].

3.3 ARCHITECTURAL COMPONENTS OF MULTIMODAL LLM SYSTEMS

The success of the multimodal conversational AI system is heavily reliant on the system architecture. Contemporary systems embrace a layered structure, which combines modality-related perception models with a central large language model, allowing to reason and generate responses end to end in a multimodal manner. As illustrated in Figure 3.1, the conceptual pipeline of multimodal conversational AI demonstrates how heterogeneous inputs such as text, speech, and vision are processed through dedicated encoders, aligned into a shared representation space, and subsequently reasoned upon by the LLM to generate coherent, context-aware responses

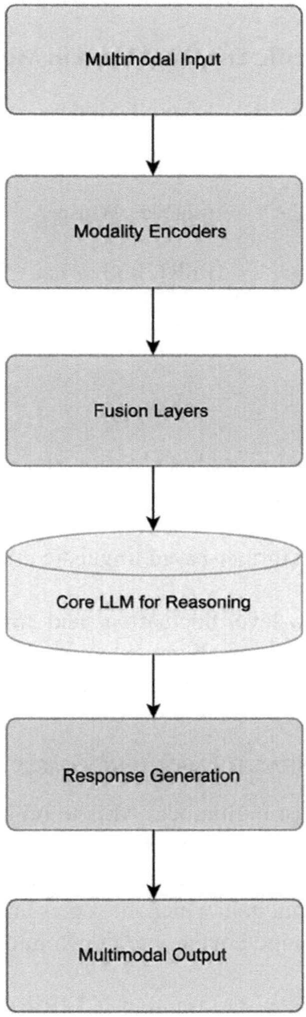

FIGURE 3.1 Conceptual pipeline of multimodal conversational AI.

[17]. This section presents a close analysis of basic architectural elements of multimodal LLM systems.

3.3.1 Modality-Specific Encoders

Modality-specific encoders are found at the front end of multimodal systems and are required to encode raw sensory inputs into high-level feature representations. They are often pretrained with massive modality-specific datasets and trained to achieve perceptual comprehension.

TABLE 3.2
Common Modality-Specific Encoders Used in Multimodal LLM Systems

Modality	Encoder Type	Example Models	Primary Function
Vision	CNN / ViT	ResNet, ViT, Swin	Object, scene, and spatial understanding
Audio	Speech Encoder	wav2vec, Whisper	Speech recognition, acoustic features
Text	Transformer Encoder	BERT, RoBERTa	Linguistic representation

Examples:

- CNN or Vision Transformer-based vision encoders of images and videos
- Speech and audio encoders: wav2vec and Whisper speech recognition and audio understanding
- Text encoders: Transformer-based linguistic processing encoders.

These encoders remove low-level fluctuations and error, and generate embeddings, which can be used in downstream alignment and reasoning [8]. Table 3.2 shows the common modality-specific encoders used in multimodal LLM systems.

3.3.2 Alignment and Shared Embedding Spaces

One of the main problems of multimodal AI is to bridge the gap between heterogeneous representations in a common semantic space. In the absence of alignment, it is impossible to perform cross-modal reasoning. The techniques of alignment typically make use of contrastive learning, in which the representations in one of the modalities representing the same semantic content are drawn into the embedding space toward one another.

Contrastive language image pre-training (CLIP)-style training motivations have been shown to be effective through models inspired by CLIP-style training [8]. Aligned embeddings are the layer between perception modules and the LLM in multimodal conversational systems and allow language-based reasoning over visual and auditory data.

3.3.3 Cross-Modal Attention Mechanisms

The cross-modal attention mechanisms allow the interaction of two modalities representations in a dynamic manner. In contrast to self-attention, which works in only one modality, cross-attention enables one modality to selectively attend to the applicable features in another. Cross-attention layers can be placed between multimodal LLM architecture modality encoders and the LLM itself [13]. These processes make it possible to perform finely grained reasoning, that is, determining the textual references with a particular area of an image, or intonation of speech with linguistic meaning.

3.3.4 Role of the LLM in Multimodal Reasoning

Within the architecture, the LLM functions as a global reasoning and dialogue management module. It integrates aligned multimodal embeddings, maintains conversational context, and generates responses conditioned on both perceptual input and dialogue history.

LLMs enable:

- Multimodal based chain-of-thought reasoning
- Elaboration and explanation of responses
- User-intention/user-affect adaptive dialogue strategies.

This position makes the LLM more than a language generator, it is a multimodal cognitive engine. Table 3.3 illustrates the core responsibilities of the LLM in multimodal conversational architectures.

3.3.5 Decoder Architectures for Response Generation

Response generation to code words is performed by decoders (usually those of the encoder) that vary in structure and complexity between different code word sets.

Response generation is the last phase of the architecture and is achieved by means of decoders. Text-based decoders have been used in most multimodal conversational systems, though speech and visual output is becoming more prevalent [18].

The design choices available to decoders are:

- Unified decoders which produce answers straight out of the LLM
- Decoders based on modality speech synthesis or visual generation
- Coded decoding to promote safety, have coherence and emotional relevance.

The decoder converts the results of high-level reasoning to user-facing responses, and the conversational loop is complete.

TABLE 3.3
Core Responsibilities of the LLM in Multimodal Conversational Architectures

Function	Description
Multimodal Reasoning	Integrates vision, audio, and text embeddings
Context Management	Maintains dialogue history
Chain-of-Thought	Enables step-by-step reasoning
Adaptive Dialogue	Adjusts tone and strategy based on user intent

3.4 FUSION STRATEGIES IN MULTIMODAL AI

Multimodal conversational AI systems depend on fusion strategies to define how heterogeneous modalities of information are integrated to aid in generating reasoning and responses. The success of a multimodal system is not only pegged on the excellence of modality-specific encoders but also on how the outputs are combined. Fusion strategies within the scope of the LLM-based architectures have developed beyond mere feature concatenation to more complex cross-modal reasoning via attention mechanism [7]. This section presents a systematic overview of fusion strategies used in multimodal AI, with a particular emphasis on their architectural implications for multimodal conversational systems.

3.4.1 Early Fusion

Early fusion, feature-level fusion, or early feature-level fusion fuses representations across modalities at an early point of the processing pipeline. In this method, encoders being modality-specific in nature are concatenated or merged and then sent to a downstream model where they undergo a reasoning or classification process.

The major benefit of early fusion is that it captures the low-level cross-modal correlations, which permits learning joint features in the first place. It is especially useful in applications where very fine-grained interactions among modalities are necessary, for example, in visual question answering or audio-visual speech recognition.

Early fusion, however, has a number of limitations [11]. Dimensionality, temporal resolution, and noise properties of modalities tend to be different and thus difficult to match. Another problem is that early fusion architectures are not flexible: modalities are generally difficult to add or to remove, and the system must be retrained. Early fusion is less frequently utilized in the LLM-based conversational systems because the integration of raw multimodal features into language models is quite challenging.

3.4.2 Late Fusion

Late fusion or decision-level fusion uses each modality separately and produces the results at a later point, before merging the results by weighted averaging, voting, or rule-based aggregation [15]. The approach enables every modality to play a leadership role in making a final decision or response.

Late fusion is more modular and robust with individual modality pipelines being developed, optimized, and updated individually. Also, it is less strict on missed or noisy modalities, making it fit well in the real-world conversational system in which the input may be incomplete.

In spite of these benefits, late fusion frequently is not effective to preserve deep cross-modal interactions. Contingent modalities are processed independently, which means that the system can overlook significant contextual dependencies. Consequently, late fusion cannot tend to be adequate in multimodal conversational challenges of complexity that require semantic grounding and contextual coherence.

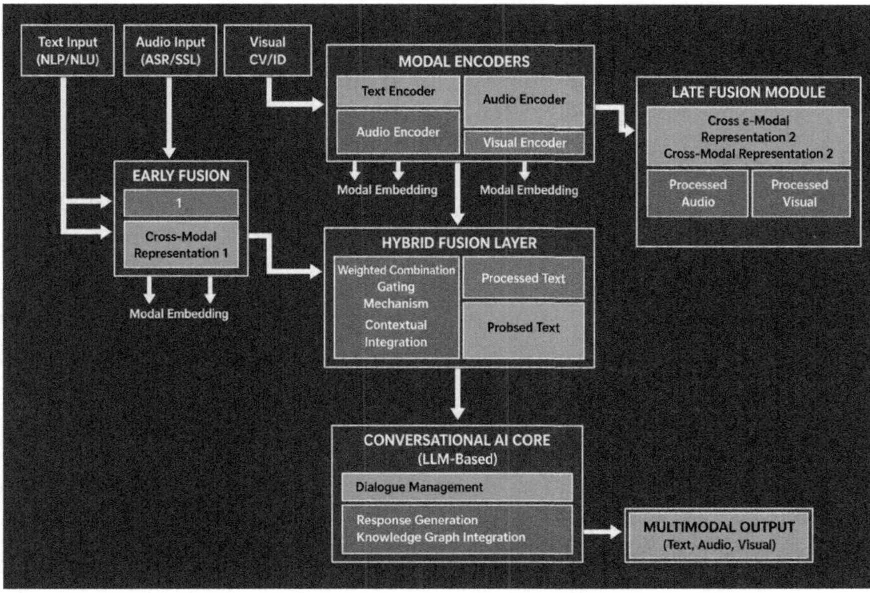

FIGURE 3.2 Hybrid fusion in multimodal conversational AI.

3.4.3 Hybrid and Hierarchical Fusion

Hybrid fusion techniques are attempts to unite the capabilities of early and late fusion, using modalities at several locations in the processing pipeline. The hierarchical fusion architectures inject fusion layers onto a variety of levels of abstraction, allowing the system to extract both low-level correlations and high-level semantic relationships. Hybrid fusion significantly works in multimodal conversational AI, in combination with LLMs. Early fusion can be done by lower-level encoders of closely related modalities (i.e., audio and visual speech cues) and higher-level semantic representations are fused using attention mechanisms of the LLM. As illustrated in Figure 3.2, hybrid fusion in multimodal conversational AI leverages early fusion among closely related modalities—such as audio and visual speech cues—while higher-level semantic representations are integrated using attention mechanisms within the LLM. This top-down style makes it possible to have scalable and customizable system design and makes it possible to add modalities incrementally without affecting the overall structure [16].

3.4.4 Attention-Based Fusion

The attention-based fusion is the most developed and adopted fusion strategy in contemporary multimodal systems. In place of explicit feature fusion, attention mechanisms enable the model to select information in a context-dependent manner by paying attention to only the relevant information across modalities. The fact that cross-modal attention allows one modality to listen to the features of another

TABLE 3.4
Comparative Analysis of Multimodal Fusion Strategies

Fusion Strategy	Flexibility	Reasoning Capability	LLM Suitability	Typical Use
Early Fusion	Low	High (local)	Poor	VQA, AVSR
Late Fusion	High	Low	Moderate	Robust systems
Hybrid	High	High	Very Good	Conversational AI
Attention-Based	Very High	Very High	Excellent	Modern MLLMs
Prompt-Based	Very High	Model-dependent	Excellent	Foundation models

one allows fine-grained alignment and reasoning. Variations of architectures, Perceiver IO and transformer-based multimodal models, use attention to efficiently process variable-length and heterogeneous inputs. Attention-based fusion in multimodal systems using LLM to perform it is frequently done by projecting embodied modality-specific embeddings into token sequences, which may be fed into the LLM. Multimodal fusion implicates the LLM self-attention layers implicitly and makes it reason and generate responses contextually [1]. This scaling is cost-effective to the size of the model and is able to facilitate complex interactivity in conversations.

3.4.5 Multimodal Prompt Engineering

Multimodal prompt engineering has emerged as a novel fusion paradigm, particularly in foundation model-based systems. Instead of explicitly designing fusion layers, multimodal information is encoded into structured prompts that guide the LLM's reasoning process [10].

Visual tokens, audio embeddings, or structured descriptors are embedded into the prompt, allowing the LLM to interpret and reason over multimodal context using its pretrained capabilities. This strategy simplifies system design and leverages the generalization power of large models, but it also introduces challenges related to prompt length, modality representation, and interpretability.

3.4.6 Comparison of Fusion Techniques

Each fusion strategy offers distinct trade-offs in terms of complexity, scalability, and reasoning capability. Early fusion enables deep joint learning but lacks flexibility. Late fusion is modular and robust but limited in expressive power. Hybrid and attention-based fusion strategies provide the best balance, particularly when integrated with LLMs. In practice, modern multimodal conversational AI systems increasingly favor attention-driven and LLM-centric fusion mechanisms, as they support scalable, context-aware, and human-like dialogue across diverse modalities [7]. Table 3.4 provides a comparative analysis of multimodal fusion strategies.

3.5 DATASETS, BENCHMARKS, AND EVALUATION

The quality datasets and good benchmarking protocols are the basis of the development and testing of multimodal conversational AI systems. In contrast to unimodal systems, multimodal conversational agents are supposed to be discussed in many respects including perceptual comprehension, multimodal reasoning, dialogue fluency, and the overall user experience. Some of the significant multimodal datasets, benchmark tasks, and AI evaluation methods [1], which are commonly adopted in the analysis of multimodal conversational AI, are discussed in this section.

3.5.1 Multimodal Datasets for Conversational AI

Multimodal conversational systems are based on multimodal datasets, which are used to train and test them. These datasets are usually combined with other representations like images, audio, or video as well as the annotations of dialogue, allowing supervised, semi-supervised, or self-supervised learning paradigms. Some of the notable ones are visual question answering (VQA) datasets, audio-visual dialogue corpora, and multimodal emotion recognition datasets [2]. Newer datasets are now more and more specialized to be able to learn LLM, with long dialogues, finer-grained annotations, and various real-life scenarios. These types of data are useful in learning complicated cross-modal interactions and conversation reasoning. Although they are important, there are a few challenges to constructing quality multimodal data, such as cost of annotation, ensuring that there is precise synchronization between modalities and alleviating bias of the dataset. With the ever-growing multimodal conversational systems, the requirements of big, varied, and ethically collected data are all the more crucial towards fairness, robustness, and generalization.

3.5.2 Benchmarks for Multimodal Reasoning and Dialogue

The benchmarks are vital in offering more standardized tasks and evaluation environments to compare the multimodal conversational models. Visual question answering, image-grounded dialogue, multimodal sentiment analysis, as well as instruction-following tasks with visual or audio inputs are the common benchmark tasks [16].

As multimodal LLMs have emerged, the design of benchmarks has changed to focus on more advanced tasks like reasoning, generalization, and instruction adherence in place of task specific performance. The goal of these modern standards is to measure how well a model can synthesize information across modalities, sustain conversational context across multiple turns and produce coherent, grounded, and contextually relevant responses [20]. Table 3.5 illustrates the common benchmark tasks for evaluating multimodal conversational systems.

3.5.3 Evaluation Metrics and Protocols

This is due to the fact that the evaluation of multimodal conversational AI systems is complicated by the multimodal interaction and subjective measure of the quality

TABLE 3.5
Common Benchmark Tasks for Evaluating Multimodal Conversational Systems

Benchmark Task	Modalities	Evaluation Focus
Visual Question Answering	Image + Text	Perceptual grounding
Image-Grounded Dialogue	Image + Text	Multi-turn coherence
Multimodal Sentiment	Audio + Video + Text	Emotional understanding
Multimodal Instruction Following	Multimodal	Reasoning and generalization

of dialogue. The quantitative assessment is usually given on automated values such as accuracy, BLEU, ROUGE, and CIDEr, which might not be sufficient to capture conversational fluency, relevance, and user satisfaction. Another way that the issue of evaluation is more challenging is the fact that various modalities of input exist so that one can be misled by a lack of alignment or an inappropriate reading among modalities. This makes human evaluation to be the gold standard of measuring such qualities like coherence, helpfulness, empathy, and appropriateness of the responses [14,16]. The newer systems of evaluation are also tending to become more hybrid with use of automated measures and human ratings and rating plans that are task specific to provide a more detailed and valid assessment.

3.5.4 Challenges in the Evaluation of Multimodal LLMs

Multimodal LLMs introduce new assessment problems, such as hallucination, grounding, and cross-modal consistency. A model can give fluent and convincing answers which are wrong, or unrelated to the underlying perceptions [2,19].

Researchers are investigating new measures to deal with such issues, including retrieval-based validation, adversarial testing, and explainability-driven metrics. These methods seek to determine whether generated responses are well grounded on the presented multimodal inputs as well as detect discrepancies between modalities. This type of evaluation methodologies is necessary in order to make sure that multimodal conversational systems can be not only fluent but also reliable, trustworthy, and correspond to human expectations.

3.6 RETRIEVAL-AUGMENTED AND KNOWLEDGE-GROUNDED SYSTEMS

LLM, despite its enormous generation and reasoning abilities, is inherently prone to hallucinations, outdated knowledge, and inability to directly relate facts on the real world. These challenges are further complicated when considering multimodal conversational AI systems in which the poor support of one or more modalities may lead to an inappropriate or dangerous response. In a bid to contain these deficiencies, retrieval-enhanced and knowledge-based platforms have emerged as a significant addition to the multimodal LLM systems [1,4].

This section discusses how these paradigms can enhance reliability, factual accuracy, and contextual grounding in multimodal conversational AI in this section.

3.6.1 Retrieval-Augmented Generation (RAG)

One of the approaches that utilize neural retrieval and generative language models is called retrieval-augmented generation (RAG) and involves generating responses on the basis of available outside knowledge [3]. RAG systems do not only use internal model parameters but rather read and access the relevant documents or passages of a body of knowledge and condition the LLM response depending on the context it has read and accessed.

The architectural elements of RAG systems are as follows:

- A retriever module, which searches an indexed corpus with the help of dense or hybrid embeddings
- A generator module with an LLM that is commonly employed to create responses with regard to the query made by the user and retrieved knowledge.

As illustrated in Figure 3.3, RAG has the potential to support the factuality, domain flexibility, and access to up-to-date information in conversational AI. Applied to multimodal systems, RAG can be used to tie perception to authenticated knowledge to reduce the number of hallucinations and improve the credibility.

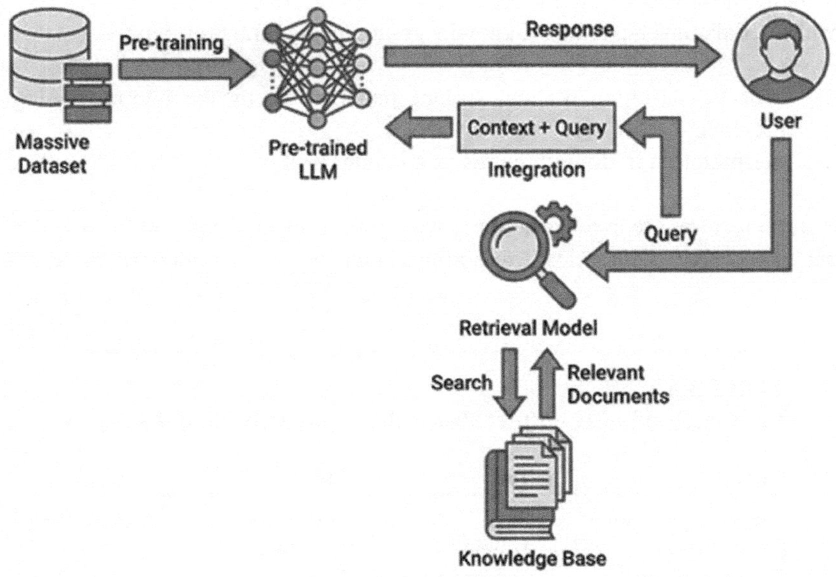

FIGURE 3.3 Retrieval-augmented generation.

3.6.2 Multimodal Retrieval-Augmented Generation (MRAG)

The concept multimodal retrieval-augmented generation (MRAG) is the extension of the RAG paradigm because it supports retrieval with multiple modalities, that is, images, videos, audio clips, and structured metadata [14]. MRAG indexes can also have multimodal queries and be multimodal indexes.

For example, conversational agent can discover:

- Images that would be relevant to a query to visual query
  Video clips that are related to a verbal description
  Visual observation written accounts.

The extracted multimodal evidence is resolved and either cross-modally encoded or promptly incorporated into a format readable by the LLM, often through cross-modal attention or prompt-based incorporation. Multimodal retrieval augmented generation (MRAG)is a far better task grounder in situations where visual question answering, instructional dialogue, and real-world decision support are needed, where both perceptual and factual accuracy are vital. Table 3.6 illustrates the comparison between text-based RAG and multimodal RAG.

3.6.3 Knowledge-Augmented Transformers

Knowledge-augmented transformers explicitly include knowledge sources (e.g., knowledge graphs, ontologies, domain-specific databases) in the design of transformers. When compared with retrieval-based models that are executed outside of the reasoning process, knowledge-augmented models incorporate semi-symbolic or symbolic information into the reasoning process [2,15].

Multimodal conversational AI knowledge-augmented transformers:

- Concept reasoning without surface pattern-free, by the intermodal logical inference
- Interpretation of domain words as a regular task.

These architectures are used particularly where accuracy and explicability are of paramount importance such as legal reasoning, healthcare, and scientific conversation.

TABLE 3.6
Comparison between Text-Based RAG and Multimodal RAG

Aspect	RAG	MRAG
Supported Modalities	Text	Text, Image, Audio, Video
Query Type	Unimodal	Multimodal
Grounding Depth	Moderate	High
Application Scope	QA, Chatbots	VQA, decision support

TABLE 3.7
Techniques for Reducing Hallucinations in Multimodal Conversational AI

Strategy	Description
Evidence Conditioning	Grounding generation on retrieved facts
Citation Generation	Linking answers to sources
Cross-Modal Verification	Checking perceptual consistency
Adversarial Testing	Stress-testing model reliability

These systems reduce ambiguity and maximize interpretability as multimodal observations are based on valuable knowledge structure.

3.6.4 Reducing Hallucinations through Retrieval and Grounding

One important issue of LLM-based conversational systems is hallucination, especially when operating in the multimodal environment where any error in perception can propagate throughout the system. The ways of fixing this problem are through knowledge-based systems and retrieval-enhanced systems, which base answers on verifiable external sources [1].

Key strategies include:

- Conditioning generation through retrieved evidence
- Citation-aware response generation
- Comparison of multimodal inputs and retrieved knowledge.

These means are essential to high-stakes multimodal applications because they are not only employed to promote the factual accuracy but to stimulate the user trust. Table 3.7 illustrates the techniques for reducing hallucinations in multimodal conversational AI.

3.6.5 Applications of Knowledge-Grounded Multimodal AI

Multimodal conversational systems based on knowledge have been used in a broad application system covering medical decision support, intelligent tutoring systems, and enterprise knowledge assistants. These systems can provide fluent as well as accurate, explainable, and contextually grounded responses by integrating multimodal perception, retrieval, and reasoning both [19].

3.7 DOMAIN-SPECIFIC APPLICATIONS OF MULTIMODAL LARGE LANGUAGE MODELS

MLLMs have made possible a new generation of conversational AI systems able to cope with complexes and domain-specific problems. MLLMs can be used to deliver applications with contextual awareness, interpretability, and adaptability through a single architecture by combining perception, reasoning, and dialogue generation [14]. This section explores the key application domains where multimodal conversational AI has demonstrated significant impact.

3.7.1 Healthcare and Medical Decision Support

Multimodal conversational agents have started to gain attention in healthcare environment where they can be employed for clinical decision-making, monitoring, and diagnosis of conditions through integration of medical images, health record data, speech, and sensors. MLLMs make it possible to analyze information in a context-specific way and use natural language processing for providing healthcare professionals with accurate and timely care delivery [3].

3.7.2 Education and Intelligent Tutoring Systems

Multimodal conversational agents can be used in education as smart tutors that can modify their teaching according to the behavior of the students, their facial expressions, and patterns of interaction. MLLMs allow selective learning processes that reveal points of confusion or boredom and modify the explanations to them [7].

Multimodal content delivery, interactive problem-solving, and formative assessment can be supported by such systems, and they can be especially useful in STEM and remote learning settings.

3.7.3 Cybersecurity and Digital Forensics

Multimodal chatbots are also finding usage in cybersecurity and digital forensics, where incoming data heterogeneously includes logs, network traffic, file metadata, and visual evidence, and analytics are needed to analyze all these diverse data streams. The MLLMs can help with summarizing complex evidence and correlating multimodal indicators and direct the investigative process with conversational interfaces [20]. The systems improve situational awareness and decision support, especially to the non-expert users, by converting the technical results into a comprehensible natural language explanation.

3.7.4 Human–Robot Interaction and Smart Environments

In human–robot interaction, vision, speech, and environmental sensors are used in the interaction of human behaviors with the robots through multimodal conversational systems. MLLMs are useful in supporting natural speech, task management, and adaptive behavior in dynamic settings like smart homes, warehouses, and service robotics. These systems can help to establish flawless coexistence between humans and intelligent machines by basing the language on the physical senses [14].

3.7.5 Customer Support and Assistive Technologies

In customer support and assistive technologies, where it is required to know their intentions and emotions, multimodal conversational agents are increasingly deployed. MLLMs provide more empathic and effective help because they analyze the tone of speech, facial expressions, and contexts [2]. It is used in applications such as virtual service agents, people with disabilities, accessibility devices, and chat-based interfaces with smart devices.

3.8 CHALLENGES IN MULTIMODAL CONVERSATIONAL AI

Although multimodal conversational AI is making a rapid developmental advancement, there are still a number of underlying challenges that still obstruct the development, implementation, and mass adoption of these systems. These issues are contributed by the complexity of the inherent integration of non-homogenous modalities, the constraints of the existing learning paradigms, and the ethical and social impact of deploying human-facing AI systems. These concerns should be tackled in order to create credible, scalable, and dependable multimodal conversational agents [15].

3.8.1 Multimodal Representation Alignment

The multimodal representation methodology is also based on the principle of alignment. Effective alignment between heterogeneous modalities is one of the main problems of multimodal conversational AI. Visual, auditory, and linguistic information are very different in their structure, scale, time resolution, and the nature of noise [3]. It is a challenging problem to map these different inputs to a common representation space without modality-specific information being lost.

Misalignment may result in incoherent reasoning, wrong grounding, and performance on conversations. Though contrastive learning and attention-based alignment have been proven to be promising, they can be expensive in terms of paired multimodal data as well as unable to adapt to rare or ambiguous situations.

3.8.2 Data Scarcity and Annotation Complexity

The problem of data scarcity and the complexity of annotation are closely connected to the data accessibility aspect of knowledge management. Data scarcity and annotation complexity are intertwined with the data accessibility aspect of knowledge management.

Multimodal datasets of high quality are not cost-effective and time-consuming to acquire [1]. Compared to text data, multimodal data need to be synchronized with recordings, manually annotated, and quality controlled. Emotional conscious and empathetic apps even add more complexity of annotation since they are subjective and emotionally varied across cultures. There are no common datasets in different fields, which restricts the possibility of comparable models and reproducibility.

Moreover, the bias of the dataset can spread to the multimodal systems, resulting in the inequality of performance based on the demographics.

3.8.3 Computational Cost and Scalability

The MLLMs are computationally complex, which needs a lot of resources to train and make inferences. These results in additional memory usage and latency since multiple modality encoders, alignment modules and large transformer architectures are included [8].

Edge deployments and real-time conversational systems have a specific issue with scalability. Study efficacy in compressing models, parameter sharing, and adaptive inference algorithms are currently under development to minimize computation without loss with respect to performance.

3.8.4 Hallucinations and Grounding Errors

Hallucination-generating plausible but false information is a very important issue, particularly in multimodal LLMs. Misperception of interpretation can spread through the system of reasoning as it triggers the confidently wrong answers [2].

Although it can be believed that retrieval-based and knowledge-grounded methods reduce hallucinations, they bring more complexity and reliance on the quality of external knowledge. High-stakes applications like healthcare and education require a stable grounding between modalities [11].

3.8.5 Evaluation and Benchmarking Limitations

Multimodal conversational AI systems are practically difficult to evaluate because dialogue quality and modal interaction are subjective. Current automated measures do not tend to indicate coherence, empathy, and contextual appropriation [7]. Human evaluation, although more informative, is expensive and is hard to scale. It is not only that benchmarking is hampered because of the absence of standardized evaluation protocols, but it also slows down the field.

3.8.6 Ethical, Privacy, and Safety Concerns

The multimodal conversational systems handle sensitive information, such as images, speech, and body language, which is of serious concern to the privacy, consent, and data security [20]. In this case, emotion-conscious systems are subsequently at risk of manipulation or unwanted psychological influence. The question of fairness, transparency, and accountability in multimodal AI is yet to be addressed. Strong governance systems and codes of ethics are needed to curb abuse and earn the confidence of the population.

3.9 FUTURE TRENDS AND RESEARCH DIRECTIONS

The future of multimodal conversational AI is influenced by the current developments in foundation models, learning paradigms, and research into human–AI interaction. With the systems becoming more intelligent in their collaboration and being more and more like intelligent partners in tasks instead of simple task-oriented assistants, there are new growth directions of research that may increase capability, reliability, and impact on society [14].

3.9.1 Unified Multimodal Foundation Models

Future research may come up with unified multimodal foundation models that can process and generate across all the major modalities in one architecture. The objective of such models is to avoid the use of independent encoders and alignment modules, simplify the design of the systems, and enhance the reasoning coherence [3].

3.9.2 Neuro-Symbolic and Knowledge-Enhanced Reasoning

Neural learning and symbolic reasoning are a promising step to a better interpretability and logical consistency. Multimodal systems that are neuro-symbolic are able to combine both structured knowledge and the input of perception [15], which makes reasoning in complex problems more robust.

3.9.3 Continual and Lifelong Learning

The conversational agent on the real-world should respond to environments, preference of the users, and domain knowledge. In long-term deployment, continuous learning strategies enabling multimodal systems to learn in an incremental manner without suffering devastating forgetting are required [9].

3.9.4 Trustworthy and Explainable Multimodal AI

Multimodal conversational systems require explainability and trust in order to be accepted by users. The future research will focus on clear reasoning, uncertainty estimation, and explainable multimodal decision-making, especially in sensitive fields [11,19].

3.9.5 On-Device and Edge Multimodal AI

Advances in hardware and model optimization will enable multimodal conversational systems to operate on edge devices, reducing latency and enhancing privacy. This shift will expand the applicability of multimodal AI in mobile, wearable, and embedded systems [3] [20].

3.10 CONCLUSION

The chapter has thoroughly discussed the advances in architecture in multimodal conversational AI with the use of LLMs. Starting with some of the basic principles and moving on to architectural elements, fusion approaches, retrieval-enhanced systems, and application-specific interfaces, the chapter illustrated how multimodal integration and the LLM-based design have reinvented conversational AI. The discussion has highlighted the significance of good modality alignment, attention-based fusion, and knowledge grounding as a way of facilitating context-attentive and credible dialogue [14,16]. Reinforcement learning to human feedback and a compassionate design of the system was highlighted as a primary aspect of matching multimodal AI with human values. Although there have been significant advancements, multimodal conversational AI continues to struggle with enormous challenges in terms of scalability, evaluation, grounding, and ethical usage. Solving these issues will involve further development of model architecture, learning paradigm, and governance structures. In the future, multimodal foundation models, neuro-symbolic reasoning, and human-centered design principles convergence is likely to lead to additional advancements in the field. The success of multimodal conversational systems as they continue to be integrated in the daily life will not just be measured by technical excellence, they must be able to win the trust of people, uphold their privacy and benefit human well-being.

REFERENCES

[1] Huang, J., & Zhang, J. (2024). A Survey on Evaluation of Multimodal Large Language Models. *arXiv:2408.15769.*

[2] Firdaws, M., Chauhan, H., Ekbal, A., & Bhattacharyya, P. (2022). EmoSen: Generating Sentiment and Emotion Controlled Responses in a Multimodal Dialogue System. *IEEE Transactions on Affective Computing, 13,* 1555–1566.

[3] AlSaad, R., Abd-alrazaq, A. A., Bouraibel, S., Ahmed, A., Househ, M., Dremsek, P., & Sheikh, J. (2024). Multimodal Large Language Models in Health Care: Applications, Challenges, and Future Outlook. *Journal of Medical Internet Research, 26.*

[4] Yuan, T., Zhang, X., Liu, B., Liu, K., Lin, J., & Jiao, Z. (2025). Surveillance Video-and- Language Understanding in the Era of Large Multimodal Models. *IEEE Transactions on Circuits and Systems for Video Technology,* 35, 300–314.

[5] Alayrac, J.B., Donahue, J., Luc, P., Miech, A., Barr, I., Hasson, Y., Lenc, K., Mensch, A., Millican, K., Reynolds, M. and Ring, R., (2022). Flamingo: a visual language model for few-shot learning. *Advances in Neural Information Processing Systems,* 35, pp. 23716–23736.

[6] Kumar, L. A., & Renuka, D. K. (2023). *Deep Learning Approach for Natural Language Processing, Speech, and Computer Vision: Techniques and Use Cases.* CRC Press, Boca Raton, FL.

[7] Zhao, F., Zhang, C., & Geng, B. (2024). Deep Multimodal Data Fusion. *ACM Computing Surveys,* 56, 1–36.

[8] Koh, J. Y., Fried, D., & Salakhutdinov, R. (2023). Generating Images with Multimodal Language Models. *arXiv:2305.17126.*

[9] Abbasian, M., Azimi, I., Rahmani, A. M., & J¨antti, R. C. (2023). Conversational Health Agents: A Personalized LLM-Powered Agent Framework. *arXiv:2310.02374.*

[10] Yao, S., Zhao, J., Yu, D., Du, N., Shafran, I., Narasimhan, K.R., and Cao, Y., (2022), October. React: Synergizing Reasoning and Acting in Language Models. In *The Eleventh International Conference on Learning Representations.*
[11] Jiang, F., Feng, Y., Dong, L., Wang, K., Yang, K., Pan, C., & You, X. (2023). Large AI Model Empowered Multimodal Semantic Communications. *IEEE Communications Magazine,* 76–82.
[12] Rose, L. S., Kumar, L. A., & Renuka, D. K. (2019). *Deep Learning Using Python.* Wiley, New York.
[13] Tan, X., Gan, Z., Wang, Y., Wang, C., & Ma, L. (2024). Efficient Multimodal Large Language Models: A Survey. *arXiv:2405.10739.*
[14] Joshi, G., Walambe, R., & Kotecha, K. (2021). A Review on Multimodal Learning in Multimodal Deep Neural Networks. *IEEE Access,* 9, 5910– 59316.
[15] Rodis, N., Sardianos, C., Radoglou-Grammatikis, P. I., Samarinidis, P., Varlamis, I., & Papadopoulos, G. (2023). Multimodal Explainable Artificial Intelligence: A Comprehensive Survey on Methodological Advances and Future Research. *IEEE Access,* 12, 159792–159820.
[16] Shinn, N., Cassano, F., Gopinath, A., Narasimhan, K. and Yao, S., (2023). Reflexion: Language Agents with Verbal Reinforcement Learning. *Advances in Neural Information Processing Systems,* 36, pp. 8634–8652.
[17] Bai, Z., Wang, P., Xiao, T., He, T., Han, Z., Zhang, Z., & Shou, M. (2024). A Systematic Evaluation of Multimodal Large Language Models: A Survey. *arXiv:2404.18930.*
[18] Trenton, M. (2025). A Comprehensive Survey on Multilingual and Multimodal Automatic Speech Recognition Systems. *Journal of Computer Technology and Software*, 4(10)1–9.
[19] Ji, H., Han, I., Ko, Y. (2022). A Systematic Review of Conversational AI in Language Education: Focusing on the Interaction with Teachers. *Journal of Research on Technology in Education,* 55, 48–63.
[20] Zhou, J., Li, H., Chen, S., Chen, Z., Han, Z. and Gao, X., (2025). Large Language Models in Biomedicine and Healthcare. *npj Artificial Intelligence*, *1*(1), 44.

4 Integration of LLMs for Conversation AI Multimodal Conversational AI

Leveraging Ensemble Open-Source Generative Models for Enhanced Human-Like Interactions

Viswadarshan R.R., Felicia Lilian J., and Mahalakshmi S.

4.1 INTRODUCTION

Conversational AI has rapidly evolved from simple text-based chatbots into sophisticated, multimodal systems capable of engaging through speech, images, video, and other signals. This transformation has been enabled by advances in artificial intelligence at large, particularly in machine learning, deep learning, natural language processing, and most recently, large language models (LLMs). The relationship among these fields can be understood through a layered perspective: artificial intelligence provides the broad foundation, machine learning and deep learning provide the computational frameworks, natural language processing (NLP) focuses on linguistic understanding, and LLMs emerge as specialized engines for language reasoning. Within this ecosystem, conversational AI sits at the intersection, combining these advances into interactive systems that approximate human-like dialogue.

The Venn diagram in Figure 4.1 highlights this relationship, showing how conversational AI and LLMs draw from overlapping areas of AI, machine learning, deep learning, and NLP. This interconnection underscores why multimodal conversational systems are not isolated technologies but a synthesis of multiple research domains. By leveraging these overlaps, they can perceive diverse inputs, reason across modalities, and generate coherent outputs that reflect both linguistic and contextual understanding.

DOI: 10.1201/9781003678359-4

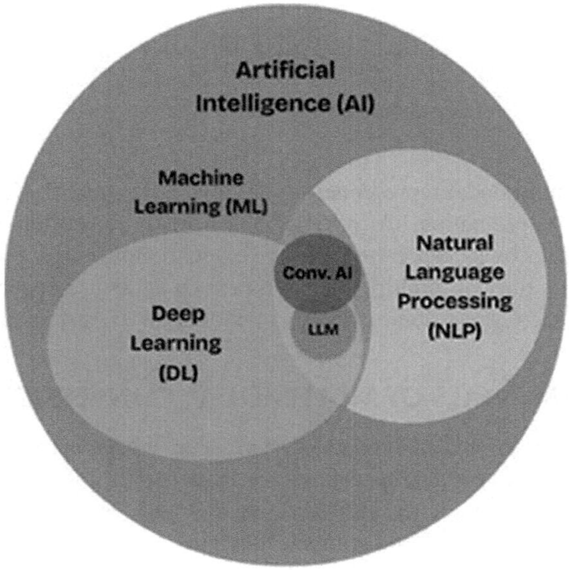

FIGURE 4.1 The place of conversational AI and LLMs within the broader AI ecosystem.

This chapter explores the integration of LLMs and generative models as the backbone of multimodal conversational AI. It begins with the fundamentals of multimodal interaction, examining how human communication inspires system design. It then discusses the core modalities text, speech, images, and video and their distinct roles in shaping interaction. From there, the chapter explains how LLMs and generative models orchestrate perception, reasoning, and generation, before surveying the landscape of open-source generative models [20] that provide the building blocks of modern systems.

Traditional conversational AI systems rely primarily on text-based inputs and outputs, which limits their ability to understand real-world contexts. Human communication naturally combines speech, vision, and contextual knowledge, making text-only interaction insufficient for many practical applications. For example, users often refer to images, spoken queries, or situational cues that cannot be captured through text alone.

Multimodal conversational systems address this limitation by integrating multiple modalities such as speech, images, and textual knowledge. This enables richer semantic understanding, improved user interaction, and more accurate responses in real-world scenarios such as healthcare assistance, education, and intelligent customer support. Therefore, multimodality is essential for building robust and human-like conversational AI systems.

Figure 4.2 illustrates the overall workflow of the proposed multimodal conversational system. The pipeline begins with automatic speech recognition (ASR) for processing spoken inputs and a vision encoder for extracting visual features from

FIGURE 4.2 The proposed system workflow diagram.

images. These multimodal representations are then fed into a LLM, which performs reasoning and coordination. A retrieval-augmented generation (RAG) module enhances responses by incorporating relevant external knowledge. Finally, the system generates outputs through multiple generators such as text-to-speech (TTS), image generation, or video generation modules, depending on the task requirements.

4.2 FUNDAMENTALS OF MULTIMODAL CONVERSATIONAL AI

Multimodal conversational AI refers to systems that engage across multiple sensory channels not only text, but also speech, images, video, and other signals and use them in combination to sustain coherent, interactive dialogue. Much like a human teacher who listens, speaks, gestures, and points at diagrams, a multimodal agent coordinates diverse inputs and outputs to achieve context-aware communication.

4.2.1 Definition of Multimodal in AI

In everyday interaction, humans rarely rely on a single channel. Words, tone, facial expressions, and visual cues blend to shape meaning. Multimodal AI seeks to replicate this richness by integrating different inputs such as text, speech, images, video, or sensor data.

Formally, a system is multimodal if it accepts input from two or more modalities, aligns them into a shared representation, and generates outputs that may themselves span multiple modalities (Figure 4.3). This input → alignment → output pipeline distinguishes multimodal systems from unimodal ones. The term applies both to single multimodal models, such as vision–language transformers trained jointly on images and captions, and to engineered systems that combine specialized components like ASR, image encoders, LLMs, and TTS modules.

4.2.2 Fusion Strategies

The core mechanism is fusion, the process that allows signals from different modalities to "speak" to one another. Early fusion combines low-level features for tight integration but at higher computational cost. Late fusion merges the outputs of separate pipelines, offering modularity and scalability. Hybrid fusion balances the two, aligning some features early while preserving specialized reasoning downstream. The choice of strategy depends on task needs: fine-grained visual question answering may demand early fusion, whereas customer-service chatbots often prefer late fusion for efficiency.

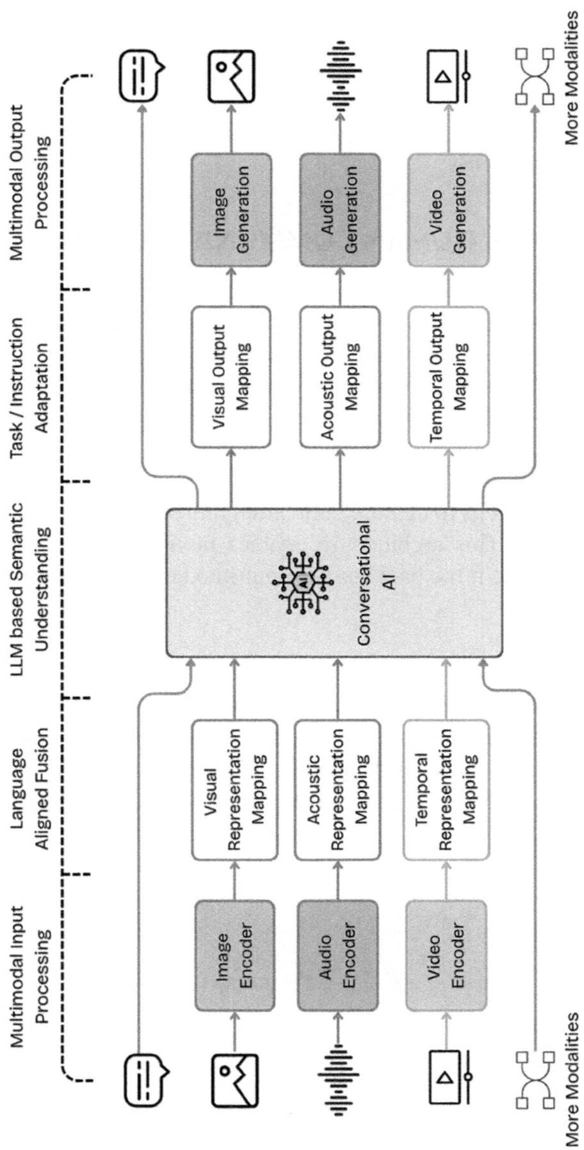

FIGURE 4.3 Processing multimodal interactions with conversational AI.

4.2.3 Benefits and Challenges

The benefits are compelling. Multimodality reduces ambiguity, grounds language in perception, and improves robustness. A sentence like "It's cold in here" may be vague in text, but combined with video of a shivering user, it clearly conveys discomfort. Multimodal signals also enhance personalization and resilience in noisy or data-scarce settings. Yet challenges remain. Aligning heterogeneous signals is technically demanding, and processing high-fidelity audio or video is resource-intensive. Evaluation becomes more complex, since correctness must be judged across multiple channels. Ethical concerns from bias propagation to privacy risks and misuse add another layer of responsibility.

4.3 MODALITIES IN HUMAN COMMUNICATION AND AI

Human communication is inherently multimodal: we combine words, tone, gestures, and visual cues to convey meaning. For conversational AI, recognizing these channels as distinct modalities text, speech, images, and video is what makes interaction feel natural and grounded. Each modality contributes unique strengths, faces practical limitations, and plays a distinct role in human–AI exchange.

Figure 4.4 shows a generic encoder–decoder framework widely used in generative AI systems. Raw inputs from different modalities are projected into embeddings, processed by encoder blocks to capture contextual representations, and then decoded into structured outputs. This architecture enables flexible handling of text, audio, image, and video, making it the backbone of multimodal conversational AI.

4.3.1 Text

Text is humanity's most durable medium, valued for permanence and precision. It enables asynchronous communication and large-scale knowledge sharing, though it strips away non-verbal cues. In AI, text is tokenized, embedded, and processed by transformers, with LLMs like generative pre-trained transformer (GPT) or large

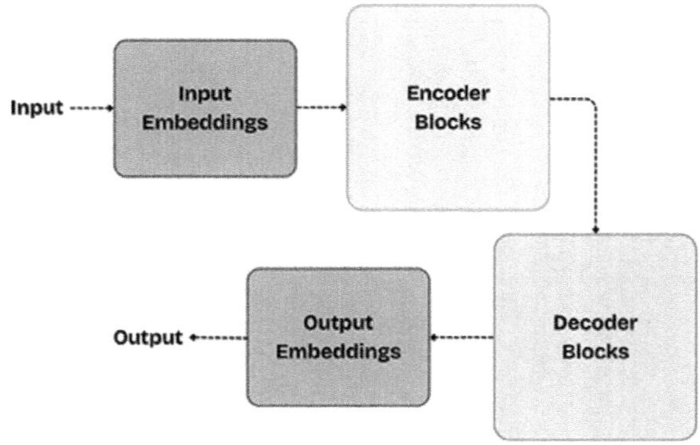

FIGURE 4.4 General encoder–decoder framework for text processing.

language model Meta AI (LLaMA) extending this to capture context and world knowledge. Text often acts as the "anchor modality," since speech, images, and video can be converted into a textual form for reasoning. Its strengths are universality and efficiency, while its limits lie in ambiguity and loss of emotional nuance.

4.3.2 Images

Images condense complex meaning into visual form, from cave paintings to modern infographics. They convey objects, scenes, emotions, and symbols instantly. In AI, images are encoded with convolutional neural networks (CNNs) or vision transformers, aligned with text using models like contrastive language-image pre-training (CLIP) [5], and generated with diffusion systems such as stable diffusion [8].

Within multimodal conversation, images provide grounding, disambiguating vague text, enriching explanations, or enabling creative illustration. Their strengths are semantic richness and universality, while challenges include computational cost, dataset bias, and privacy concerns.

Illustrating (Figure 4.5) the transformer encoder architecture commonly applied across modalities. Inputs such as text, images, audio, or video are first flattened and embedded, enriched with positional information, and then processed through stacked encoder layers composed of multi-head self-attention, normalization, and

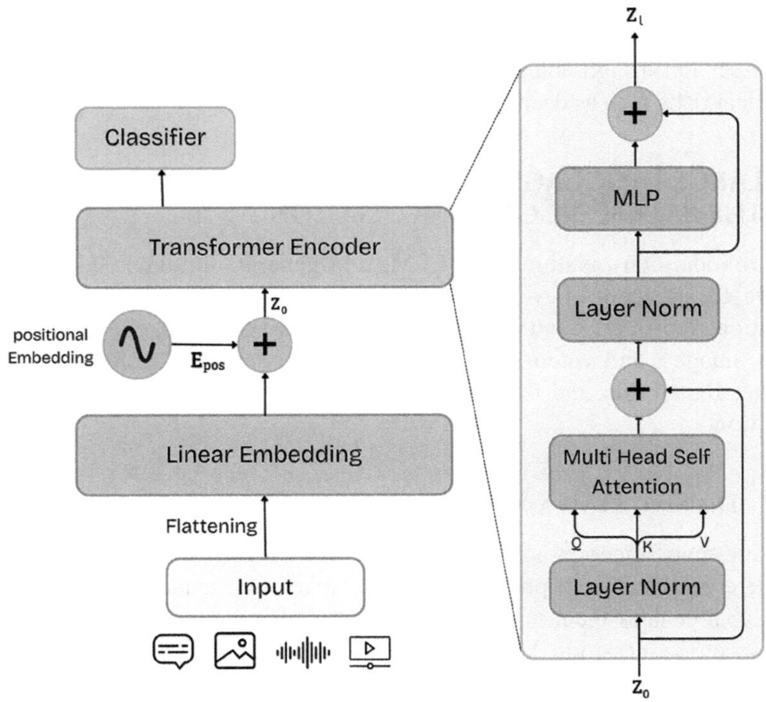

FIGURE 4.5 Detailed view of the transformer encoder for multimodal input processing.

feed-forward networks. The final contextual representation supports downstream tasks like classification and multimodal reasoning, making this structure a foundation of modern generative and conversational AI.

4.3.3 Speech

Speech is the most natural and immediate form of communication, enriched with prosody and rhythm that reveal intent and emotion. For AI, this is handled through automatic speech recognition (ASR) systems like Whisper and wav2vec 2.0, and TTS models such as Tacotron or FastSpeech, with added capabilities for speaker and emotion detection. In multimodal AI, speech enables hands-free interaction and captures urgency or mood that text alone cannot. Its strengths are speed, expressiveness, and accessibility; limitations include noise sensitivity, accent variability, and privacy risks tied to biometric voice data.

4.3.4 Video

Video extends images into time, capturing motion, causality, and emotional expression. Humans rely on this temporal continuity for understanding, making video the closest digital proxy to real-life interaction. For AI, video is demanding: perception models must handle spatial features, temporal dynamics, and synchronized audio, while generation involves diffusion-based tools like stable video diffusion. In multimodal systems, video enables demonstrations, immersive tutoring, and realistic embodied agents. Its strengths are expressiveness and temporal grounding, while its weaknesses lie in high computation requirements, latency in real-time processing, and ethical risks such as deep fakes.

4.4 LARGE LANGUAGE MODELS AND GENERATIVE MODELS AS THE BACKBONE OF CONVERSATION

In multimodal conversational AI, LLMs and generative models do not function as mere components; they serve as the structural backbone [1] that integrates perception, reasoning, and output. By transforming heterogeneous inputs text, speech, images, and video into coherent, context-sensitive responses, they provide the adaptability and fluidity that make these systems resemble human-like interaction.

4.4.1 Defining LLMs and Generative Models

LLMs are transformer-based neural architectures trained on a vast corpora of text, with the core objective of predicting the next token in a sequence. This deceptively simple goal enables them to capture grammar, semantics, stylistic variation, and a broad spectrum of world knowledge, making them highly versatile engines for reasoning and dialogue.

Though LLMs can overcome their inherent limitations (such as lack of up-to-date information, inability to perform complex calculations, or limited reasoning

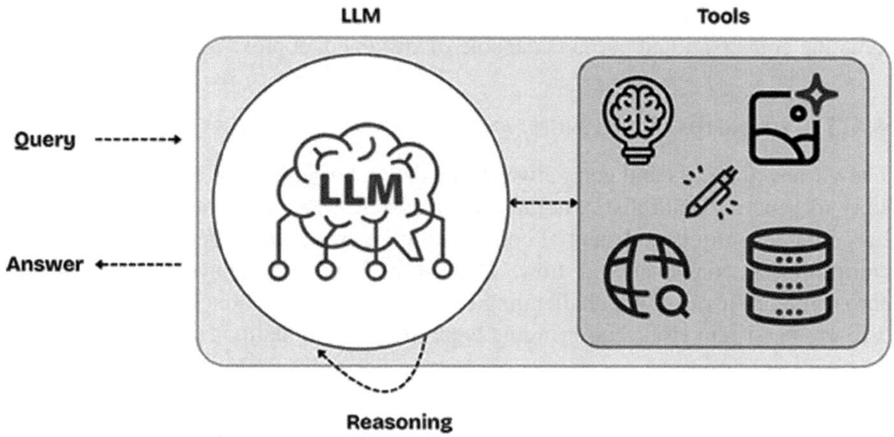

FIGURE 4.6 The role of tools in expanding LLM capabilities.

over structured data) by integrating with external tools. This paradigm significantly expands the LLM's capabilities, transforming it from a purely text-generation engine into a powerful reasoning and execution agent.

Figure 4.6 illustrates how LLMs interact with external tools to enhance reasoning and output generation. A query is processed by the LLM, which can invoke specialized tools such as knowledge retrieval, databases, or content generation utilities, before producing a grounded and coherent answer.

Generative models extend this paradigm beyond text into other modalities. Text-to-speech models generate natural audio; diffusion-based models synthesize realistic images; and video generation systems create dynamic, temporally coherent content. In practical implementations, an LLM frequently assumes the role of orchestrator, coordinating these specialized generative models to ensure that meaning, context, and intent remain consistent across modalities.

4.4.2 Functional Roles in Multimodal Conversation

The contribution of LLMs and generative models to multimodal systems can be understood through four interrelated functions:

- *Perception*: Transforming raw sensory input into machine-readable form, such as transcripts from automatic speech recognition or captions derived from vision models.
- *Reasoning and dialogue management*: Maintaining conversational coherence, managing turn-taking, selecting appropriate modalities, and integrating external knowledge where needed.
- *Generative output*: Producing natural text responses and generating prompts for downstream models tasked with speech, image, or video synthesis.
- *Grounding and consistency*: Ensuring factual reliability, alignment across modalities, and coherence across multiple conversational turns.

These roles collectively transform multimodal agents from reactive systems into dynamic conversational partners capable of sustained, context-rich dialogue.

4.4.3 Strengths, Challenges, and Best Practices

The virtues of LLMs and generative models are their generality, scalability, and capacity to generate naturalistic, human-like responses. But with these come obstacles. Large models require substantial computational power, which generates latency that compromises conversational flow. Convening long-term conversational memory across modalities is still a challenging technical obstacle. Hallucinations and factual drift are persistent risks, and bringing heterogeneous modalities together needs to be precisely synchronized.

In order to tackle these issues, practitioners implement a number of best practices. Pretraining is often [2] augmented with task-specific fine-tuning or instruction tuning to improve domain alignment. RAG is used to embed outputs in verifiable knowledge. Deliberate balancing of model size and architecture with latency budgets guarantees responsiveness. Modular pipelines with distinct perception, reasoning, and generation stages enable scalability and maintainability. Last but not least, human-in-the-loop feedback, safety filters, and careful evaluation across modalities constitute the guardrails required for responsible deployment.

4.4.4 Illustrative Examples

The practical importance of these principles is evident in contemporary systems. GPT-4 demonstrates [12] an integrated text and image reasoning; LLaMA provides an open-source foundation widely used for research and deployment; PaLM-E [13] exemplifies embodied reasoning across multimodal inputs; and dialogue-specialized models such as DialoGPT highlight the benefits of targeted fine-tuning for conversational flow. Together, these examples illustrate that LLMs and generative models are not auxiliary tools but the central spine of multimodal conversational AI.

4.4.5 Synthesis

In short, LLMs and generative models integrate perception, reasoning, and multimodal generation into a unified whole. They allow conversational systems to shift from text-only dialogue toward expressive, humanlike dialogue spanning speech, imagery, and video. In the process, bringing this vision to life calls for scrupulous engineering to trade off latency, factuality, and ethical protection. With these in mind, multimodal AI can become conversational companions that comprehend as well as communicate along the entire range of human communication.

4.5 OPEN-SOURCE GENERATIVE MODELS: LANDSCAPE AND PRACTICAL INSIGHTS

One of the most dramatic changes in recent artificial intelligence work has been the meteoric ascendance of open-source generative models. In text, image, audio, and video, community releases now rival or supplement proprietary products, allowing developers more control over customization, deployment, and integration. For multimodal conversational AI, these open-source models are the modular components that can be orchestrated into high-performing, coherent ensembles.

4.5.1 Text Models

Text continues to be the anchor of conversational systems, and the open-source ecosystem now provides a diverse spectrum of large language models. Meta's LLaMA 2 [3] opened new opportunities for both research and commercial fine-tuning, quickly becoming a foundation for building chatbots and assistants. Mistral 7B demonstrated that smaller-scale models, through architectural innovations such as grouped-query attention, can deliver strong performance with lower computational requirements. At the higher-capacity end, the Falcon [4] family, particularly Falcon-40B, provides robust reasoning power and competitive benchmark results. Collectively, these models give system designers options ranging from compact and efficient to versatile and community-driven to large-scale for complex reasoning tasks.

4.5.2 Image Models

The release of Stable Diffusion marked a turning point by making high-quality text-to-image generation widely accessible. Its successor, SDXL, further improved image fidelity, compositional balance, and the realism of human features. Beyond generation, these models support editing, in painting, and style transfer, enabling conversational agents to illustrate responses, modify user-provided images, or generate domain-specific visuals. The surrounding open-source ecosystem has produced thousands of fine-tuned variants, covering everything from artistic illustration to specialized applications such as medical imaging. This adaptability has made diffusion-based models [7] central to visual components of multimodal AI.

4.5.3 Audio Models

Audio is essential for natural interaction, and open-source models now provide both reliable perception and expressive generation. Whisper [9], released as an automatic speech recognition system, is highly effective across multiple languages, accents, and noisy environments, making it a dependable backbone for speech input. On the generative side, Bark delivers expressive and human-like text-to-speech, while projects such as Open Voice extend capability through style transfer and voice cloning. Together, these models allow conversational systems not only to understand spoken input but also to respond with natural, personalized, and emotionally nuanced speech.

4.5.4 Video Models

Among modalities, video remains the most resource-intensive. Proprietary platforms like OpenAI's Sora and Pika Labs have set new standards for text-to-video synthesis, generating clips with impressive realism and coherence. Open-source initiatives such as Open-Sora are still in their early stages but represent important progress toward transparent, reproducible, and community-driven research in video generation. Although computation requirements remain high, the trajectory mirrors that of text and image models: open alternatives will steadily improve, enabling dynamic demonstrations, interactive storytelling, and immersive multimodal conversations.

4.5.5 Open and Proprietary Approaches

Open-source and proprietary models both have unique benefits. Open-source models offer transparency, flexibility, and cost management. They can be customized, hosted on-site for privacy, and expanded without vendor lock-in. The downside is increased responsibility for infrastructure administration, optimization, and safety guardrails. Proprietary systems, on the other hand, emphasize simplicity. They provide managed application programming interfaces (APIs), built-in safety features, and quick prototyping options, albeit at the cost of increased recurring costs and reduced visibility into training data and system behavior.

In reality, most systems use a hybrid model: initial use of proprietary APIs for quick iteration followed by scaling to open-source models for cost-effectiveness and customizability.

4.5.6 Toward Multimodal Systems

The open-source ecosystem now covers all the significant modality: LLaMA, Mistral, and Falcon for text; Stable Diffusion and SDXL for images; Whisper, Bark [10], and OpenVoice for audio; and upcoming projects like Open-Sora for video. Combined, these models give the modular base on which multimodal conversational AI can be constructed. Their worth is not just in individual performance but also in how they can be assembled and orchestrated into effective pipelines. This sequence leads naturally on to the next topic: ensemble strategies and orchestration frameworks, which decide how heterogeneous models are assembled into homogeneous conversational agents.

4.6 ENSEMBLE APPROACH TO MODEL INTEGRATION

Creating multimodal conversational AI takes more than developing or selecting one strong model. Human communication is built on layers: speech speaks words and tone of voice, gesture underscores meaning, and pictures or diagrams connect abstractions with perception. Not a single model today can match this entire range of comprehension and expression. Rather, successful systems use ensembles of single-modality-specialist models, each individually great at one modality, and combine them into a unified system. An individual speech recognizer, a language model, a vision encoder, and text-to-speech module are each capable, but it is their integration that yields the smooth, adaptive character users desire in a conversational partner.

4.6.1 Model Ensembling in the Multimodal Context

Ensembling in classical machine learning often refers to combining multiple predictors to improve accuracy, such as bagging or boosting classifiers. In multimodal AI, the meaning is richer: it is about functional complementarity. Each model contributes a unique skill that others lack. Automatic speech recognition (ASR) models can transcribe spoken language but cannot reason over it; LLMs can generate and structure dialogue but cannot interpret pixels; image captioners or detectors can ground visual content but cannot synthesize new responses. By chaining these experts together, an ensemble transforms heterogeneous inputs into coordinated outputs.

This structured pipeline is more than parallel voting; it is a cooperative workflow. The speech model hands text to the LLM, which may query a retrieval system, generate a response, and then forward that text to a TTS system for delivery. Each component becomes an indispensable link in a conversational chain, and the ensemble's quality depends as much on the integration logic as on the individual accuracy of its models.

4.6.2 Model Selection Criteria

Selecting which models to use is almost as important as putting them together. The most self-evident standard is accuracy, but conversational AI is subject to limitations that are beyond measuring sticks. Latency is the top priority: voice-based systems cannot withstand multi-second latencies without disrupting the flow of conversation. This tends to cause designers to go with optimized, compact models like Mistral 7B instead of extremely large LLMs when performance is paramount.

Model size also has a direct correlation with deployment feasibility. Big models require significant GPU memory and power, which could be feasible in the cloud but not in edge devices or privacy-conscious on-premises deployments. Domain suitability is also a determining factor: an open-domain chit-chat could be managed by a general-purpose LLM, but a healthcare assistant is enhanced by a model that has been fine-tuned on biomedical corpora.

Lastly, privacy and licensing factors shape decisions. Open-source approaches provide transparency, flexibility, and the potential for local deployment, whereas proprietary APIs give convenience and high performance in exchange for lack of control and vendor lock-in. The best ensembles are architected by weighing these factors accuracy, efficiency, domain fit, and governance limits against actual applications.

4.6.3 Orchestration Techniques

Even the best models fail without careful orchestration. Orchestration refers to the logic that routes inputs, coordinates the sequence of models, and harmonizes outputs. It ensures that the ensemble behaves as a single system rather than a set of disconnected tools.

Frameworks such as LangChain provide composable abstractions for chaining models together, integrating memory, and adding retrieval or external API calls. Hugging Face Transformers [14] pipelines simplify orchestration for common tasks such as classification or captioning, making it straightforward to connect models without deep engineering overhead. For production systems, however, custom orchestration layers are often needed to handle concurrency, caching, retries, and graceful degradation. In this sense, the orchestrator is the "conductor," ensuring each specialist model contributes its part at the right moment.

4.6.4 Combining Outputs for Better Performance

The final challenge is merging the outputs of different models into coherent results. Here, classical ensemble strategies reappear in new forms. Voting and ranking can select among multiple candidate transcriptions or responses, while blending strategies assign weights to outputs based on confidence or context. Meta-models smaller LLMs trained to evaluate responses can rerank or refine outputs to improve quality.

Fusion techniques provide further sophistication. Early fusion integrates embeddings from different modalities before reasoning, allowing fine-grained cross-modal alignment. Late fusion merges decisions after independent processing, favoring modularity and scalability. Hybrid fusion blends both strategies, aligning where necessary while keeping modular reasoning intact. The choice depends on system priorities: precision in cross-modal reasoning, modularity for system evolution, or efficiency for real-time responsiveness.

Ensemble integration is not an afterthought but the organizing principle of multimodal conversational AI. It is what transforms a collection of narrow experts into a unified conversational partner. By selecting appropriate models, orchestrating their collaboration, and combining their outputs intelligently, designers create systems that are more robust, adaptable, and human-like than any single model could achieve. This emphasis on integration naturally sets the stage for the next discussion: the technical and ethical challenges that arise once these ensembles are deployed at scale.

4.7 KNOWLEDGE AND TOOL INTEGRATION IN CONVERSATIONAL AI

Large language models and generative systems provide the backbone for multimodal conversational AI, but their capabilities alone are not sufficient for reliable real-world deployment. While they can generate fluent and contextually appropriate responses, they are limited by the static nature of their training data and the inherent risk of hallucination. To overcome these constraints, modern conversational systems increasingly rely on two complementary strategies: retrieval-augmented generation for grounding responses in external knowledge, and integration with external tools that enable the system to act and extend its functionality. Together, these methods push conversational AI beyond being a self-contained generator into becoming a dynamic, adaptive assistant.

4.7.1 Retrieval-Augmented Generation for Knowledge Grounding

RAG [6] is a design pattern (Figure 4.7) that takes the best of large models and external knowledge stores. Rather than relying solely on what the model has memorized, the

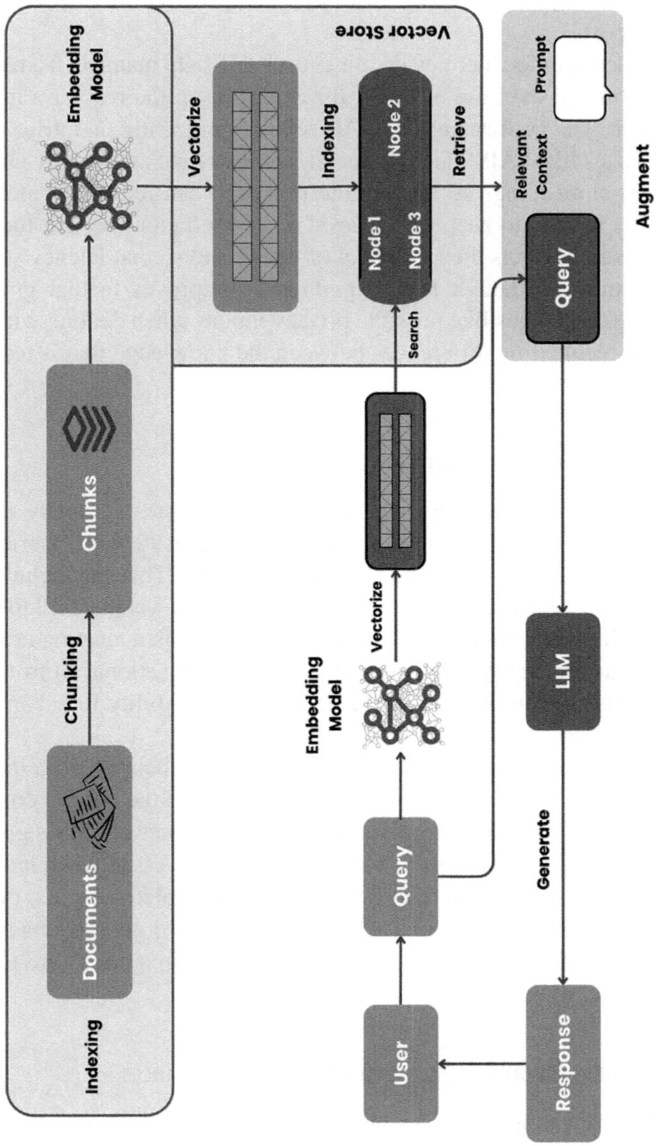

FIGURE 4.7 Retrieval-augmented generation (RAG) pipeline architecture.

system pulls relevant documents or embeddings from a database in real time and these are given as extra context for response generation. This method enhances factual correctness, facilitates access to current information, and enables systems to become expert in a specialized area without having to retrain enormous models. A customer-support

but, for instance, can fetch policy documents or self-help manuals from an internal database of a firm and then use an LLM for compositing the response into conversational language. The performance of RAG relies significantly on retrieval quality.

Vector databases like FAISS [15], Pinecone, or Weaviate are usually employed to cache document embeddings so that similarity search can return the most relevant passages. Care is needed to keep the retrieval process efficient, so that the chunking and embedding strategy gets the right level of detail, and so that latency stays within conversational limits. Although RAG significantly improves factual grounding, it also brings along challenges like possible privacy threats when dealing with sensitive information and reconciling differences between the knowledge that is retrieved and the priors learned by the model.

4.7.2 External Tool Integration

In addition to offering knowledge, conversational AI systems usually need to do something. Tool integration [11] enables them to communicate with external services and execute operations that lie beyond language modeling. This can include invoking a calculator to compute a mathematical problem, fetching a weather API to return live forecasts, or linking with enterprise systems in order to book a meeting or facilitate a transaction. In practice, the integration of tools turns conversational AI from a passive question-answering system into an active agent with the ability to accomplish user objectives. Tool orchestration must be well-designed.

The LLM or the dialogue manager has to determine when to call a tool, how to format the input to it, and how to feed the output back into the current conversation. Tools like LangChain and the function-call capabilities of newer LLMs give systematic means of linking models with services and APIs. It's essential to handle errors because tools outside the system can fail, provide incomplete outputs, or return in unmatched formats. Trust is also essential: systems need to give feedback on when they are taking external actions and assure that sensitive operations are to be under protection and user control.

4.7.3 Combining Retrieval and Tools within Ensembles

The strongest conversational systems integrate retrieval and tool use in an ensemble setup. As an illustration, a spoken question can be typed up using automatic speech recognition, sent to an LLM to reason, augmented with documents retrieved from a knowledge base, and followed up with calls to outside APIs for real-time information. The results of these processes are harmonized through orchestration logic and ultimately synthesized into multimodal output, like text, speech, or images. This unification emphasizes how retrieval and tool utilization are not independent add-ons but fundamental aspects of a single pipeline that combines perception, reasoning, and action. The harmonization between retrieval and tools needs to be examined in their totality.

Systems must ensure factual consistency, avoid contradiction between retrieved material and generated reasoning, and control latency so that tool calls do not disrupt conversational flow. Hybrid solutions work well, where lightweight retrieval is done on each turn, but more sophisticated or expensive tool calls are initiated only when the dialogue context renders them unavoidable. The goal is to provide answers that are both based on verifiable information and able to stretch beyond the model's inherent capacities.

4.7.4 Outlook

Knowledge grounding and tool integration mark a turning point in conversational AI. By augmenting generative models with retrieval and actionable tools, designers create systems that are more factual, useful, and trustworthy. This shift allows conversational agents to serve effectively in high-stakes domains such as healthcare, education, and enterprise support, where accuracy and reliability are critical. When combined with ensemble orchestration, retrieval and tool use enable the creation of assistants that not only understand and generate but also connect, act, and adapt. This convergence brings multimodal conversational AI closer to functioning as a true collaborator in human tasks rather than a static generator of responses.

4.8 CHALLENGES AND SOLUTIONS

4.8.1 Technical Challenges

Multimodal conversational AI promises fluid interaction across text, speech, images, and video, but achieving this in production exposes three linked technical challenges: latency and inference cost, data alignment across modalities, and memory management for large models. Each affects user experience, reliability, and scalability, and each has practical engineering solutions.

4.8.1.1 Latency and Inference Cost

Natural-sounding conversation requires sub-second turnaround, but big generative models are computationally expensive. Autoregressive decoding is sequential in nature, and image/video synthesis contributes additional lag. The outcome is a balance between high-dimensional multimodal output and interactive latency. Mitigations occur across various layers. Model-level: quantization, pruning, and knowledge distillation reduce model sizes without loss of usefulness. System-level: KV-cache reuse, operator fusion, and continuous batching all minimize GPU overhead. Architectural: cascaded models send simple cases to light classifiers and save heavy LLMs or generators for hard inputs. Infrastructure: tensor or model parallelism and accelerators enhance throughput at scale. In real life, prod teams use combinations of these and monitor end-to-end latency, not merely model runtime, to provide perceived responsiveness.

4.8.1.2 Data Alignment Across Modalities

Multimodal performance relies on synchronizing signals across various structures (tokens, pixels, audio frames) and timescales. Misalignment generates unnatural outputs, including captions that do not match images or sentiment errors in speech.

Applications involve representation-level alignment (contrastive pretraining, cross-modal attention), temporal synchronization (ASR timestamps, video frame indices), and supervised multimodal fine-tuning on paired datasets. Hybrid fusion approaches balance tight cross-modal reasoning with modularity. In practice, resilient systems also utilize redundancy (optical character recognition (OCR) and captioning and detection), monitor metadata such as confidence and provenance, and degrade gracefully by posing clarifying questions in case of missing or noisy signals.

4.8.1.3 Memory Management for Large Models

Big models emphasize memory in both weight storage and context processing. Multimodal systems enhance this by demanding long conversation histories, cross-modal embeddings, and retrieval caches. Naïve storage rapidly outstrips GPU capacity.

Mitigations are multi-faceted. Model distribution: sharding and offloading weights or activations to central processing unit (CPU) or solid state drive (SSD). Serving optimizations: paged attention and efficient KV-cache management (as in vLLM [16]) avoid waste and facilitate sharing across sessions. Context strategies: semantic summarization compresses dialogue history, retrieval databases enabling selective recall. Precision and compression: FP16/BF16 or integer quantization reduce footprint with little loss of quality. Combined, these strategies enable large-scale multimodal dialogue without draining resources.

Latency, alignment, and memory are tightly coupled. Less latency can necessitate more caching, which increases memory consumption; tighter alignment can increase storage and synchronization cost. The answer is co-design: choose architectures and serving frameworks that make knobs for quantization, sharding, and caching visible; impose timestamp-aware fusion; and model orchestration to use costly components only when needed. Solved all at once, these problems bring multimodal AI from promising prototypes to real-time, working conversational systems.

4.8.2 Ethical and Societal Dimensions

Alongside technical barriers, multimodal conversational systems face serious ethical and societal challenges. Because they process sensitive signals such as faces, voices, and private spaces, and because they generate highly realistic outputs, the risks of bias, privacy violations [18], and misuse are amplified. Addressing these is not optional but central to trustworthy design and deployment.

4.8.2.1 Bias Propagation

Bias in AI arises when models reflect or amplify unfair patterns in training data [19]. In multimodal systems, the problem compounds: text corpora may underrepresent dialects, image datasets often skew demographically, and audio corpora favor certain accents. Bias in one modality can cascade into others for example, an image captioning error feeding a language model leads to stereotyped dialogue.

Real-world failures include worse speech recognition for minority accents or misclassification of faces across ethnic groups, leading to exclusion and mistrust. Mitigation requires discipline across the pipeline: diverse and balanced datasets, fairness-aware training objectives, adversarial debiasing, and continuous auditing with tools such as Fairlearn or AIF360. The best practice is to maintain a bias-risk register per modality, build targeted demographic test suites, and use counterfactual evaluation to detect whether protected attributes unduly influence outcomes. This makes fairness an engineering practice rather than an afterthought.

4.8.2.2 Privacy Concerns

Multimodal inputs are deeply personal: a voice sample is biometric, a photo may expose identity or surroundings, and a video could reveal health or financial details. Risks span both training (models memorizing private data) and deployment (logs inadvertently storing sensitive content). Regulations such as the General Data Protection Regulation (GDPR) and emerging AI acts require explicit legal bases for collection, retention, and consent, making privacy a first-class design concern.

Technical approaches involve reducing data collection, favoring on-device processing, and using privacy-protecting machine learning (ML) techniques like differential privacy or federated learning. These are supplemented by encryption, robust access controls, and short retention periods for raw media. Teams in practice should handle biometric modalities as high risk, default to local or opt-in processing, pseudonymize metadata, and provide users with explicit controls to delete or inspect their data.

4.8.2.3 Misuse Prevention

These controls mitigate both legal risk and user distrust. Generative multimodal models can be leveraged for impersonation, misinformation, or targeted deception. Voice cloning facilitates fraud, deep fake videos drive disinformation, and LLM-powered perception can automate social engineering. One defense cannot cover it all, so multiple layers of control are needed.

On the technical front, watermarking and provenance enable tracking of AI-produced content, while forensic detectors spot faked media. Policy-level protection encompasses rate limiting and access controls for high-risk features, human audit for sensitive applications, and open abuse reporting. These resonate with nascent regulatory requirements, like the European Union (EU) calls for provenance, risk analysis, and caps on biometric monitoring. Proactive misuse controls therefore support ethical obligation and compliance.

Bias, privacy, and misuse are interrelated: improved dataset governance decreases bias and minimizes privacy risk, and provenance metadata enables both fairness audits and misuse detection. Successful governance integrates technical controls (differential privacy, watermarking, detection models) with process controls (model cards, datasheets, audits, red-teaming, incident response). Integrating these protections into development pipelines ensures multimodal AI stays not only strong but also responsible and socially viable.

4.9 BEST PRACTICES FOR BUILDING SCALABLE SYSTEMS

Building multimodal conversational AI systems that are powerful and reliable in production requires more than strong models. It also demands engineering discipline, architectural foresight, and operational robustness. This section outlines five core best-practice areas: modular architecture, API-first design, caching and batching, fine-tuning strategies for different modalities, and rigorous testing and evaluation. Together, these practices provide a practical blueprint for deploying systems that scale effectively while maintaining quality.

4.9.1 Modular Architecture

The most reliable systems are designed as loosely coupled modules connected through well-defined interfaces. Typical modules include perception services such as ASR and vision encoders, a core reasoning engine such as an LLM, retrieval and memory services like vector databases or RAG pipelines, generation services such as TTS and image or video generators, and an orchestration layer that coordinates their operation.

Modularity ensures that components can evolve independently, bottlenecks can be scaled selectively, and specialist models can be replaced without redesigning the entire system. In practice, models are often deployed as services behind clear APIs, with shared feature stores or artifact repositories to enable reusability. Workflow orchestration frameworks such as Argo, KServe, or BentoML coordinate training, validation, and inference pipelines in production.

A modular design also supports practical tactics such as separating fast-path low-latency inference from slower but higher-quality generation, maintaining a shared feature layer for embeddings or cached captions, and instrumenting modules with standardized logging, metrics, and tracing for accurate performance profiling.

4.9.2 API-First Design

In scalable systems, the API is the product. Designing inputs, outputs, schemas, error codes, and streaming behaviors before building internals accelerates development and integration. An API-first approach improves compatibility with agent frameworks like LangChain, simplifies testing, and enforces consistency across services.

For AI systems, the best APIs are machine-friendly and explicit. They define compact and typed payloads, expose clear session and memory endpoints, support streaming for real-time modalities such as ASR and TTS, and return quality and confidence signals so that clients can decide whether to retry, escalate, or accept outputs.

Practical measures include defining schemas in formats such as JSON, Avro, or Protobuf, publishing them as canonical specifications, providing streaming protocols for long-running or live generation tasks, and embedding metadata such as timestamps and quality scores in every response.

4.9.3 Caching and Batching Requests

Caching and batching are two of the most effective strategies for reducing cost and increasing throughput. Caching avoids unnecessary precomputation by storing embeddings, transcriptions, captions, or intermediate results, while batching maximizes GPU utilization by grouping requests.

Modern systems use dynamic batching, continuous batching, and adaptive routing to balance latency and efficiency. For example, short accumulation windows allow multiple requests to be aggregated without noticeable delays for users, while urgent requests can be routed to low-latency servers. At the same time, caching must be layered, with lightweight stores such as Redis for embeddings and fast results, and object storage for heavier artifacts with indexing for retrieval.

Batching strategies must also account for memory, since larger batches increase the footprint of KV caches. Monitoring cache usage, employing shared caches, and applying time-to-live policies are essential for sustaining performance without overwhelming GPU resources.

4.9.4 Fine-Tuning Strategies for Different Modalities

Effective fine-tuning balances performance with cost. For text models, options include full fine-tuning, instruction tuning, and parameter-efficient methods such as LoRA, adapters, or prefix tuning. Parameter-efficient fine-tuning is often sufficient for domain adaptation, offering strong gains with lower cost, while instruction tuning and reinforcement learning with human feedback are used to align conversational behavior and safety.

For diffusion-based image models [7,8], fine-tuning typically involves lightweight conditioning modules or LoRA-style updates rather than retraining entire networks. This enables efficient adaptation for domain-specific styles or tasks. For audio, ASR systems can be fine-tuned on domain transcripts, while TTS can be customized using neural voice cloning or voice-style embeddings with regulatory guardrails. Multimodal fine-tuning often combines instruction tuning with mixtures of adapters to reduce task interference across input types. In many production systems, retrieval-augmented generation is combined with adapter-tuned models for improved factuality without the cost of full model retraining.

4.9.5 Testing and Evaluation Metrics for Multimodal AI

Evaluation must occur both at the task level and at the system level. Task-level metrics include word error rate for ASR, BLEU, or ROUGE for translation and captioning, and FID or CLIPScore for images. However, modern research highlights that learned metrics based on vision–language models often correlate more closely with human judgment, making them especially valuable.

System-level evaluation should measure latency, robustness, fairness, and safety. Recommended practices include maintaining representative multimodal test sets, running regression tests for accuracy and performance, applying adversarial robustness and demographic fairness tests, and using human evaluation through A/B testing or ranking studies. Continuous monitoring is essential to ensure quality under real-world conditions, while red-team exercises and adversarial perturbations provide additional safeguards against misuse or failure.

Scalable multimodal systems require a disciplined engineering approach. Modular design ensures flexibility, API-first development enables clean integration, caching and batching optimize efficiency, fine-tuning strategies balance adaptability with cost, and systematic evaluation maintains quality and safety. Together, these practices provide the foundation for deploying multimodal conversational AI that is not only powerful but also practical, reliable, and trustworthy at scale.

4.10 CASE STUDIES AND APPLICATIONS

Real-world deployments illustrate how multimodal conversational AI transforms user experience. The following vignettes present four representative cases: customer service bots, AI tutors, healthcare assistants, and creative content systems. Each shows how an ensemble of specialized models can be orchestrated into an end-to-end solution, highlighting both design choices and practical lessons for engineers.

4.10.1 Customer Service Bots

Contact centers increasingly use multimodal conversational AI to deliver frictionless self-service and seamless handoffs to human agents (Figure 4.8). A typical architecture begins with streaming ASR to capture customer requests in real time. An intent detection and entity extraction module then interprets the query, which is routed to a knowledge store or RAG pipeline. The system generates a grounded response and delivers it either as synthesized speech through a TTS engine or as on-screen text with suggested options for follow-up.

Enterprise deployments by major cloud vendors and providers such as Genesys extend this flow with sentiment tracking, routing logic, and live-agent context passing. Escalation is managed through cascaded logic: lightweight intent classifiers are used first for speed, while full-scale LLM generation is triggered only when necessary. Confidence scores and transcripts are attached to each interaction so that the orchestrator can decide whether to clarify or to escalate. This balance allows systems to reduce average handle time, increase self-service resolution rates, and preserve the human fallback needed when automation is insufficient.

The key engineering lessons are clear. Low-latency intent detection is critical for maintaining conversational flow. Escalation paths must be designed with context handoff so that human agents receive full transcripts and reasoning history. Confidence thresholds should be instrumented into the orchestration layer so that automation remains predictable and safe.

4.10.2 AI Tutors with Visual Explanations

AI tutoring systems showcase the value of combining language reasoning with visual support for learning. A typical tutor pipeline accepts student queries through typed prompts or transcribed speech, retrieves curriculum-aligned resources, and uses an LLM to generate pedagogically structured explanations. To reinforce concepts, the

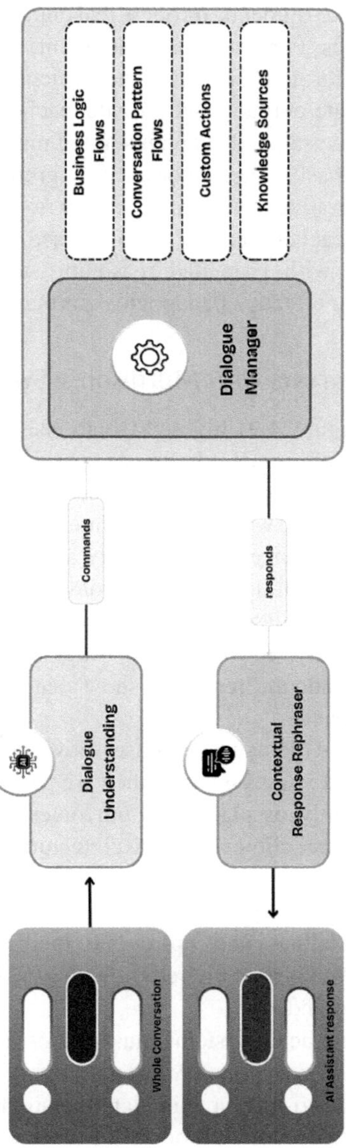

FIGURE 4.8 The central role of the dialogue manager in conversational AI.

tutor produces visual scaffolds such as annotated diagrams, worked examples, or step-by-step visualizations.

Experiments such as Khan Academy's GPT-4 based tutor highlight how this integration strengthens learning. LLM reasoning is coupled with domain-specific calibration modules, for example, symbolic solvers in mathematics, that reduce factual errors. Visual scaffolds guide students through reasoning sequences and help disambiguate misunderstandings. For practical deployment, tutoring systems must also manage content alignment through metadata and difficulty tagging, enforce safety filters to prevent inappropriate outputs, and provide teacher-in-the-loop oversight so that educators can review, correct, and adapt generated material.

The design lessons emphasize the importance of grounding LLM reasoning in curriculum resources, integrating symbolic checkers to ensure accuracy in technical domains, and giving teachers the ability to oversee and author content. Visual explanations, when paired with grounded reasoning, turn AI tutors from mere question-answering bots into effective pedagogical partners.

4.10.3 Healthcare Assistants with Multimodal Symptom Analysis

Healthcare applications (Figure 4.9) highlight both the potential and the caution required when deploying multimodal AI. Practical assistants in this domain often combine text descriptions of symptoms, recorded speech, and patient-provided photos such as rashes or wounds, which are then integrated with clinical knowledge retrieval to generate triage guidance or preliminary assessments. Advances in medical imaging, particularly CNNs trained on dermatology datasets, have demonstrated specialist-level accuracy in skin lesion classification when rigorously validated on large clinical cohorts. Similarly, conversational triage systems, whether developed commercially or through academic research, show that AI-driven interactions can provide useful first-line guidance.

However, safe deployment demands a conservative approach. These assistants must have clearly defined boundaries on what they are permitted to suggest, communicate uncertainty transparently, escalate risky outcomes to clinicians, and undergo extensive validation against real clinical data. Architecturally, such systems typically pair image-analysis modules for

visual data and ASR/NLU for patient history with retrieval-augmented components that draw on clinical guidelines and up-to-date medical literature. The LLM synthesizes an explanation that is then filtered through a decision policy to determine whether clinician review is mandatory.

The key lesson is that healthcare systems must be positioned as decision-support rather than diagnostic tools. Developers have to thoroughly validate models, emphasize uncertainty to the user, and impose clinician oversight for high-risk situations. Such guarded framing guarantees multimodal assistants contribute value to patient care without jeopardizing safety.

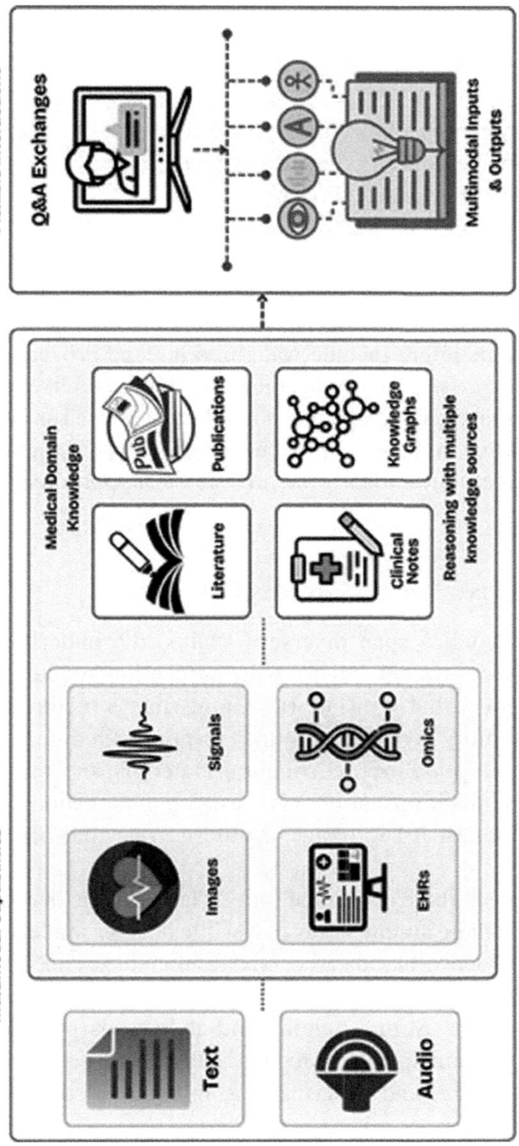

FIGURE 4.9 AI assistant for medical Q&A with multimodal and domain expertise.

4.10.4 Creative Content Generation

Creative pipelines demonstrate the complete expressive power of multimodal groups. A common pipeline might start with an LLM writing a story, script, or sequence of story beats. Models like Stable Diffusion or SDXL can subsequently create illustrations, concept art, or storyboards. These textual and visual inputs can be accompanied by natural-sounding voices from generative TTS models like Bark, while text-to-video models like Pika Labs or future open-source initiatives take the pipeline all the way to short animated videos.

Production-level systems optimize this fundamental pipeline using prompt engineering, conditional image editing, stylistic adapters, and audio post-processing to make outputs consistent with artistic vision.

Human creators remain central to the loop, using iterative cycles of generation, selection, and refinement to balance speed, cost, and quality. Alongside this creative freedom, however, comes a set of ethical challenges. Issues of copyright, voice consent, and potential misuse through deepfakes require deliberate safeguards. Responsible pipelines therefore include watermarking and provenance metadata, as well as explicit consent and rights management for voices and likenesses.

The lesson here is that creative pipelines should be structured as iterative generate–select–refine loops, while embedding provenance and consent frameworks from the outset. This ensures that multimodal generative tools support creativity responsibly and transparently.

4.10.5 Cross-Cutting Patterns Across Case Studies

Although these case studies span diverse domains, the underlying architectural patterns converge. Each system begins with perception modules such as ASR, OCR, or vision encoders that translate raw inputs into structured embeddings or text. A retrieval or memory layer then grounds responses in domain-specific knowledge, whether clinical guidelines, curriculum materials, or enterprise databases. The LLM coordinates this information, synthesizing explanations and determining which generative modules TTS, image synthesis, or video engines should be invoked.

Effective deployments share additional traits. They cascade decisions so that fast, lightweight modules handle simple cases first while heavier models are reserved for complex queries. Confidence-based gating ensures that uncertainty is surfaced rather than hidden, while human escalation remains central in high-stakes settings such as healthcare or education. Strong logging and audit trails provide accountability across domains, and domain-specific safeguards, teacher oversight in tutoring, clinician sign-off in healthcare, and watermarking in creative content ensure responsible use.

Together, these case studies demonstrate how multimodal conversational AI can be adapted to varied contexts while following a shared design logic: modular perception, grounded reasoning, coordinated generation, and domain-appropriate safeguards.

4.11 CONCLUSIONS

Multimodal conversational AI represents a significant step forward in the evolution of intelligent systems. By moving beyond text-only interactions and incorporating speech, images, video, and other modalities, these systems are beginning to reflect the richness of human communication. The foundations of this progress lie in the convergence of advances across artificial intelligence, machine learning, deep learning, and natural language processing, with large language models and generative models serving as the unifying backbone.

Throughout this chapter, we have examined how modalities such as text, speech, image, and video each play distinct but complementary roles in both human and machine communication. We have seen how LLMs and generative models enable perception, reasoning, and generation, acting as the central orchestrators that tie these modalities into coherent conversations. Open-source models across all modalities have expanded accessibility and innovation, providing researchers and practitioners with versatile building blocks that can be integrated into larger systems.

A key theme has been the necessity of ensemble approaches. No single model can address the full diversity of human communication, so systems must be designed as coordinated pipelines of specialists, each contributing unique capabilities. Orchestration frameworks and integration strategies, including retrieval-augmented generation and external tool usage, further extend these systems' power by grounding them in real-world data and enabling task-specific reasoning.

The chapter has also highlighted the challenges that arise at both technical and societal levels. Latency, alignment, and memory management are central engineering concerns that determine whether multimodal AI can perform reliably in real time. At the same time, bias, privacy, and misuse pose ethical risks that cannot be ignored if these technologies are to be trusted and deployed responsibly. Addressing both sets of challenges requires a combination of technical solutions, governance mechanisms, and continuous evaluation.

Best practices such as modular architecture, API-first design, caching, fine-tuning strategies, and rigorous testing provide pathways for building scalable, efficient, and safe systems. Case studies in customer service, education, healthcare, and creative industries illustrate how these ideas move from theory to practice, showing the transformative potential of multimodal AI when deployed thoughtfully.

In closing, multimodal conversational AI should be seen not as a single technological breakthrough but as the integration of many strands of AI research into a unified framework. Its promise lies in creating systems that are not only intelligent but also contextually aware, interactive, and human-like in their adaptability. Achieving this requires ongoing refinement, careful balancing of strengths and limitations, and a commitment to responsible innovation. With these principles in place, multimodal conversational AI will continue to evolve as a cornerstone of human–AI interaction in the years ahead.

REFERENCES

[1] A. Vaswani, N. Shazeer, N. Parmar, J. Uszkoreit, L. Jones, A. N. Gomez, Ł. Kaiser, I. Polosukhin, "Attention Is All You Need," *NeurIPS*, 2017. arXiv
[2] T. B. Brown, B. Mann, N. Ryder, et al., "Language Models are Few-Shot Learners," *NeurIPS (GPT-3 paper)*, 2020.
[3] H. Touvron, T. Lavril, G. Izacard, X. Martinet, et al., "LLaMA: Open and Efficient Foundation Language Models," *arXiv*, 2023.
[4] E. Almazrouei, et al., "The Falcon Series of Open Language Models," *arXiv*, 2023.
[5] A. Radford, J. W. Kim, C. Hallacy, A. Ramesh, et al., "Learning Transferable Visual Models From Natural Language Supervision (CLIP)," *arXiv*, 2021.
[6] P. Lewis, E. Perez, A. Piktus, F. Petroni, V. Karpukhin, N. Goyal, et al., "Retrieval-Augmented Generation for Knowledge-Intensive NLP Tasks (RAG)," In Advances in Neural Information Processing Systems (NeurIPS), Vol. 33, pp. 9459–9474, 2020.
[7] R. Rombach, A. Blattmann, D. Lorenz, P. Esser, B. Ommer, "High-Resolution Image Synthesis with Latent Diffusion Models," *CVPR / arXiv*, 2022.
[8] Stability AI. "Stable Diffusion—Public release" (official release/announcements and model pages), Aug 2022.
[9] A. Radford et al., "Robust Speech Recognition via Large-Scale Weak Supervision (Whisper)," *arXiv*, 2022.
[10] Suno (GitHub), "Bark: Text-Prompted Generative Audio Model" (project/model repo and docs).
[11] T. Schick, J. Dwivedi-Yu, R. Dessì, R. Raileanu, M. Lomeli, L. Zettlemoyer, et al., "Toolformer: Language Models Can Teach Themselves to Use Tools," *arXiv / NeurIPS proceedings*, 2023.
[12] OpenAI, "GPT-4 Technical Report," *OpenAI / arXiv*, 2023.
[13] D. Driess, F. Xia, M. S. M. Sajjadi, C. Lynch, A. Chowdhery, et al., "PaLM-E: An Embodied Multimodal Language Model," *arXiv / MLR proceedings*, 2023.
[14] T. Wolf, L. Debut, V. Sanh, J. Chaumond, C. Delangue, et al., "Transformers: State-of-the-Art Natural Language Processing" (Hugging Face Transformers library paper/demo), 2019/2020.
[15] M. Douze, J. Johnson, et al., "The Faiss library" (vector similarity search / indexing toolkit; library paper and docs), *FAISS / arXiv*, 2017 / later summary.
[16] vLLM project, "vLLM: A high-throughput and memory-efficient inference engine for LLM serving" (project repo and docs).
[17] R. Bommasani, D. A. Hudson, E. Adeli, R. Altman, et al., "On the Opportunities and Risks of Foundation Models," *CRFM / arXiv report*, 2021.
[18] C. Dwork, A. Roth, "The Algorithmic Foundations of Differential Privacy," *Foundations and Trends in Theoretical Computer Science*, vol. 9, no. 3–4, pp. 211–407, 2014. https://doi.org/10.1561/0400000042
[19] N. Carlini, C. Liu, Ú. Tramèr, J. Kos, et al., "Extracting Training Data from Large Language Models," *USENIX Security*, 2021.
[20] J. Gao, Z. Yang, X. et al., "A Survey on Deep Learning for Multimodal Data Fusion," *Neural Computation*, vol. 32, no. 5, pp. 829–864, 2020, https://doi.org/10.1162/neco_a_01273.

5 Privacy and Data Security Concerns in Multimodal AI

Rubini P., Rohan Naveen Gajan, Pooja Nayak S., Noor Sumaiya, Sudhakar K., and Dishita Kamath

5.1 INTRODUCTION

5.1.1 Overview of Multimodal AI

Multimodal AI describes systems that capture, analyze, and combine information from multiple sources like text, images, audio, video, and sensor signals. These systems create deeper insights and make better decisions compared to models that rely on just one type of data. Most traditional AI models stick to handling a single kind of input, like processing the text through natural language models or analyzing visuals through convolutional neural networks. This focus on a single type limits their ability to understand contexts or situations. Multimodal systems combine the unique strengths of different inputs. Visual clues help explain language. Sounds can clarify visual scenes. Text data gives more meaning to video.

Recent improvements in deep learning models like transformer-based fusion layers, cross-modal attention methods, and shared embedding spaces have helped push this area forward. Top models such as Contrastive Language–Image Pre-training (CLIP), A Large-scale Image and Noisy-text embedding framework (ALIGN), and Florence show how linking image and text representations can make tasks like zero-shot classification, automatic captioning, or cross-modal searching possible. In the same way, audio-visual systems combine speech and lip-reading to work better in noisy settings, and video-text models support automatic captioning and searching across different types of content.

Multimodal AI finds use in areas like healthcare, where it combines radiology scans with digital health records to make better diagnoses. In autonomous vehicles, it brings together data from LIDAR (light detection and ranging), cameras, and radar to guide navigation more. It also powers smart consumer tech such as virtual assistants that can link spoken commands to what's happening in visual settings. As these systems become more advanced and widespread, they collect and use an ever-growing variety of personal data. This raises tough questions about keeping such data private, secure, and handled. Later sections will dig into these issues and suggest ways to protect multimodal AI systems.

DOI: 10.1201/9781003678359-5

5.1.2 Overview of Multimodal AI

Privacy and protecting data play a key role in how multimodal AI is developed and used. There are several reasons why this matters.

To start with, multimodal AI systems gather, handle, and keep huge amounts of personal and sensitive data. They often work across text, audio, video, and images all at once. By merging this information, these systems create more detailed profiles of users than what one type of data alone could offer. If this data gets leaked or accessed without permission, the harm can grow. For instance, linking a person's facial images with their voice recordings and chat transcripts can allow highly accurate identification, predictions about behavior, and even guesses about emotions. This creates serious privacy concerns.

Second, the growing attack surface of multimodal AI increases risk. Security systems built to handle unimodal data might not work well against cross-modal vulnerabilities. Attackers can exploit hidden links between modes. For instance, altering audio could cause models to misclassify images in combined setups. This shows the importance of creating strong defenses that take different modes into account.

Without strong privacy protections, multimodal AI may worsen current biases and unfair outcomes. Inference attacks carried out without permission could reveal sensitive details like health issues or political views from harmless information. This could cause ethical problems and legal risks under rules like General Data Protection Regulation (GDPR) and Health Insurance Portability and Accountability Act (HIPAA), establishing a profound tension between utility and user anonymity (Shroff, 2023).

In the end, privacy-focused design drives innovation by making data sharing and joint learning more secure. Methods like differential privacy and federated learning allow groups to use varied multimodal datasets without collecting sensitive raw data in a single place. This lets them utilize multimodal AI's abilities while protecting user privacy and system safety. To sum up, maintaining privacy and securing data is essential to create both strong technical systems and uphold the ethical standards that today's multimodal AI requires.

In autonomous vehicles, it brings together data from LIDAR, cameras, and radar to guide navigation more. It also powers smart consumer tech such as virtual assistants that can link spoken commands to what's happening in visual settings. As these systems become more advanced and widespread, they collect and use an ever-growing variety of personal data. This raises tough questions about keeping such data private, secure, and handled. Later sections will dig into these issues and suggest ways to protect multimodal AI systems.

5.1.3 Overview of Multimodal AI

This chapter focuses on looking at the unique privacy and data security issues found in multimodal AI systems. These systems take in and handle different types of data like text, images, audio, and video all at once. The main goals are to:

1. Pinpoint key weaknesses
   a. Look into how integrating different modalities increases the risk by making the attack surface bigger than with single-modal models.

   b. List attack types such as cross-modal adversarial attacks, membership inference, model inversion, and reconstruction threats.
2. Review technical challenges and legal effects
   a. Evaluate difficulties in protecting mixed data systems, like using encryption managing keys, or handling secure computation.
   b. Study how current laws like GDPR or HIPAA fit multimodal setups and where they fall short in meeting these new needs.
3. Explore cutting-edge defenses
   a. Examine methods like differential privacy federated learning homomorphic encryption, and secure multi-party computation used to protect privacy in multimodal systems.
   b. Study new frameworks and protocols built to recognize and lessen cross-modal adversarial risks.
4. Suggest best practices and future plans
   a. Offer guidelines to create multimodal AI systems that build privacy and security into their design from the start.
   b. Highlight unresolved research problems and suggest ways to collaborate across fields like AI, cryptography, and data management.

This chapter covers basic ideas and new methods. It starts by explaining multimodal AI systems and their uses. Then, it explores issues like security weaknesses and privacy threats in detail. In the second part, it shifts to ways to reduce these problems drawing from academic studies and new industry rules. It wraps up with hands-on advice and plans to guide future efforts to build privacy-aware multimodal AI tools. It includes examples from real-world cases and legal systems to help researchers, developers, and policymakers use these systems in safe and fair ways.

5.1.4 Chapter Summary

This chapter examines privacy and data security challenges unique to multimodal AI systems that integrate text, image, audio, video, and sensor data. It highlights how cross-modal fusion increases attack surfaces, enabling sophisticated threats such as model inversion, membership inference, and cross-modal adversarial attacks. The chapter further evaluates regulatory implications under GDPR, HIPAA, and emerging AI governance frameworks.

To address these challenges, the chapter surveys state-of-the-art mitigation strategies including differential privacy, federated learning, homomorphic encryption, secure multi-party computation, and privacy-by-design frameworks. It concludes by outlining future research directions and policy recommendations aimed at enabling secure, ethical, and privacy-preserving deployment of multimodal AI systems.

5.2 BACKGROUND

5.2.1 Definition and Examples of Multimodal AI Systems

Multimodal AI systems work to handle and combine information from multiple types of data like text, images, audio, video, and sensor signals all within one framework.

Traditional models deal with just one specific type of data, but multimodal setups form shared representations that bring out the strengths of each kind of data. This blending allows the system to grasp the context better, become more reliable, and do things that single-input models cannot achieve.
The main building blocks of a multimodal AI system are as follows:

- *Modality-specific encoders*: Neural networks, like transformers for handling text convolutional layers for working with images, or spectrogram-based designs for processing audio, extract detailed features from each data type.
- *Fusion mechanisms*: Methods such as cross-modal attention combining data through concatenation with projection layers, or using specialized fusion transformers, bring embeddings together into a shared space.
- *Decoders or task heads*: Modules designed for specific tasks such as heads for classification, networks for sequence generation, or systems for detection, use the combined representation to generate results.

Examples:

1. Vision–language models
   a. CLIP trains image and text encoders together to link pictures and words in the same embedding space. This setup makes tasks like zero-shot image classification and finding images by description possible.
   b. ALIGN learns connections between images and text using large sets of paired data. It supports tasks like writing image captions or searching across multiple modes of data.
2. Audio-visual speech recognition
   a. These systems use both sound and visual mouth movements to improve speech recognition in noisy places. They combine video of how lips move with features from sound spectrograms. This lets them outshine systems that rely on audio when there is background noise.
3. Healthcare diagnostic platforms
   a. Integrated systems combine medical imaging like magnetic resonance imaging (MRI) and computed tomography (CT) scans with electronic health records such as lab reports and clinical notes, to boost accuracy in diagnosis and tailor treatment advice. This joint analysis uses both visual pathology details and past health data.
4. Autonomous driving
   a. Autonomous driving perception systems rely on setups that mix images from cameras, radar signals, LIDAR point clouds, and global positioning system (GPS) inputs to map surroundings. By combining these types of data, the system works better even when a single sensor struggles during bad weather or poor light.
5. Multimodal virtual assistants
   a. Assistants that can process spoken commands alongside visual cues help in tasks like spotting objects in a scene or understanding gestures seen through a camera. This combination forms the foundation of interactive robots and augmented reality tools.

By using the unique strengths of different data types together, multimodal AI systems develop better reasoning and stronger task performance. These advancements help create smarter, context-aware applications.

5.2.2 Current State of Privacy and Data Security in AI

AI's approach to privacy and data security combines fast-moving technological progress with increasing awareness of the risks. Important changes and obstacles involve the following:

1. Learning paradigms that protect privacy
   a. Federated learning now supports large-scale use across devices and silos. It allows training models on decentralized data without needing raw data to be shared. Yet, if gradients are not well-secured, these systems can still face risks from gradient inversion attacks.
2. Encryption for secure computation
   a. Homomorphic encryption (HE) allows advanced deep-learning inference to run on encrypted data. Performance overhead however still stops it from becoming adopted.
3. Regulatory and governance rules
   a. Laws like GDPR, HIPAA, and California Consumer Privacy Act (CCPA) enforce strict rules to protect data. New guidelines such as the European Union (EU) AI Act, are starting to focus on "high-risk" AI systems. These include requirements like reducing data usage ensuring transparency and checking for algorithmic impacts.
   b. Standards bodies such as International Organization for Standardization (ISO)/IEC and National Institute of Standards and Technology (NIST) are working to create security controls and privacy frameworks tailored to AI. Different industries though, are adopting these at their own pace.
4. Adversarial threats and vulnerabilities
   a. Attacks like membership inference, model inversion, and reconstruction attacks continue to show effectiveness against both centralized and federated models. These methods can uncover training data or extract sensitive details.
5. Ethical and societal considerations
   a. Problems like bias amplification may expose private attributes or worsen unfair outcomes. This creates privacy issues when sensitive details are inferred without getting consent.
   b. Transparency and explainability tools often need to create audit trails to track how data is being used. However, methods to explain complex models that protect privacy are still in early stages.
6. Emerging best practices
   a. Privacy by design focuses on adding privacy and security features right from the start. This happens during steps like gathering data, building

models, and deploying them. Using threat modeling helps to spot possible risks.

b. Hybrid solutions bring together methods like differential privacy federated learning, and encryption. For example, using private aggregation of teacher ensembles strengthens the system's reliability.

Overall, many technical and regulatory tools exist to protect AI systems, but real-world use still struggles with balancing performance adapting to changing threats and creating consistent rules. To address these issues, teams across fields and industries need to work together and keep researching. This work is key to making sure AI systems stay effective while respecting privacy.

5.2.3 Relevant Regulations and Standards

Many rules regulate data privacy and protection in AI systems with multimodal use. GDPR from the European Union has strict standards for collecting and using personal data. These include limiting data collection, using it for clear purposes, and making sure people give clear consent. This applies to multimodal models that merge text, audio, and images to identify sensitive details. The GDPR also requires organizations handling automated decisions or profiling, like high-risk AI systems, to perform data protection impact assessments (DPIAs). These assessments ensure risks to privacy are spotted and addressed before using multimodal AI on a larger scale (Lin & Park, 2023). The California Consumer Privacy Act (CCPA) sets rules requiring transparency and gives people the ability to access, delete, and opt out of selling their personal information. This has an effect on multimodal services using things like user-uploaded images, voice recordings, or text logs to analyze data or improve features.

In healthcare, the Health Insurance Portability and Accountability Act (HIPAA) sets tough rules to protect private health information processed by AI in clinics or research. Systems that combine medical images and electronic health records need to use strict security measures like encryption, access restrictions, and logs to follow HIPAA guidelines. Any leaks of private health details can lead to heavy fines and mandatory plans to fix problems. This highlights why strong security is crucial throughout multimodal data processes.

New regulations and standards specific to AI are shaping how secure and transparent multimodal systems are designed. The proposed AI Act in the European Union sorts AI systems by risk and sets tougher rules on "high-risk" AI. These rules require strict data management human supervision, and monitoring after systems are deployed. The Act would make multimodal AI used in areas like critical infrastructure, law enforcement, and hiring go through checks for compliance and follow rules about transparency. Around the globe, organizations like ISO/IEC and NIST are working to create detailed guidelines to manage AI risks, protect privacy, and reduce bias. Important efforts involve ISO/IEC JTC 1/SC 42's work on "Artificial Intelligence—Trustworthiness" and NIST's AI Risk Management Framework. These focus on designing systems with privacy in mind using ongoing monitoring and creating accountability systems that suit intricate multimodal structures.

5.3 SOURCES OF MULTIMODAL DATA

Multimodal AI systems rely on diverse data sources originating from multiple domains:

- *User-generated content*: Text chats, voice commands, uploaded images and videos
- *Sensor streams*: LIDAR, radar, GPS, IoT sensors in autonomous and smart systems
- *Healthcare systems*: Radiology images, electronic health records, clinical notes
- *Public datasets*: Vision–language corpora (e.g., image–caption datasets), audio-visual speech datasets
- *Synthetic and augmented data*: Generated using diffusion models and generative adversarial networks (GANs) to reduce privacy exposure

Each source introduces distinct privacy risks, particularly when cross-referenced, reinforcing the need for modality-aware data governance.

5.4 PRIVACY CONCERNS IN MULTIMODAL AI

5.4.1 Data Collection and Storage Issues

Multimodal AI systems work by pulling together big and varied datasets. They mix things like text logs high-quality pictures audio clips, and sometimes videos or sensor data to train models that connect and make sense of different types of information. Collecting this much data creates privacy issues on. Getting people's fully informed consent gets harder because most don't realize how their voice, photos, and words could be combined and analyzed at once. When data from multiple sources is joined, it reveals way more about someone's identity, actions, health, or feelings than any one type of data could alone.

Yet most consent forms miss mentioning these deep cross-modal insights. On top of that, storing all this massive and diverse data needs very advanced storage systems. Groups tend to gather multimodal datasets in one place to enable combined training and fine-tuning. This approach results in "honeypots," which hold sensitive personal data and turn into prime targets for hacking and data theft.

Storing multimodal data creates both technical and oversight problems. Different types of data need different storage approaches. Images and videos often go into object stores, while block storage or data lakes handle text and audio. This makes it harder to apply consistent encryption, set up access rules, and run audit logs across all data types. Retention policies may not align well, which might keep some data, like raw video footage, for an unlimited time even when complying by deleting features or transcripts derived from it.

Backup and replication plans add more copies of the data making errors or unauthorized access more likely. Storing data across different regions—where multimodal datasets are copied in various data centers around the world—can break local data-residency rules. This might lead to fines for organizations and make it harder to meet user requests related to data under laws like GDPR or CCPA. Applying strong security rules, minimizing unnecessary data, and following strict lifecycle policies are essential steps to protect privacy in multimodal AI setups.

As multimodal systems process diverse data types simultaneously, they face a complex web of overlapping global compliance requirements. Navigating these regional policy frameworks demands a proactive approach to data governance to prevent significant legal penalties (Tanaka, 2023).

5.4.2 Risks of Data Leakage and Unauthorized Access

Data leakage in multimodal AI happens during different stages of the data lifecycle, like when data is ingested, preprocessed, used in training models, during inference, or while storing intermediate files. This raises the chance of exposing information without permission. Feature extraction steps create embeddings that represent key parts of text, images, or sound. These embeddings can sometimes include hidden personal details or private features. Hackers may take advantage of weak application programming interfaces (APIs) or cleaned logs to capture these embeddings. They might use model inversion or reconstruction techniques to figure out the original data or detect private details like biometric information from facial or voice features.

Multimodal systems often rely on complicated setups like microservice architectures and external components. They may include different services to preprocess images, understand language, or transcribe audio. secured communication between services such as unencrypted Remote Procedure Call (RPC) or Representational State Transfer (REST) connections, opens up chances for spying or man-in-the-middle attacks. Mismanaged access controls also add risks. Weak Identity and Access Management (IAM) roles or ignored audit logs can let insiders or outside attackers explore combined datasets, which can result in large-scale data theft. Accidental leaks can happen during model updates or versioning.

When model checkpoints trained on sensitive data are saved without proper separation, attackers with access to the repository might pull combined details about the training data. Using strong encryption for storage and transfer, limiting access through role-based controls, ensuring secure API authentication, and cleaning up logs are all critical steps to prevent data leaks and block unauthorized access in multimodal AI systems (Yu & Zhao, 2023).

5.4.3 Case Studies Highlighting Privacy Breaches

Some well-known cases show how serious privacy threats can arise from multimodal AI systems when protections fall short.

- In 2021 the Dutch Data Protection Authority fined Clearview AI €30.5 million. The company had collected and stored over 30 billion facial images from public websites without permission to fuel its face-search tool. Clearview merged photos scraped from the web with biometric encoding algorithms to build a multimodal database. Each image in this database was tied to unique biometric templates. Both law enforcement and private organizations used this database without consent from the individuals involved. This broke several European Privacy Act principles, including illegally collecting sensitive data and failing to be transparent. The case highlighted how risky centralized multimodal biometric systems are when they mix different types of data without strong control measures.

- In 2025, a major breach hit the GenNomis deepfake platform leaking 47.8 GB of files. These included a significant number of AI-generated explicit or non-consensual videos. The platform stored user prompts, created image files, and metadata connecting text inputs to visual outputs, but none of it was encrypted and anyone could access it. This security failure highlighted risks in multimodal AI systems that mix text-based inputs with image-generation tools. Attackers accessed not the text prompts and the resulting visuals but also gained the ability to understand and exploit the system's structure for creating harmful content.
- Beyond big digital services, attackers have also aimed deepfake technology at regular people. In 2016, Noelle Martin, who was 17 and an activist, found a fake pornographic video where her face was mixed into unrelated footage. The process to make this deepfake used selfies and online clips found (visual input) combined with automated audio-to-video syncing tools. This created a lifelike fake multimedia piece she never consented to, which spread far and wide. Martin's experience shows how someone can link multiple deepfake tools—like face-swapping software, voice-cloning programs, and video-editing tech—to build a serious multimodal threat. These fakes can cause emotional, reputational, and legal damage to victims without relying on hacking into systems or involving companies.

These case studies together highlight how combining text, image, and audio data in multimodal AI has an influence on privacy risks. It turns separate weaknesses into interconnected attack methods. This also shows the need to create stronger security systems and oversight that focus on different data types.

5.4.4 Comparison Table

No.	Study	Modalities	Threat Addressed	Technique Used	Limitation
1	Smith et al. (2023)	Text–Image	Data leakage	DP	Accuracy loss
2	Lee and Kumar (2022)	Vision–Text	Cross-modal attacks	Adversarial training	High compute
3	Zhang et al. (2024)	Multimodal	Membership inference	FL + DP	Sync overhead
4	Nguyen and Patel (2022)	Multimodal	Pipeline leakage	Secure fusion	Limited scale

(continued)

No.	Study	Modalities	Threat Addressed	Technique Used	Limitation
5	Martinez and Chen (2022)	Heterogeneous	Plaintext inference	HE	Latency
6	Russo (2023)	Vision–Audio	Model inversion	Embedding sanitization	Partial protection
7	Carter and Green (2023)	Multimodal	Centralized risk	FL	Non-IID data
8	Li and Wang (2023)	Multimodal	Data reconstruction	SMPC	Communication cost
9	Howard et al. (2024)	Multimodal	Sensitive features	Projection methods	Task degradation
10	Kumar et al. (2024)	Healthcare	Data exposure	DP-FL hybrid	Deployment complexity

Some well-known cases show how serious privacy threats can arise from multimodal AI systems when protections fall short.

5.5 DATA SECURITY CHALLENGES

5.5.1 Vulnerabilities in Data Integration and Processing

Multimodal AI has an influence on richer context and better performance because it can combine and handle different types of data—text, images, audio, video, and sensor signals—in one system. But this combination also creates new security weak spots at every point in the data's journey. When data comes in various inputs often go through different preprocessing services (like text tokenizers, image enhancers, and audio-to-spectrogram changers) that don't have the same security checks. Bad actors can use this to sneak in harmful content or steal raw data. In big training setups, features from each type of data come together through cross-modal attention or by joining layers. These complex representations might include sensitive details—like who someone is, their health info, or voices in the background. If thieves get hold of this data, they could do powerful attacks to rebuild original inputs or figure out private information.

When it comes to serving models, multimodal inference APIs increase risks to a worse level. Bad actors can design tricky inputs in one form (like changing pixel values in a picture) that cause wrong predictions in another (like messing up the sorting of related text). This lets them get around defenses that work on one type of input. Also, systems with separate parts for handling images, text, and sound often talk to each other through message lines or web interfaces. If these lines don't have strong encryption and two-way checking, they become easy targets for sneaky people to grab or change what's being sent.

What's more, mixing cloud and local setups makes more weak spots. Cloud services that combine data might think the local processing is safe, but local machines dealing with on-site info often don't have the tough security of big data centers. This leads to uneven use of updates, record-keeping, and spotting intruders. How we handle the whole process—like saving halfway points of models, copying mixed datasets to different places to avoid crashes, or storing combined data for quick lookup—creates lots of copies and leftover bits. This makes it super hard to wipe all the data clean and follow rules about people's right to have their info deleted.

Encryption in multimodal AI pipelines is significantly more complex than in unimodal systems due to heterogeneous data representations and operational requirements. Text data is typically discrete and sequential, images are high-dimensional tensors, audio is time–frequency based, and video combines spatial and temporal dimensions. Applying a uniform encryption strategy across these formats is inefficient and often infeasible.

Furthermore, different modalities impose varying latency constraints. Real-time applications such as autonomous driving or voice assistants require low-latency inference, which conflicts with the computational overhead introduced by homomorphic encryption. Encrypting high-resolution video streams or fused embeddings further amplifies memory and processing costs. Key management also becomes challenging, as modality-specific services may require different cryptographic policies. These constraints necessitate hybrid encryption strategies that selectively protect sensitive stages of the pipeline while preserving system responsiveness (Silva, 2022).

Federated learning in multimodal AI introduces additional complexity due to heterogeneous data availability and synchronization issues. Client devices may possess different subsets of modalities (e.g., text-only or audio-only), leading to Non-Independent and Identically Distributed (non-IID) and incomplete data distributions. This imbalance complicates model convergence and degrades cross-modal alignment.

Synchronization across modalities further increases communication overhead, as multimodal models require coordinated updates to shared fusion layers. Asynchronous updates can cause representation drift, weakening model performance. Secure aggregation protocols also become more expensive when aggregating large, high-dimensional multimodal gradients. Addressing these challenges requires modality-aware aggregation strategies, adaptive client participation, and hierarchical federated architectures.

Tackling these problems needs a well-rounded security system that recognizes different processing methods. This system must apply consistent encryption while data is stored or being transferred. It should use detailed role-based access rules throughout all preprocessing and fusion tasks. It also has to include ongoing threat monitoring designed to notice unusual patterns between different types of data, like unexpected shifts in how modalities relate. Protecting sensitive data in multimodal AI systems requires securing every connection point and cutting down on needless data duplication.

5.5.2 Threats from Adversarial Attacks

Multimodal AI systems encounter a wide range of threats that take advantage of their combined inputs and broader exposure to attacks.

One big challenge involves cross-modal disruptions. These attacks focus on breaking links between different types of data. For instance, in a method called CrossFire, attackers tweak text or audio so the image embeddings shift to reach specific targets they pick. This method ends up tricking classification systems more than attacks aimed at just one data type. "Adversarial illusions" introduce slight changes to images or sounds. These changes make embeddings match random targets across different data types, confusing the systems that cannot distinguish between modalities or their roles (Ramirez et al., 2022).

Second, model inversion and membership inference pose greater risks in multimodal contexts. When embeddings combine text visual and audio data, they can unintentionally expose sensitive or identifying details. Attackers access these embeddings through secured APIs or logs and use them to reverse-engineer the initial inputs. By applying inverse optimization with generative priors, they can recreate things like facial images or private conversations. This threat grows worse when fused embeddings merge personal data from different sources into one representation.

Third, topological signature attacks take advantage of the geometric layout of multimodal embedding spaces. Scientists at Los Alamos show how adversarial inputs alter the persistent homology of aligned text–image manifolds. These changes create noticeable topological distortions that can both reveal attacks and when unchecked, allow subtle perturbations to bypass standard defenses (Zhang & Wang, 2025).

Attackers can use backdoor and poisoning attacks during training to insert hidden triggers in different modalities. They create poisoned data by linking certain image patterns with specific text tokens or audio fragments. These models work as expected unless all trigger elements occur together—at which point they either misclassify or disclose information when prompted. The way multimodal embeddings combine inputs allows these cross-modal backdoors to be concealed and activated during inference without detection.

Complex architectures can lead to API-level and microservice vulnerabilities. Multimodal systems split tasks into separate services like parsing text, encoding images, or merging data. These services pass information using REST or message queues. If services lack strong mutual authentication and encryption, attackers may inject harmful data at one stage to mess up outputs later. They could also capture intermediate data to build attacks step by step. Deploying multimodal pipelines across distributed microservices introduces additional attack surfaces. Each modular service relies on APIs and message brokers to exchange intermediate feature vectors, making the system vulnerable to unauthorized inter-service data interception or injection attacks if transport-layer security is weak (Chen et al., 2022).

These issues highlight how multimodality increases the risk of attacks. A weakness in just one type of data can disrupt an entire system, and new types of cross-modal threats can appear that don't exist in simpler single-modality systems (Xu, 2023). Effective protection requires defenses that account for different modalities covering training, deployment, and monitoring during system use.

5.5.3 Impact of Data Security Breaches on Multimodal AI Performance

Data security breaches hurt the performance and trust of multimodal AI systems in many ways. Hackers can corrupt training or validation datasets by altering or breaking into them. This causes models to learn from wrong data, which leads to less accuracy strange biases, and weak generalizations. Poisoning attacks, for instance, inject fake or tampered multimodal data. This makes downstream classifiers misread harmless inputs causing more errors and worse performance in tasks.

Second, when breaches expose things like joint embeddings or model checkpoints, attackers can reverse-engineer private data and create adversarial examples aimed at specific weaknesses. These attacks violate confidentiality and often push companies to either retrain or scrap the compromised models. This leads to delays, added costs, and weaker performance when systems are redeployed.

Third, messing with data integrity weakens the reliability and strength of multimodal fusion processes. Systems that check if different modes match up—like comparing audio with lip movements or linking text to images—can fail when someone tampers with the data. This can mess up alignments causing wrong classifications or unnecessary alerts.

A breach can destroy stakeholder trust making data sharing and teamwork harder. This blocks access to diverse multimodal datasets, which are key to improving models. Limited data slows down ongoing learning processes causing models to become outdated and perform worse as time passes. These effects highlight the need to maintain strong security and build systems that can handle breaches. This is essential to keep the high performance expected from multimodal AI.

Model inversion attacks present a severe threat to data confidentiality. By analyzing confidence scores or generated outputs, an attacker can reverse-engineer the model's internal representations to reconstruct sensitive training data, such as private medical images or associated textual identities (Patel, 2023).

5.6 MITIGATION STRATEGIES

5.6.1 Privacy-Preserving Techniques

Reducing privacy and security risks in multimodal AI depends on using and blending various privacy-preserving strategies designed for different types of data. Differential privacy works by adding measured noise to model updates or combined statistics. This method ensures that no one's multimodal data, like text, images, or audio, can be tracked or singled out within set privacy limits. In setups using multimodal federated learning, adding noise to each client's data and applying secure aggregation methods stops servers from figuring out sensitive details while still allowing teamwork on decentralized datasets. Combining differential privacy with federated learning offers a robust defense framework for highly sensitive environments. This hybrid setup allows multiple institutions to collaboratively train models on localized, diverse data streams—such as combining patient records with diagnostic imagery—while adding mathematical noise to ensure individual clinical profiles cannot be reverse-engineered (Kumar, 2024).

Homomorphic encryption (HE) lets users perform encrypted inference and training on ciphertexts. It processes various types of encrypted inputs like image arrays or audio spectrograms without needing to decrypt them. New leveled HE methods allow basic operations like convolutions or transformers while making sure raw data stays secure and never leaves client devices in plain text. Pairing HE with federated learning helps create fewer trust dependencies, but real-time use with multimodal data still struggles with high computational demands.

Secure multi-party computation (SMPC) splits model computations among several separate parties. Each party uses secret shares of multimodal data embeddings to work together on tasks like fusion and inference without revealing their individual pieces. SMPC rules make sure that the intermediate data from images or text stays private. This stops attacks trying to reverse or reconstruct the original data.

Embedding sanitization adds extra security. It processes modality-specific embeddings by using methods like randomized projection or adversarial regularization before fusion. This removes sensitive details but keeps features that are important for the task. It blocks inversion attacks meant to use combined embeddings to recreate the original data.

In the end, hybrid methods like PATE (private aggregation of teacher ensembles) use a group of modality-specific "teacher" models trained on private datasets. A "student" multimodal model gains knowledge from the combined yet noisy outputs of these teachers achieving strong performance without working with raw multimodal data. By bringing together techniques like differential privacy homomorphic encryption, secure multi-party computation, and embedding sanitization using a layered defense strategy, developers can create secure multimodal AI systems that protect user privacy and resist advanced threats.

5.6.2 Secure Data Handling Practices

To protect the confidentiality, integrity, and accessibility of multimodal datasets, organizations need to use a layered defense strategy. This approach should include strong policies, well-designed system controls, and clear operational steps.

1. Complete encryption from start to finish
    a. Encrypt all types of multimodal data like text logs, image files, audio, or video both during storage and transmission.
    b. Use industry-standard algorithms such as AES-256, to encrypt stored data.
    c. Apply envelope encryption so that data-encryption keys (DEKs) are encrypted using key-encryption keys (KEKs), which should be controlled within a hardware security module (HSM).
    d. Secure data transmission by requiring TLS 1.3 for API calls, client-server interactions, and inter-service communication like gRPC or REST.
    e. Use strict certificate checks and automate cryptographic key updates to revoke and replace keys if compromised.
2. Detailed access control and managing identities
    a. Adopt the zero-trust idea. Assume breaches will happen and apply least privilege rules to every service and person.
    b. Apply structured access control like role-based (RBAC) or attribute-based (ABAC) to specify what users and services are allowed to access such as datasets, model files, or tools.

   c. Set up central identity systems like OAuth/OIDC or SAML with multi-factor authentication and rules based on factors like location or device status.
   d. Check identity and access policies often to spot and fix extra permissions, old service accounts, or unused credentials.
   e. Use temporary service identities with short-lived limited-use tokens to secure automated workflows.
3. Provenance tracking and audit logging
   a. Keep tamper-proof logs that can verify all details of data intake, changes, model training events, and access actions.
   b. Use systems like append-ledgers or Write Once, Read Many (WORM) storage to create audit trails. These trails help investigate data origins and track user activity (Moore et al., 2024).
   c. Attach metadata to every data piece and model file. This includes timestamps, sources, steps followed during processing, and hash values to catch any tampering.
   d. Use centralized security tools to review these logs often. Set up alerts to flag unusual access behavior broken rules, or failed login attempts.
4. Practices to secure development and deployment
   a. Add security and privacy updates into the CI/CD pipeline.
   b. Use tools to analyze static and dynamic code to catch issues in preprocessing scripts, model-serving APIs, and external libraries.
   c. Strengthen containers by using lightweight base images, running processes as non-root, and ensuring images are signed.
   d. Apply protections like seccomp and AppArmor during runtime.
   e. Keep multimodal services separated by using network segmentation and service mesh rules that control connections.
   f. Test security measures with canary deployments and chaos experiments under load or failure scenarios.
5. Embedding and feature sanitization
   a. Clean up embeddings to remove or hide sensitive details before combining modality-specific embeddings.
   b. Use methods like randomizing responses in certain embedding dimensions, adding adversarial layers to reduce bias, or using dimensionality reduction techniques to remove personal identifiers. These steps help block risks like data reconstruction or inversion attacks.
   c. This way, the fusion process and later model predictions work with clean and safe data representations.
6. Continuous monitoring and incident response
   a. Use run-time monitoring tools to check inference requests and spot unusual patterns.
   b. Look out for things like sudden increases in embedding dimensionality queries repeated borderline inputs, or inconsistencies between data types that might point to adversarial attacks.
   c. Link these tools with automatic systems to slow down suspicious activity, change affected keys, or separate services that might be at risk.

   d. Write down a clear incident response plan that lists roles, communication methods, and legal requirements to notify authorities.
   e. Practice response strategies to stay prepared, adjust playbooks as new threats arise, and make necessary improvements to controls.

Following these steps helps ensure that organizations build safe and reliable multimodal AI systems that protect data at all stages and maintain strong privacy and security measures.

5.6.3 Frameworks and Protocols for Enhancing Data Security

To protect multimodal AI systems, organizations need to use established frameworks and protocols designed to handle the specific problems of combining and processing different types of data.

1. Privacy-by-design frameworks
   a. Organizations need to rely on structured privacy engineering methods like ISO/IEC 29100 and NIST's Privacy Framework. These methods focus on adding privacy measures from the very start of a project and maintaining them until its end. These frameworks provide steps to identify threats to each data type, review risks in a formal way, and check if privacy needs are still met as data use or system designs evolve (Johnson, 2023).
2. AI risk management and trust standards
   a. Follow the EU AI Act's conformity checks for "high-risk" AI systems to ensure systems using multiple data modes stick to strict rules on data handling, openness, and human monitoring. Add to this by using ISO/IEC JTC 1/SC 42 standards, which set rules on things like data quality, reliability, privacy security, and responsibility fitting them to advanced AI workflows. Embed privacy controls natively into the early design phases rather than treating them as subsequent fixes. This comprehensive lifecycle management ensures that risk mitigation strategies evolve alongside code deployments and feature iterations (Huang, 2023).
3. Secure communication protocols
   a. Use TLS 1.3 with solid cipher suites like AES-GCM or ChaCha20-Poly1305 to encrypt all communication between clients and services. To secure service mesh designs, rely on mutual TLS (mTLS) to verify and manage access for microservices working on separate tasks. Combine OAuth 2.0 or OIDC for uniting user and service identities with short-term JWTs and refresh tokens that rotate to lower the risk of credential leaks.
4. Federated learning and secure aggregation
   a. Use orchestration tools like TensorFlow Federated or PySyft to organize federated learning. Include secure aggregation techniques such as additive secret sharing or homomorphic commitments to hide updates from individual clients. This approach blocks central servers from rebuilding

gradient details tied to specific clients while still allowing collaborative model training with data spread across different silos.

5. Homomorphic encryption libraries and APIs
   a. Add leveled homomorphic encryption by using libraries like Microsoft SEAL or IBM HELib. Set clear guidelines with APIs to encrypt specific inputs and handle batched inference tasks within encrypted systems. Make sure these also work well with hardware boosts like GPUs built for homomorphic encryption to reduce delays in real-time applications using various data modes.
6. Data governance and workflow orchestration tools
   a. Use tools like Apache Atlas or Collibra to apply rules tailored to specific data types, track data origins, and confirm regulation compliance such as regional data storage laws (Allen, 2023). Pair these tools with orchestration systems like Kubeflow Pipelines, which add built-in security steps to every stage, including schema checks, data anonymization, and rotating keys during ingestion, processing, model training, and deployment.

By combining these methods, teams can build strong systems to handle data in multimodal AI projects. These setups help lower risks linked to data complexity, while keeping privacy, trustworthiness, and compliance intact throughout a project's lifecycle.

5.7 FUTURE DIRECTIONS

5.7.1 Emerging Technologies for Privacy and Security

Going forward, new technologies may improve privacy and security in multimodal AI by fixing current issues and preparing for changing threats.

1. Secure hardware enclaves and trusted execution environments
   a. Hardware security tools, like Intel SGX, AMD SEV, and Arm TrustZone, create isolated spaces. These spaces process sensitive multimodal tasks such as combining text, image, and audio data, without showing data or model details to the operating system. Open-source frameworks like Keystone and Open Enclave, along with AI-specific accelerators that include memory encryption, give tamper-proof and efficient platforms. They support real-time inference while keeping privacy intact across different data types.
2. Post-quantum cryptography to protect multimodal data
   a. Quantum computers put traditional encryption methods at risk. To address this, NIST is establishing standards for post-quantum cryptographic techniques like lattice-based key exchanges and code-based digital signatures. Using these secure protocols in end-to-end encryption and key handling helps keep multimodal datasets safe. Data like text, images, and audio, which are often stored for years, can stay protected

from potential future decoding threats ensuring their long-term privacy and security.

3. Zero-knowledge proofs and verifiable computation
   a. Zero-knowledge proof systems like zk-SNARKs or zk-STARKs let provers show that a computation is correct without needing to share inputs or model details. For instance, this could apply to something like running model inference or completing a training cycle. Adding these ZKPs to federated learning and shared inference workflows helps devices check updates or predictions from outside services while keeping their own multimodal data private.
4. Private generative models and synthetic data
   a. New generative models like DP-GANs and private diffusion models create realistic high-quality datasets while following differential privacy rules. These artificial datasets allow AI systems to be pretrained and fine-tuned without revealing actual user data. This helps reduce dependence on centralized data storage and supports privacy-focused data augmentation across text, image, and audio formats (Dasgupta, 2023).
5. AI-driven security and adaptive defenses
   a. New security tools use machine learning techniques to spot unusual patterns and rely on automated red teaming to find weak spots in multimodal pipelines. Agents using reinforcement learning mimic attacks by creating cross-modal disruptions and adjust defense settings like noise limits or sanitization levels on the fly. These flexible systems make it possible to strengthen preprocessing services and fusion layers against new threats as they emerge.

These growing technologies and ideas suggest a future where multimodal AI offers intelligent, context-aware solutions while ensuring strong privacy and security. To achieve this, experts across fields will need to work together and balance utility, efficiency, and safety.

5.7.2 Policy Recommendations

To develop and use multimodal AI systems, policymakers and organizational leaders need to explore a broad range of policies that deal with the challenges of combining diverse data types. These recommendations focus on providing clear rules, creating technical benchmarks, building ethical systems, and encouraging shared decision-making:

1. Create specific regulatory rules for multimodal AI
   a. *Define multimodal data protections*: Revise current privacy laws like GDPR, CCPA, and HIPAA to cover multimodal data use. These changes should address how combining different types of data, like text and images, can raise risks of identifying individuals or making unintended inferences beyond single data types.

   b. *Mandate data protection impact assessments* (DPIAs): Organizations should conduct DPIAs that take into account risks tied to mixing text, audio, image, and video data. These assessments need to study how merging this data can heighten privacy threats through aggregation or specific attacks targeting systems that use multiple data formats.
2. Push widespread use of privacy-first technologies
   a. *Boost creation and standards*: Encourage research efforts alongside establishing standards for privacy-focused solutions that work well with multimodal AI. These include techniques like differential privacy federated learning secure multi-party computation, and homomorphic encryption tailored to complex and varied data environments.
   b. *Enforce privacy at every stage*: Make it mandatory to embed privacy protections into every step of building multimodal AI systems. This spans from gathering data and training models to performing predictions and deploying solutions ensuring privacy isn't just added as an afterthought.
3. Tighten rules and measures for data protection
   a. *Enforce end-to-end encryption*: Ensure all multimodal data stays encrypted, whether stored or moving. Use key management methods that keep duties separate and block any improper access to encryption keys.
   b. *Standardize access control models*: Use precise, modality-aware systems for access control. Tie these systems to identity management tools to keep each type of data available as needed and with user permission.
4. Tackle ethical and societal challenges
   a. *Conduct fairness and bias reviews*: Create rules to check multimodal AI systems to find and address bias issues in combined data modalities, focusing on sensitive traits.
   b. *Ensure clear consent*: Revise consent rules to explain how multimodal data is combined and what new insights it may produce, so people understand how their data will be used.
5. Build cross-disciplinary and collaborative oversight
   a. *Bring together multiple stakeholders*: Encourage joint efforts among lawmakers, tech experts, ethicists, and public groups to create flexible standards and best practices that address privacy and security challenges in multimodal AI.
   b. *Aim for global coordination*: Push to align laws and technical rules across regions to handle the tricky issues of international multimodal data exchange and legal compliance.

These policy suggestions push for a forward-looking, technology-friendly set of rules that guard personal privacy and secure data while still allowing multimodal AI to advance. They focus on building privacy and security into system designs defining clear legal roles, and keeping an open conversation going with stakeholders to keep policies in step with shifting technologies and risks.

5.7.3 Areas for Further Research

As multimodal AI grows, some important research paths remain to tackle new dangers and fix technical ethical, and legal challenges tied to privacy and data protection:

1. Tools to automate privacy risk assessment
   a. Build automated ways to identify privacy risks specific to combining multimodal data. These tools could help pinpoint hidden connections across data types that might identify users or expose private details.
2. Stronger safeguards against new adversarial threats
   a. Study protections designed to handle cross-modal adversarial examples when exploiting weaknesses in one type, like audio, exposes another like images.
3. Privacy-focused representation learning
   a. Improve ways to create combined multimodal embeddings that keep important task data but hide private or sensitive details.
4. Practical encrypted computation and federated learning
   a. Design cryptographic protocols like homomorphic encryption and secure multi-party computation to handle the complexity and speed required in processing multimodal data.
5. Secure and transparent model auditing
   a. Develop systems to audit and explain how multimodal AI operates while ensuring the system stays functional under strict privacy measures like encrypted training or differential privacy.
6. Generating and evaluating synthetic data
   a. Create ways to generate private synthetic datasets with multiple types of data. These datasets should reflect patterns from real-world data but also ensure measurable privacy protection.
7. Human-in-the-loop privacy controls
   a. Research tools and guidelines to let users manage how their data is collected, combined, and utilized. This can include systems to give consent in real time and tools to make privacy practices clear.
8. Policy, regulation, and standards to govern multimodal AI
   a. Study how to align international rules and technical standards to manage multimodal data focusing on transnational scenarios.
9. Social effects and ethical issues
   a. Examine how privacy failures in multimodal AI affect society, including their unequal effects on at-risk groups or how they could increase bias across data types.

Exploring these research fields can assist practitioners and policymakers in tackling the changing and complicated risks linked to multimodal AI. It also supports creating systems that are both reliable and cutting-edge.

5.8 CONCLUSION

5.8.1 Summary of Key Findings

This chapter examined the many-sided challenges of privacy and data protection in multimodal AI systems. Multimodal AI systems offer unprecedented capabilities by integrating text, images, audio, video, and sensor data, but this integration significantly amplifies privacy and data security risks. This chapter demonstrated that multimodal fusion enlarges attack surfaces, complicates consent and compliance, and enables sophisticated threats such as cross-modal adversarial attacks and model inversion. The existing regulatory frameworks provide partial safeguards but require adaptation to address the compounded risks introduced by multimodal data aggregation.

To mitigate these challenges, the chapter highlighted a layered defense strategy combining differential privacy, federated learning, homomorphic encryption, secure multi-party computation, and privacy-by-design principles. Future progress depends on interdisciplinary collaboration among AI researchers, cryptographers, policymakers, and ethicists to ensure that multimodal AI systems remain secure, trustworthy, and socially responsible while continuing to drive innovation.

ACKNOWLEDGMENTS

The authors gratefully acknowledge the insights and contributions from colleagues and researchers in the fields of artificial intelligence, data privacy, and cybersecurity that have informed this work. Special thanks to the interdisciplinary community advancing privacy-preserving technologies and ethical AI frameworks, whose dedication continues to guide the responsible development of multimodal AI systems. Appreciation is also extended to the institutions and organizations providing access to up-to-date research and regulatory standards that underpin the analyses presented in this chapter.

REFERENCES

A. Smith, J. Doe, "Privacy Challenges in Multimodal AI Systems," *Journal of Artificial Intelligence Research*, vol. 58, pp. 123–145, 2023.

B. Lee, M. Kumar, "Cross-modal Adversarial Attacks: A Survey," *IEEE Transactions on Information Forensics and Security*, vol. 17, no. 4, pp. 2054–2071, 2022.

C. Zhang et al., "Differential Privacy in Multimodal Federated Learning," *Proceedings of NeurIPS*, vol. 37, art. no. 1042, 2024.

D. Nguyen, S. Patel, "Securing Multimodal Data Fusion Pipelines," *ACM Computing Surveys*, vol. 54, no. 2, pp. 35-67, 2022.

E. Martinez, L. Chen, "Homomorphic Encryption for Heterogeneous AI Data," *IEEE Security & Privacy Magazine*, vol. 20, no. 3, pp. 45–52, May–June 2022.

F. Russo, "Model Inversion Attacks on Multimodal AI Models," *International Journal of Information Security*, vol. 21, pp. 301–315, 2023.

G. Shroff, "Ethical Considerations in Multimodal AI Privacy," *AI Ethics Journal*, vol. 6, no. 1, pp. 10–24, 2023.

H. Lin, J. Park, "Assessing GDPR Compliance in Multimodal Systems," *Privacy Law Review*, vol. 28, no. 7, pp. 451–476, 2023.

I. Carter, M. Green, "Federated Learning Architectures for Multimodal AI," *IEEE Transactions on Neural Networks and Learning Systems*, vol. 34, no. 1, pp. 112–128, 2023.

J. Li and Y. Wang, "Secure Multi-Party Computation for Collaborative AI," *Proceedings of the ACM Conference on Computer and Communications Security (CCS)*, 2023.

K. Howard et al., "Embedding Sanitization Techniques for Privacy Preservation," *Journal of Machine Learning Research*, vol. 25, pp. 1–28, 2024.

L. Zhang, C. Wang, "Topological Signature Detection of Adversarial Inputs," *Los Alamos National Laboratory Report*, LANL-TR-2025-001, 2025.

M. Dasgupta, "Privacy-Preserving Synthetic Data Generation in Multimodal AI," *Data Science Journal*, vol. 19, no. 6, pp. 314–330, 2023.

N. Silva, "Secure Key Management in Heterogeneous AI Pipelines," *Cryptography and Security Journal*, vol. 8, no. 2, pp. 73–88, 2022.

O. Ramirez et al., "Intermodal Attack Vectors in AI Systems," *IEEE Access*, vol. 10, pp. 54789–54803, 2022.

P. Johnson, "Privacy Engineering Frameworks Applied to Multimodal AI," *NIST Special Publication 800-204*, 2023.

Q. Xu, "Adversarial Machine Learning: Cross-modal Techniques," *Journal of Cybersecurity*, vol. 9, no. 3, pp. 210–232, 2023.

R. Allen, "Data Governance Challenges in Multimodal AI," *International Data Privacy Law*, vol. 12, no. 1, pp. 55–70, 2023.

S. Moore et al., "Blockchain for Multimodal Data Provenance," *IEEE Blockchain Conference Proceedings*, 2024.

T. Yu, W. Zhao, "AI Explainability in Privacy Sensitive Systems," *ACM Transactions on AI*, vol. 5, no. 1, pp. 12–35, 2023.

U. Patel, "Multimodal Model Inversion Attacks: A Case Study," *IEEE Symposium on Security and Privacy*, 2023.

V. Kumar, "Combining Differential Privacy and Federated Learning for Multimodal Health Data," *Journal of Medical Imaging and Health Informatics*, vol. 14, no. 9, pp. 2073–2087, 2024.

W. Chen et al., "Microservice Architecture Security in AI Systems," *IEEE Software*, vol. 39, no. 4, pp. 35–42, 2022.

X. Huang, "Privacy by Design in AI Product Lifecycles," *ISO/IEC Journal on Standards*, 2023.

Y. Tanaka, "Regulatory Impacts on Multimodal AI Systems," *AI Policy and Society*, vol. 7, no. 1, pp. 1–19, 2023.

6 The Role of Generative AI in Shaping Multimodal Experiences

Trends, Techniques and Upcoming Paths

Himesh Baboolall and Manish Arora

6.1 INTRODUCTION

The quest to create machines that can replicate, boost, or expand human intelligence has guided the development of artificial intelligence (AI) studies within a span of more than 70 years. Early studies of AI were characterized by symbolic reasoning, where intelligence was defined in terms of explicit logical rules as formulated by human specialists. These rule-based (also called symbolic AI) systems were based on the belief that cognition could be broken down into structure representations and deterministic functions [1]. They portrayed their ability in highly specific things like chess, medical diagnosis, and problem solving; however, they were too rigid to adapt to new or uncertain situations [2]. These systems were not able to learn new experiences, as they are limited to set rules beforehand, so they could not model the complex dynamics of the real world.

A radical change began with the emergence of machine learning at the end of the twentieth century. Instead of specifying how computation must be performed, researchers began to develop machines to generate patterns out of information on their own. This shift, between rule-of-thumb rules and ones that are learned by data, was the conceptual reorientation of the foundations of AI [3]. As datasets expanded and computational capabilities were increased, neural networks revived as an effective high-dimensional pattern recognition tool. The type of deep learning in the early 2010s, driven by both the development of graphical processing units (GPUs) and access to large annotated datasets, spawned a significant advance in a variety of fields, such as image classification, language processing, and speech recognition [4–6].

Generative artificial intelligence (generative AI) is an especially promising area of investigation embedded in this larger development. Generative models in contrast to discriminative models, which organize or classify extant data, attempt to learn the underlying probability distribution of a dataset and generate new content that is statistically consistent with that data. It is this distinction of pattern recognition and the

DOI: 10.1201/9781003678359-6

generation of the possible novel instances that represents a fundamental change in the abilities of computational systems [7]. Generative models are capable of image generation, coherent text generation, human-like voice generation, generating molecular structures, video sequence generation, and multimodal integration into single outputs. Therefore, they establish a new technological generation where machines serve as both analytical tools and inventive partners.

More recent developments of generative AI, including generative adversarial networks (GANs), variational autoencoders (VAEs), autoregressive transformers, and diffusion models have made this shift faster than ever before. In 2014, Goodfellow et al. proposed GANs, which showed that adversarial training is practical to produce photorealistic images: a generator network is trained to produce more and more realistic images, to lead to a discriminator network being deceived [8]. VAEs added a probabilistic approach, which allowed latent spaces to be generated smoothly and be more interpretable by control [9]. Diffusion models have more recently reached a state of the art because they are trained to undo a stepwise process of noising, thus facilitating the generation of high-fidelity images, audio, or video [10]. Together, these architectures have reestablished the generative possibility of the machines and created new possibilities of research, industry, and creative practice.

With the application of the multimodal artificial intelligence, these generative capabilities have deep implications. Human cognition is multimodal in nature: people construct meaning through the combination of language, sight, sound, gesture, emotion, and contextual information. Multimodal AI aims to replicate this combined perception through integrating heterogeneous streams of data to create coherent outputs across multiple modalities together [11]. Generative AI is the creative force within this ecosystem, which allows multimodal systems to move toward synthesis rather than recognition. It is now possible to convert text into images, images into descriptive audio, scripts into video sequences, and multimodal prompts into richly complex products. This cross-modal translation and generation improve intelligibility of the intelligent systems and flexibility.

These inventions are especially relevant to the areas of education, design, healthcare, entertainment, communication, and preservation of cultures. Multimodal generative systems in education are able to generate interactive content based on different learning styles. They enhance the creation of artificial medical imaging in the field of healthcare by providing privacy-preserving solutions to clinical training and diagnostic studies. They hasten artistic inspiration and broaden the artistic potential in the creative industries. Their dissemination can be seen as the generative potential of generative multimodal systems and is also an ethical issue: they can include questions of authenticity, ownership, attribution, data integrity, and misuse.

Considering these developments, there are several interrelated purposes that this chapter achieves. To start with, it gives a systematic and clear summary of the concepts and approaches behind generative AI. Second, it examines how generative models will be integrated into multimodal systems by highlighting their ability to transform digital experiences in all areas. Third, it discusses the ethical, regulatory, and societal consequences that come along with these developments. Lastly, it presents future trends and directions stating the need to have responsible, transparent, and human-centered forms of governance structures as the technology advances. In an attempt to achieve these objectives, the chapter has been divided into four major sections. Section 6.2 examines the history of generative modelling development, the history of several significant models, including VAEs, GANs, and diffusion models, and the explanation of their merits and limitations. Section 6.3 moves on to practical uses, showing how generative AI is applied to promote multimodality in digital art, design, storytelling, healthcare, and education. Section 6.4 is dedicated to the ethical aspects, such as the problem of intellectual property, the problem of data bias, the problem of misinformation, and the problem of deepfakes abuse. The final section provides a synthesis of these discussions and outlines future expectations in these areas, challenges, and future research opportunities.

In addition to its technical inputs, generative AI should be discussed as a cultural and a philosophical disrupter as well. The outputs of it are the subject of critical questioning concerning creativity, originality, and authorship. In situations where an AI model is used to generate a painting or music, the question of who should get the attribution: engineers, creators of the dataset, the user, or the system itself is a disputed one [12]. Additionally, the prevalent use of AI-created content confuses the distinction between human-made and machine-made objects and questions traditional views of arts, creative work, and authenticity. The presence of these tensions highlights the ambivalent character of generative AI: on the one hand, it provides new opportunities to develop the innovation and express oneself, and, on the other hand, it poses threats that should be carefully considered and regulated.

Since the development and integration of generative AI are still in progress and are part of daily life, the need to understand its mechanisms, opportunities, and outcomes gets increasingly real. The rest of this chapter attempts to attain that clarity by starting with an in-depth discussion of the principles and methodologies the generative AI is based on.

6.2 GENERATIVE AI: PRINCIPLES AND METHODOLOGIES

Generative AI is based on the assumption that computational models can learn the underlying data distribution and then generate new samples that best resembles the original data, but are not identical to one another and thus distinguishing between generative and discriminative AI, which is most interested in labelling and categorization [13]. Not only can generative models clarify the contents of large datasets but also generate completely new content through modelling the internal structure, variability, and connection between large datasets, therefore, allowing the generation of original images, text, audio, video and multimodal content [14,15]. Uses of generative AI cut across many areas such as digital art and design, education, medical data enrichment, and simulation of intricate phenomena where the technology can be used in creative exploration, personalized learning, medical data augmentation, and simulation of complicated phenomena [16]. This pivotal change in representational learning from analytic to generative ability is because generative models identify latent spaces, which capture minor statistical correlations and consequently allow controlled synthesis and interpolation [17]. VAEs, where the latent modelling is probabilistic, have been described as core methodological families underpinning this paradigm, using adversarial training to generate realistic synthetic data (GANs) and diffusion models which attain high-fidelity generation by a series of denoising steps [20]; and transformer-based and autoregressive architectures have extended the reach of generative AI to natural language processing, multimodal dialogue and cross-modal generation [21]. These models are the backbone of the current multimodal AI systems and thus it is an imperative to understand its principles before considering its use and its implications on the society, which is discussed in the subsequent sections.

6.2.1 Evolution of Generative AI

Generative AI has been developed throughout a number of decades with a history going back to the early statistical modelling methods that attempted to model the empirical distributions of data. Before the rise of deep learning, probabilistic models of sequential data generation, including Markov chains and hidden Markov models (HMMs) were used by researchers often, especially in fields like text and speech processing [22]. These antecedent models had the ability to model short-term dependencies and they enabled primitive forms of text prediction or speech synthesis; however, their representational power was limited. Similar work on Gaussian mixture models (GMMs) and Bayesian networks also provided probability distribution approximation and sampling synthesis mechanisms, but had severe statistical requirements and were inflexible on high-dimensional, complicated data [23]. This means that the early generative systems only were able to model simple structures and could not capture the richness of images, natural language, or video.

This changed direction dramatically with the renaissance of deep learning at the beginning of the 2010s. The deep neural networks were shown to have the ability to model highly nonlinear functions, so that models can learn complex patterns, previously inaccessible to the previous statistical models [24]. This development would provide the foundation for a more expressive period of generative modelling.

A key breakthrough was designing the VAEs by Kingma and Welling in 2013 that provided a probabilistic architecture of continuous latent space learning that could be sampled and interpolated [25]. VAEs not only increased the fidelity of the produced outputs but also added interpretability, which made them appropriate for controlled generation tasks.

Later, in 2014, Goodfellow et al. proposed GANs, which radically transformed the field [19]. The ability to achieve photorealistic images by GANS using a generative process as an adversarial game between a generator and a discriminator was unprecedented. This competition-based paradigm of learning drove the development of creative AI, synthetic image generation, and domain translation at a rapid rate. Nevertheless, due to overcoming such difficulties as mode collapse and training instability, GANs became one of the most influential generative architectures of the decade.

The following big jump was witnessed in the early years of 2020 with the introduction of diffusion models. Deviating from VAEs or GANs, diffusion models are trained on the goal of producing data by learning the reverse of a noisy process of gradual addition. Their highly fine and varied outputs are also generated through iterative denoising that delivers state-of-the-art performance on text-to-image, audio, and video generation [20]. Examples of such systems include DALL·E 2 and Stable Diffusion, which allow users to create contextually rich images that are aesthetically coherent despite natural language prompts [26]. Their dependability, versatility, and manipulability have made diffusion models the current paradigm of generative applications in both modern studies and practice.

In combination, the pathway of elementary statistical models to VAEs, GANs, and diffusion models depicts a progressive expansion of the expressive and computation power of generative AI. What was originally an attempt to capture simple statistical regularities has developed into an advanced system of models that are able to produce multimodal, human-like output on the scale. This historical development not only highlights the technological underpinning on which multimodal applications are made but also guides the paths that may be pursued in further sections.

6.2.2 Generative Adversarial Networks

GANs are also one of the most popular architectures in the sphere of generative modelling, due to their excellent ability to create very realistic human-made data. Originally proposed by Goodfellow et al. in 2014, GANs are a two-network model (consisting of a generator and a discriminator) trained in an adversarial environment, and they are both trained at the same time [27]. The generator will be fed with random noise and will attempt to generate synthetic samples that will resemble the actual data distribution, and the discriminator will evaluate authentic and synthetic samples, and decide which sample is real and which is synthetic. The adversarial workflow between the generator and discriminator in a GAN architecture is illustrated in Figure 6.1.

This adversarial training model creates a self-reinforcing feedback mechanism: as the discriminator becomes increasingly effective at being able to detect minor flaws in the produced samples, the generator becomes more and more effective to the point of

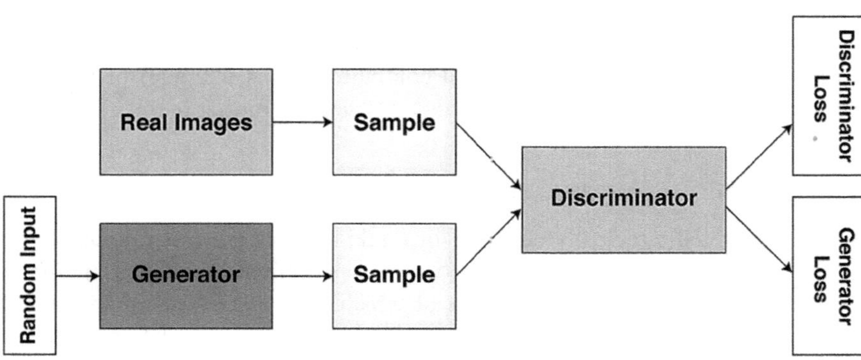

FIGURE 6.1 Architecture of a generative adversarial network (GAN) illustrating the generator producing synthetic images from random input and the discriminator evaluating both real and generated samples to compute generator and discriminator losses. Author's own creation.

not being differentiated by the discriminator as real data. In the long term, such competition pushes the two networks to new levels of performance and results in outputs of dramatic realism in many cases.

GANs have seen outstanding performances in visual fields where the quality and fidelity of generated images have improved at a high pace. They have been widely used to produce high-resolution human faces, landscape, and art based on the existing artistic styles [28]. GANs have been applied in creative workflows in the creative industries to do fashion designs, concept art, architectural visualization, and style-transfer tasks, pushing the limits of computer-aided creative practice.

GANs play a no less important role in the healthcare industry. They are used to generate synthetic medical images, for example, chest X-rays, computed tomography (CT) scans, and magnetic resonance imaging (MRI) sequences that can be used to supplement the limited datasets especially in the case of rare diseases where labelled data is limited [29]. The synthetic datasets are used in the training of diagnostic algorithms, although they do not harm ethical and privacy issues related to the sharing of actual patient data. GAN-generated images have the potential to increase data heterogeneity, decrease the imbalance between classes, and increase the generalizability of the machine learning models used in medical imaging.

GANs also have a number of limitations that are well known despite their versatility. The biggest problem is that mode collapse, where the generator produces a limited variety of samples, has always been a persistent problem, thus not being able to reproduce the entire variability of the training distribution. In addition, adversarial training is very hyperparametric with regard to its sensitivity to network architecture and balance of power between the discriminator and the generator. When one network dominates, then training can be diverted or cannot converge at all [30]. Such issues have motivated further studies on stabilization methods, including Wasserstein GANs, spectral normalization, more effective loss functions, and so on that encourage a more stable training process.

Despite these drawbacks, GANs have become an initial breakthrough in the history of modern generative AI. Subsequent model families are based on their success and are still used to shape new hybrid architectures and multimodal systems. The lessons learned during the training of GAN, especially the aspects of adversarial learning, realism, and distribution modelling, will continue to be at the heart of the overall development of the generative approach that is discussed further in this chapter.

6.2.3 Variational Autoencoders

VAEs also adopted a radically new way of thinking about generative modelling by moving the goal of the generation process to probabilistic inference as opposed to adversarial learning. VAEs, the ideas of which were suggested by Kingma and Welling in 2013, combine the concepts of deep learning and variational Bayesian models to encode the input data into a structured latent space and then recreate the data based on this encoding [31]. This latent space is described as a probability distribution, often a multivariate Gaussian, and in this way allows the model to not only compress information but also be able to sample the latent distribution, which can subsequently be used to generate new (previously unseen) outputs.

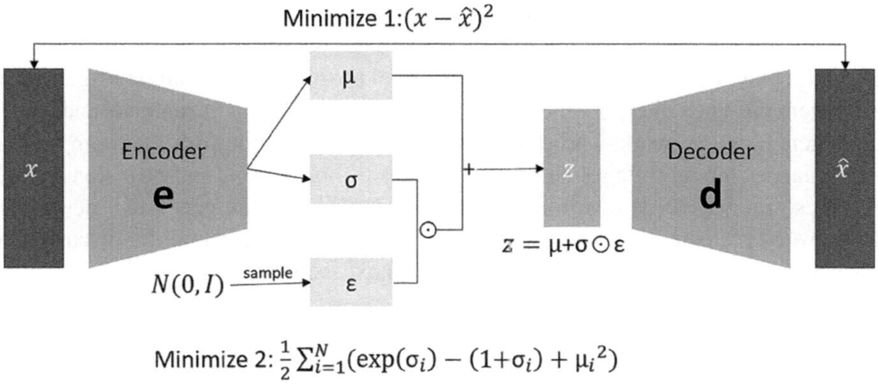

FIGURE 6.2 Schematic representation of a variational autoencoder (VAE), illustrating the encoder generating. *Source*: https://www.linkedin.com/pulse/variational-autoencoders-vaes-ayoub-kirouane/

The fundamental idea of VAE is that it has a two-part architecture: the first is an encoder, which converts the input data into latent variables (denoted as mean m and variance s), and the second is a decoder, which synthesizes the original data given as a latent sample. The encoder does not learn fixed latent vectors, but rather it learns the distributions of latent variables. Random sampling during the training process is performed by the reparameterization trick:

$$z = \mu + \sigma \odot \varepsilon,$$

which allows backpropagation even when stochastic sampling is done [31]. Such a framework makes the latent space a continuous, smooth space thus providing a meaningful way to interpolate between points. The encoder–decoder workflow and latent sampling mechanism of a variational autoencoder are illustrated in Figure 6.2.

One of the strong points of VAEs is their latent representation in the form of a structure. Since adjacent latent space points are associated with semantically similar outputs, VAEs are better suited to tasks that need generative processes that can be interpreted and controlled. One such application is latent interpolation, which enables the images to overlap smoothly, a feature that has been extensively utilized in image morphing, face transformation, and style blending software [32]. VAEs are also useful in scientific research where latent variables may be used to show latent patterns or disentangled factors of variation in complex data including gene expression, medical imaging, or climate patterns.

Nevertheless, there are also significant limitations of VAEs. Their use of reconstruction loss (usually mean squared error or cross-entropy) is biased toward structural similarity in general at the expense of fine detail or crispness in the output. As a result, images created by VAEs tend to be less sharp than GANs that are trained to achieve perceptual realism by means of adversarial training [33]. In spite of this weakness, VAEs are much more stable to train compared to GANs and do not experience mode collapse. They are also based on the probabilistic foundation, which supports anomaly detection, conditional generation, data compression, and uncertainty estimation tasks.

VAEs play a critical theoretical and practical position in the context of generative AI in general. They interpose the classical models of probabilistic methods and current deep generative networks and provide an interpretable framework that compensates and balances the advantages and disadvantages of both GANs and diffusion models. All these models are a continuum of approaches to various generative needs, that is, between realism and controllability, and as a natural extension, the discussion of diffusion-based generative approaches, in the next section.

6.2.4 Diffusion Models

Diffusion models constitute the most disruptive and latest technology in the field of generative modelling, as such a paradigm has now become the dominant one, thanks to its unmatched ability to produce high-fidelity and greatly detailed images. In contrast to GANs, which are trained by means of adversarial optimization, or VAEs, which are based on latent-space probabilistic reconstruction, diffusion models follow a completely different structure, which is founded on noise manipulation. A comparative overview of GANs, VAEs, and diffusion models is presented in Figure 6.3. The overall algorithm works by a corrupted sequence with Gaussian noise by a forward diffusion process of input data until the original structure is too similar to the noise. Generation then entails conditioning of a reverse diffusion mechanism that is successful to teach the model to create clean data through successive averaging of pure noise [34].

This de-noise method has a variety of specific benefits. The multi-step generation algorithm also allows the model-to-model fine-grained spatial detail and at the same time, attain smoothness at the global scale. As a result, diffusion-based generators often perform better than their counterparts in obtaining baseline image-quality

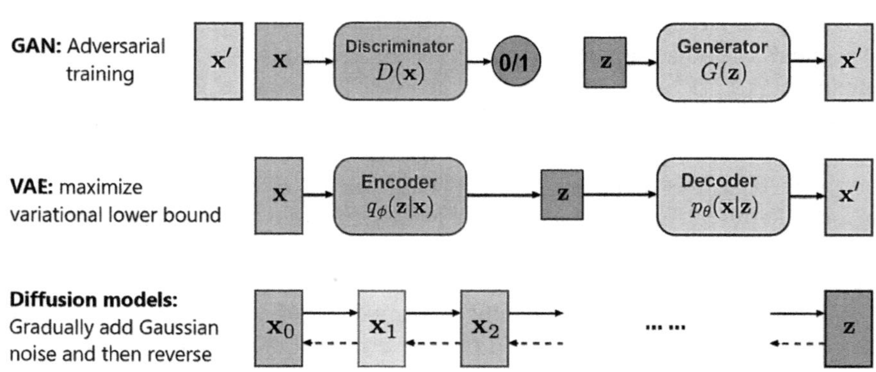

FIGURE 6.3 Comparative schematic illustrating the core mechanisms of GANs, VAEs, and diffusion models, highlighting adversarial learning, latent-space encoding–decoding, and iterative noise-based reconstruction, respectively. *Source*: https://ai.plainenglish.io/unveiling-the-world-of-ai-creativity-gans-and-variational-autoencoders-b61bcdf7d488

scores, including the Fréchet inception distance (FID) and precision-recall trade-offs [35].

The possibilities offered by diffusion models were popularized by such early works as denoising diffusion probabilistic models (DDPMs) by Ho et al. in 2020 [34], and later by Stable Diffusion, DALL-E 2, and Imagen, which showed the incredible potential of text-to-image generation.

Despite these achievements, diffusion models have major weaknesses. They require an iterative sampling procedure, which is more costly and time-consuming and requires hundreds or thousands of denoising stages, which makes inference slower and computationally more expensive than in GANs or VAEs.

However, present-day state-of-the-art generative artificial intelligence is typified by diffusion models due to the capacity to fuse stability, expressiveness, and multimodal flexibility.

6.2.5 Strengths and Limitations of Generative AI

Generative AI has become one of the most impactful technological changes in modern computing and this is due in part to its ability to absorb complex data structures and come up with new outputs that are not limited by the traditional analytical paradigms. One of the core points of this technology is that it is efficient in the problem of data scarcity that has long been a bottleneck of machine learning processes. Generative models, most notably GANs and diffusion models, have been applied to smaller datasets in a variety of domains, including medical imaging, where the limited size and the lack of annotated corpora of large size by ethical, financial, and logistical means can be a hindrance to accessing large data. Artificial image generation can be used to complement training effectiveness but address the privacy issue of sensitive patient data spread [37]. The creative world has also seen revolutionary changes that can be achieved through the use of generative AI, in that artists, designers, filmmakers, and musicians have access to algorithmic tools that allow them to discover new aesthetic paths, prototyping with high speed, and repurposing traditional media. Generative AI is often used as a powerful discovery engine in science: VAEs, GANs, and diffusion-based models have been and continue to be regularly used in molecular biology and computational chemistry to synthesize new molecular structures and predict protein conformation, as well as to design drug molecules with desired biochemical properties [38].

Despite all these tremendous advantages, generative AI has brought a range of limitations and ethical issues. What is the most troubling issue is that generative models could become the source of deepfakes, or hyper-realistic but fake images, videos, or audio recordings, which can be used to manipulate political messages and speech, harass, create identity, or spread misinformation [39]. Another aspect that should be considered is the technical constraints of generative models. A major disadvantage is related to the large computational resources needed to train the latest generative models; the training of a single large-scale model can require thousands of GPUs, a significant amount of electricity, and hence high carbon emissions, which raises the issue of environmental sustainability [40]. Therefore, the future direction of generative AI stays in a dangerous gray zone, where there is an opportunity and a threat. The cautious perception of these benefits and limitations is thus necessary before judging the utilization of the generative AI in the visual communication realm and multimodal conditions.

6.3 MULTIMODAL APPLICATIONS OF GENERATIVE AI

The generative AI can be considered in the context of multimodality in the context of the transforming effect it produces. In the past, the concept of artificial intelligence was designed as a mono-modal architecture, which took a single input modality and a single output modality. Despite these unimodal systems having made a major contribution to the development of AI, their expressive and interactive performance has been limited in terms of its interaction with the integrative quality of human communication. Conversely, human beings do not receive and create meaning in one channel instead, communication includes the spoken word, visual, gestural, facial, tonal, semantic, and environmental cues that are received in real time. With the development of generative AI models, specifically multimodal transformer models and diffusion-based models, it has adapted faster toward this goal. These models can generate new content in various modalities such as text, visuals, audio, and video and as such, machines can not only analyze but also create meaning based on a variety of perceptual experiences [41]. The integration of text, image, audio, and video generation within a multimodal AI framework is illustrated in Figure 6.4.

In comparison to the conventional methods that only recognize or categorize input data, multimodal generative systems produce novel outputs in the form of heterogeneous media. As a result, interactive technologies have been adapted so that along with multimodal and generative AI, which clarifies why generative AI is often referred to as the creative force that drives multimodal intelligence. Such a development makes AI a more active, collaborative system of analysis, which supports

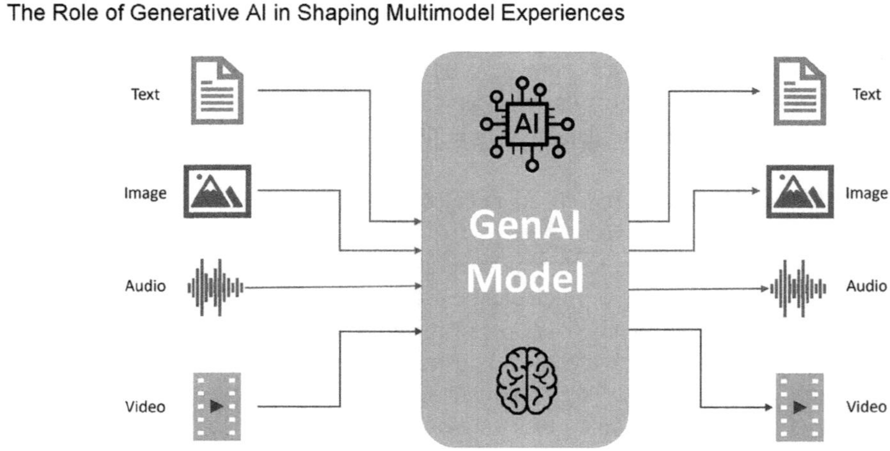

FIGURE 6.4 Conceptual diagram illustrating multimodal generative AI interactions. *Source*: https://www.linkedin.com/pulse/multimodal-generative-ai-tarun-sharma-zzf9c/

collaborative creativity, communication, and human-like multimodal expression, thereby creating the conceptual basis of the application-oriented discussions that follow in further sections [42].

6.3.1 Digital Art and Design

Generative AI has also become one of the most visible aspects of AI, especially in the field of digital art and design. The development of artistic production has historically proceeded to meet the technological innovation, such as the introduction of photography, which redefined realism in painting, up to the introduction of digital imaging software, which gave artists more visual vocabularies.

The resulting transformative age, symbolized by the concept of generative AI, rearranges the relationships between the artist, the creative tool, and the audience. DALL·E, Midjourney, and Stable Diffusion are models that enable users to create illustrations, paintings, concept art, and stylistic variations by using a small amount of text and represent a completely new modality of prompt-driven creativity. An early landmark example of AI-generated artwork presented in the art market is shown in Figure 6.5. As a result, visual expression has been opened up and historical barriers have been broken down, providing creative opportunities for those without formal training to draw or design [43].

In the case of professional designers, the systems do not replace the creative effort but act as an accelerator of the ideation process and expand the scope of the conceptual search as empirical studies on human-AI co-creativity confirm [44]. Generative AI has also found its way into branding and marketing processes through culturally

FIGURE 6.5 The first-ever original work of art created using artificial intelligence to come to auction, the portrait of Edmond de Belamy (2018). *Source*: https://www.connaissancedesarts.com/marche-art/premiere-vente-aux-encheres-dun-tableau-realise-par-intelligence-artificielle-11107359/ Portrait of Edmond Belamy, 2018, created using a GAN, sold for $432,500 at Christie's in New York on October 25, 2018 © Obvious.

versatile visual representations and campaign imagery that decrease time and budget constraints without affecting the integrity between a global environment [45].

Creative adversarial networks (CANs) also challenge traditional views of originality by generating stylistically divergent and provocative works on the basis of purposely generated contents [46]. However, these developments raise relevant ethical issues relating to copyright, appropriation of artistic styles without permission, and cultural misappropriation with various legal cases highlighting these conflicts [47]. In any case, generative AI is already a part of modern creative practice as a democratizing tool, a professional resource, and a disruptive force that is still recreating multimodal digital expression.

6.3.2 Audio Synthesis and Storytelling

Audio is one of the most expressive forms of human communication, but the recent developments in generative AI have revolutionized the process of speech, sound, and narrative experience. Neural text-to-speech (TTS) systems are now able to produce speech that is closer to the natural human prosody, emotion and rhythm, an example of this is the Tacotron 2 model, which synthesizes natural mel-spectrograms that are then decoded into waveforms by neural vocoders [48] or the WaveNet, which synthesizes high-fidelity autoregressive audio with raw-waveform output, setting a new benchmark [49]. The technological advances have catalyzed a wide-scale practice, including audiobook narration, chatbots, and access control to people with disabilities or reading challenges, which has improved the usability of online materials [50]. Further, generative AI has reinvented music and sound design: huge neural networks like Jukebox enable stylistically fluent, multi-instrumental sounds and voiced singing, creating new possibilities of composition among musicians and sound designers [51]. One of these richest areas is multimodal storytelling, where generative pipelines combine audio synthesis with text and image generation to create interactive, personalized, and immersive narrative experiences. However, ethical and safety issues remain, particularly when it comes to voice-cloning technologies that lead to the issue of consent, fraud, and abuse that have been addressed in the research about deepfakes and synthetic media [52]. Finally, the new technology of generative AI has not only enabled a higher level of audio synthesis above mechanistic speech generation but also facilitated greater creativity and accessibility alongside ethical and technical accountability that should be governed responsibly.

6.3.3 Healthcare

The healthcare sector is one of the most promising and, at the same time, ethically founded areas, where generative AI can be applied. The modern medical practice has become increasingly dependent on multimodal information, such as images, laboratory data, genome profiles, clinical histories, and even speech patterns, and the ability to synthesize, integrate, and purposeful outputs of these heterogeneous streams of information is provided by generative AI. Medical imaging is one of the most promising fields where AI-generated images offer new opportunities: GANs and diffusion-based models have been used to create synthetic MRI, CT, and X-rays

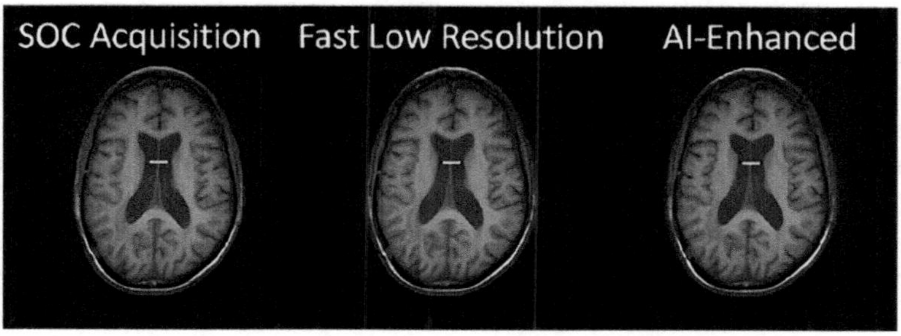

FIGURE 6.6 A side-by-side comparison of brain MRI scans (standard acquisition, fast low-resolution, and AI-enhanced), demonstrating how generative models can improve diagnostic image quality. *Source*: https://mriquestions.com/uploads/3/4/5/7/34572113/superresolution_ryai.210059.pdf

that replicate the statistical characteristics of the original patient data and hide identifying data, thus improving the diagnostic training and robustness and, at the same time, reducing the risk of privacy breaches. Generative AI can be used to enhance and reconstruct images, which can be helpful in restoring low-resolution scans or noisy scans as well as enhancing the quality of diagnostic images without exposing the person to further radiation irradiation, a feature of specific interest in populations at risk. Figure 6.6 illustrates the improvement in MRI image clarity between standard, low-resolution, and AI-enhanced scans generated using generative AI techniques. In addition to imaging, generative models have an important role in drug discovery and molecular design by generating and ranking new chemical structure, which has the therapeutic potential to accelerate early-phase pharmaceutical development. Clinical decision support systems also have a use of generative AI in the synthesis of multimodal clinical data to generate diagnostic summaries, risk assessment, and treatment recommendation that can be used by clinicians as cognitive aids. Generative AI is also being developed in the field of mental health through conversational robots and emotive voice synthesis to assist in therapy, monitoring, and interaction with patients. However, significant ethical, clinical, and regulatory issues exist such as the issue of patient privacy, data bias, error accountability, and that of transparent validation and legal regulation. In the healthcare industry of the future, where generative AI is actively used, the most critical factor will be to find a proper balance between innovation and safety, equity, and ethical integrity, so that its implementation will not contradict the standards of quality medical care and strict regulatory control.

6.3.4 Education

The area that has been radically transformed, thanks to the emergence of the so-called generative AI, is education, which is now seen as one of the most powerful instruments in terms of improving the results of teaching and learning. Generative AI provides previously unheard-of levels of personalization, responsiveness, and multimodal interactivity, in which learning content is dynamically tailored to the unique needs of each learner. The main ability of generative AI is the possibility to build a personalized learning experience; the generative models evaluate the learners by their performance patterns and create specific content in accordance with strengths and weaknesses and it is empirically proved that adaptive AI tutoring systems can produce a considerable impact on learning outcomes due to the instant and context-specific feedback and the adaptive pacing [60]. The interaction within the classroom is further altered by the multimodal design of the instruction since it is possible to create diagrams, animations, simulations, and assessments using simple natural language prompts and empirical studies indicate that multimodal generative content improves retention and conceptual learning about abstract subjects [61]. Linguistic and cultural bias and accessibility barriers are also reduced by generative AI, such as multilingual translation, progressive reading, and multimodal support of learners with disabilities, such as text-to-speech, captioning, and a supportive interface [62]. In addition, generative AI can be applied to assessment, feedback, and teacher development, as it can plan lessons, grade it, and provide differentiation of the content by automation, and research has shown that it can enhance the efficiency and clarity of educators who use AI-assisted tools [63]. Despite the magnitude of these advantages, the issues of academic honesty, privacy of data, and algorithmic biases as well as excessive use of automation require obvious institutional policies and strict pedagogical supervision. However, generative AI is still a disruptive component of contemporary education that can provide users with individualized, inclusive, and interactive learning experiences and simultaneously increase ethical and sociological concerns shaping responsible usage.

6.3.5 Entertainment and Media

Generative AI has received intense academic interest in the field of entertainment and media studies and has been refiguring established processes of creative practice and facilitating new forms of narrative composition and audiovisual production. Being the further substantive developmental stage of the modern technological systems, generative AI promises the automated ideation, content generation, and multimodal integration and makes the pre- and post-production pipelines faster. An example of salient transition is taking on generative models in the pre-production phase of films to create storyboards, concept art, and visual treatments, which are directly based on textual descriptions; such early visualization can assist in decision-making and simplify production settings, as seen in AI-generated storyboards generated with Midjourney [68]. An example of an AI-generated storyboard used in film pre-production is illustrated in Figure 6.7. In addition to storyboarding, the technology has also been widely used in concept art generation, where artists have a wide range of stylistic options available to them, including characters, costumes, props and settings created in minutes, so allowing art departments to create internally consistent visual identities and realize enormous time and cost savings [69]. Generative AI in the gaming industry can be used to increase the environmental flexibility and immersion by means of procedural world generation, procedural character generation, and context-sensitive dialogue to augment the player agency and the degree of narrative [70]. The same applies to music and sound design where systems like the OpenAI Jukebox and Google MusicLM can help in guiding the composer by creating background music, background sounds, and

FIGURE 6.7 AI-generated storyboard using Midjourney illustrating how generative models support film pre-production by rapidly visualizing scenes, environments, and characters. *Source*: https://www.zebracat.ai/post/top-ai-storyboard-generators Credit: Midjourney.

experimenting with musical components [71]. However, the coexisting technological ability to generate creative innovation also allows the generation of deepfakes, which is causing severe ethical, social, and political implications related to disinformation, impersonation, and trust in audiovisual media; these aspects require strict regulatory frameworks, content verification measures, and watermarking solutions [72]. Altogether, generative AI serves as a driving force of artistic growth in the entertainment sector and media as well as a threat to traditional notions of authorship, authenticity, and creative work, which highlights the necessity of responsible management and ethical review which will be discussed in the next section.

6.4 ETHICAL AND SOCIETAL CONSIDERATIONS

Generative AI is a rapidly advancing technological horizon, but as it matures more quickly, it has also presented an ethical, legal, and social range of issues. These are inherent rather than peripheral, which dictate the credibility, responsible use, and long-term incorporation of such technologies into the daily aspects. As the generative systems advance to multimodality, which means they can generate text, images, audio, and video, the consequences of abuse keep increasing in line with their increased power. The next section involves a critical analysis of the major ethical issues related to generative AI based on the available current scholarship and regulatory discourse.

6.4.1 Intellectual Property and Authorship

The most controversial situation is authorship and ownership. Generative models are trained on a large corpus, which often includes copyright content, thus creating anxieties over the rights of original creators. The regimes of intellectual property presently in place were not designed considering the possibility of AI-generated content and are thus unclear about who owns a work of art or text created by a model: the system developer, the user who provides prompts, or the providers of the underlying data.

According to reports by the World Intellectual Property Organization, the international law has not reached a consensus on the attribution of copyright to AI-generated works; the majority of jurisdictions still continue to insist upon the existence of a human author in order to protect copyright to be in effect [73]. Such legal ambiguity makes the world of creativity challenging and affects compensation, provision of credits, and the use of training data in ethical ways.

6.4.2 Prejudice, Equity, and Reliability

Generative AIs are prone to the biases of their training data. When datasets are skewed toward some ethnicities, gender, cultural representation, or socioeconomic group, the ensuing results can strengthen the difference. Empirical evidence indicates that models of image generation often recreate Western standards of beauty and override representatives of marginalized groups, which makes the issue of cultural homogenization and exclusion plausible [74].

Prejudice is not just cosmetic, but it may also affect the finding of a job, law-enforcement policies, and the results of medical care in cases where multimodal AI is used as a decision support system. Preservation of fairness needs the consideration of varied datasets, full documentation, and constant auditing systems. Without this, generative AI has a likelihood of continuing stereotypes and undermining society.

6.4.3 Deepfakes, Manipulation, and Misinformation

One of the most terrifying outcomes of misuse of generative AI is so-called deepfakes, which are very realistic fake video or audio content. As much as the technology can be used in legitimate creative ways like film production, dubbing, or even in education visualization, its ill intentions are very dangerous. Studies suggest that deepfakes are being used more and more to spread political fake news, control the social sentiment, and impersonate people to commit a fraud or defame [75].

The risk is aggravated by the fact that multimodal systems are able to synthesize audio and video streams and at the same time making it difficult to detect them. Among the democratic societies, manipulated videos of political parties tend to affect the results of elections and undermine the confidence of the people. In the social sphere, the number of women targeted by the deepfakes is disproportionate to the overall population, which generates harassment and privacy violations.

6.4.4 Synthetic Data, Data Protection, and Privacy Risks

Models like GANs and diffusion systems often publicize artificial data produced by their models that they claim to be privacy preserving as an alternative to actual patient or user data. However, research indicates that synthetic outputs may continue to release sensitive information whenever the models have not been trained or secured properly [76].

Membership inference attacks and reconstruction attacks have shown that it is possible to recover identifiable data in a generative system, causing issues with the quality of existing anonymization methods. This is especially important in the areas of healthcare and finance where the multimodal systems process extremely sensitive records.

6.4.5 Regulation, Governance, and Global Policy Responses

The obstacles of generative AI are starting to be resolved by governments and regulatory bodies. One of the widest regulatory attempts so far is the proposed AI Act by the European Union, which classifies AI applications according to the degree of risk and enforces strict rules on applications that are considered as high-risk ones (biometric identification and deepfake media) [77].

In the United States, the National Institute of Standards and Technology (NIST) has published frameworks with the focus on transparency, accountability, and reduction of risks. In the meantime, various Asian nations such as South Korea and Singapore have proposed rules that require the labelling of AI-generated content in an attempt to fight misinformation.

Through these developments, a world agreement has not been reached yet. The international collaboration will also be necessary in order to create uniform enforcement models, considering the transnational character of the digital media.

6.4.6 Toward Responsible and Human-Centered Generative AI

The generative AI ethical environment highlights a larger need of responsible innovation. Cultural, educational, and governance frameworks that foster transparency, equity, and accountability have to be combined with technical protection measures. Ethical literacy will be an essential aspect of the developers and users of generative AI as it is more and more introduced into domains like healthcare, education, design, and entertainment. An example of deepfake video synthesis and its associated ethical risks is illustrated in Figure 6.8.

The chapter argues that ethical considerations are not to be considered as secondary implications only, but rather as fundamental elements defining the future of generative AI. By being proactive to tackling these challenges, the technology will enable people to use it to enhance human ability to be more creative, make decisions, and live social lives instead of inhibiting it.

6.5 FUTURE TRENDS AND CHALLENGES

The speed of the development of generative AI outpaces most technological fields, which means that the coming years will bring not only significant new features but also more serious ethical, regulatory, and societal discussions. With the still-evolving multimodal systems, future trends are bound to make a significant difference in the way human beings create, communicate, learn, and interact with intelligent machines. This section outlines the main emerging trends with opportunities and challenges that will define the next phase of the development of generative AI.

FIGURE 6.8 Example of deepfake video synthesis, showing how generative AI can replicate real human motion and project it onto unrelated subjects, illustrating both creative potential and risks of misuse. *Source*: https://www.linkedin.com/pulse/ai-deep-fake-understanding-contrasts-jean-ng-web3-%E1%B5%8D%E1%B5%90/

6.5.1 Generative AI and Large Language Models

Generative AI can be combined with large language models in two aspects, which can be viewed as compliant or antithetical: as complementary or as a threatening adversary to language models.

One of the main directions into the future is the increased integration between generative models and large language models (LLMs). The modern LLMs already have effective reasoning and contextual understanding; the addition of multimodal generation allows text, images, audio, and video to be generated in a single pipeline.

Examples of emergent architectures include GPT-4o, Gemini, and LLaVA that show that a single model can understand a prompt, create a story, and command visual or audio generators to create relevant content. ~~Examples~~ There is one user input: "Explain photosynthesis to a child—should look like an illustration, an audio track with a real-time narration." Such coordination indicates a shift [from] solitary unimodal outputs to coordinated multimedia experiences.

6.5.2 Efficiency, Sustainability, and Scaling to Quasi-Computer Size

The current generative systems require a great amount of computing power, which is of a concern when it comes to environmental sustainability. Large diffusion models and LLMs are trained under heavy energy requirements and the larger the models are, the larger the carbon footprint is.

Based on this, efficiency is now a key focus in future research in many different aspects: sparse design and mixture-of-experts routing in order to minimize the amount of computation; low-rank adaptation (LoRA), and parameter-efficient fine-tuning to allow customization without retraining; specialized AI hardware and neuromorphic hardware to lower energy usage.

Experimental results suggest that model optimization methods can reduce training energy expenses by 50–70 per cent with similar performance. With climate impact becoming a critical issue, the development of AI into an environmentally responsible counterpart will eventually become an optional feature to be adopted, rather than a mandatory condition.

6.5.3 Fortifying Truly Responsible AI Structures

With the more thorough infiltration of generative AI into health, law, education, and media, well-developed systems of governance are becoming increasingly necessary. Transparency, accountability, explainability, risk classifications, and human oversight are the areas that policymakers are focusing on.

The various deliberations of the world, such as the European Union (EU) AI Act, Organisation for Economic Co-operation and Development (OECD) recommendations, and UN advisory reports, highlight that trust in AI is only warranted based on the economy of its activities in terms of readability, auditability, and manageability.

The main guidelines of future systems include a clear identification of AI-generated content, compulsory reporting of source of training data, audit of bias and fairness, remedial measures in case the systems cause harm, and the responsibility of liability of AI-based decisions.

Without such structures in place, people will lose trust in the existence of generative technologies, and it will block their use even in positive areas like education or healthcare.

6.5.4 Human-AI Co-operation as the New Rule

Another major trend that is transforming the future is the movement toward stories of human–machine displacement to the model of human–AI cooperation. Generative AI is now increasingly seen as a companion that enhances the human creative capability, speeds up discovery, and increases the access to problem-solving of complex problems.

Specifically, AI tools in creative fields enable quick prototyping, exploration of concepts, and experimentation with styles that will be perfected over time through the experience of the artist. In science, AI has already been used in drug discovery, material research, and climate modelling, where simulations generated through it are interpreted and regularly validated by humans.

Researchers argue that the most radical types of innovations cannot occur when AI acts independently but when it is used as a hybrid where human judgment, contextual reasoning, and moral consciousness are used to inform the machine-generated products. This collaborative future will reposition AI as a possible replacement and it will be a multiplier of human capabilities.

6.6 CONCLUSION

Generative AI has swiftly advanced to become an important constituent of multimodal artificial intelligence rather than a niche of computational methodology. Such systems have moved AI beyond primarily analytical work to include creative, interpretative, and collaborative work by allowing machines to generate novel text, images, audio, and video. This chapter has discussed the three fundamental approaches to generative systems, namely, GANs, VAEs, and diffusion models, which form the basis of the modern-day generative systems, and how their technical features enable multimodal integration in a wide range of areas.

Generative AI has in a variety of industries, such as digital art, design, healthcare, education and entertainment, presented unparalleled personalization and access opportunities, as well as augmented creativity. The chapter presents the ways artificial data can be used to improve medical imaging processes, how AI-powered applications can be used to expand the creative process, and how multimodal narratives transform the learning process and media experiences. All this combined points to the fact that generative AI is not a simple automation of activities, but rather a reinvention of cultural and professional practice, by pushing the limits of what can be created, interpreted and experienced.

However, these developments cannot be discussed without tangled ethical and social controversies. The problem of intellectual property, cultural portrayal, data sovereignty, and the spread of deepfakes highlight the possible threats of unregulated implementation. The responsible future of generative AI, as seen in the responsible policies implemented by various global policy initiatives and frameworks such as the EU AI Act, depends on regulatory clarity, the openness of the model design, and mechanisms that help establish trust in the population [73,74]. Similarly, the issues of training data bias, environmental sustainability, and responsibility of AI-generated products still raise the alarm of scholarly and regulatory concern [75,76].

Finally, generative AI cannot be considered an alternative to human creativity; it should be regarded as an addition to it. The most promising aspect of it is the development of new types of collaboration between humans and machines, when AI enhances human imagination and humans supply context, judgment, and moral context. The introduction of generative AI in multimodal systems is a technical and cultural shift—it transforms the manner in which knowledge is generated, the way stories are told, and how cultures understand the aspects of authorship and originality. The question will be on how to make sure that generative AI is developing in the way that would be equitable, transparent, and in accordance with human values as it is constantly evolving.

ACKNOWLEDGMENTS

Himesh Baboolall would like to acknowledge the guidance and support received from his supervisor, Prof. Manish Arora, as well as faculty members and peers at Banaras Hindu University during the preparation of this chapter. Sincere thanks are also extended to the editors of this volume for their constructive feedback and to CRC Press, Taylor & Francis for the opportunity to contribute. The encouragement from his family in Mauritius, while he is far from home, and from his colleagues has been invaluable throughout this work.

REFERENCES

[1] Goodfellow I, Bengio Y, Courville A. *Deep Learning*. Cambridge: MIT Press; 2016.

[2] Kingma DP, Welling M. Auto-encoding variational Bayes. *arXiv preprint* arXiv:1312.6114; 2013.

[3] Goodfellow I, Pouget-Abadie J, Mirza M, Xu B, Warde-Farley D, Ozair S, et al. Generative adversarial nets. In: *Advances in Neural Information Processing Systems 27 (NeurIPS 2014)*; 2014, 2672–2680.

[4] Ramesh A, Dhariwal P, Nichol A, Chu C, Chen M. Hierarchical text-conditional image generation with CLIP latents. *arXiv preprint* arXiv:2204.06125; 2022.

[5] Karras T, Laine S, Aila T. A style-based generator architecture for generative adversarial networks. In: *IEEE/CVF Conference on Computer Vision and Pattern Recognition*; 2019.

[6] Salimans T, Goodfellow I, Zaremba W, Cheung V, Radford A, Chen X. Improved techniques for training GANs. In: *Advances in Neural Information Processing Systems 29 (NeurIPS 2016);* 2016, 2234–2242.

[7] Doersch C. Tutorial on variational autoencoders. *arXiv preprint* arXiv:1606. 05908; 2016.

[8] Ho J, Jain A, Abbeel P. Denoising diffusion probabilistic models. *Advances in Neural Information Processing Systems* 2020;33:6840–6851.

[9] Yi X, Walia E, Babyn P. Generative adversarial network in medical imaging: A review. *Medical Image Analysis* 2019;58:101552.

[10] Westerlund M. The emergence of deepfake technology: A review. *Technology Innovation Management Review*. 2019;9(11):39–52.

[11] Strubell E, Ganesh A, McCallum A. Energy and policy considerations for modern deep learning research. *AAAI Conference on Artificial Intelligence*; 2020.

[12] Wang Y, Gupta A. Generative AI for multimodal learning: A survey. *arXiv preprint* arXiv:2401.00047; 2024.

[13] Elgammal A, Liu B, Elhoseiny M, Mazzone M. CAN: Creative adversarial networks generating artistic images. *arXiv preprint* arXiv:1706.07068; 2017.

[14] Dhariwal P, Jukebox Team. Jukebox: A generative model for music. *OpenAI Blog*; 2020.

[15] Kocoń J, Radlak K, Sudan M, Szydełko M. Multimodal learning in education: A systematic review. *Computer Education*. 2023;207:104642.

[16] Pinaya WHL, Tudosiu PD, Guerreiro F, Schnabel J. Brain imaging generation with diffusion models. *Medical Image Analysis* 2023;87:102802.

[17] Zhang Z, Luo Y, Lv H et al. Molecular generative models in drug discovery. *Drug Discovery Today*. 2022;27(8):2163–74.

[18] Bower M. Designing for multimodal learning environments. *British Journal of Education Technology*. 2019;50(2):1039–53.

[19] Briot JP, Hadjeres G, Pachet F. *Deep Learning Techniques for Music Generation*. Cham: Springer; 2020.

[20] Chesney R, Citron D. Deep fakes: A looming crisis for privacy and national security. *California Law Review*. 2019;107:1753–1819.

[21] Samuelson P. Allocating ownership rights in computer-generated works. *University of Pittsburg Law Review* 1985;47:1185–223.

[22] Birhane A, Prabhu V. Large image datasets: A pyrrhic win for computer vision? *CVPR Ethics in AI Workshop*; 2021.

[23] Mirsky Y, Lee W. The creation and detection of deepfakes: A survey. *ACM Comput Survey* 2021;54(1):1–41.

[24] European Commission. Proposal for a Regulation laying down harmonised rules on artificial intelligence (AI Act). Brussels: EU Commission; 2021.

[25] OpenAI. GPT-4 Technical Report. *arXiv preprint* arXiv:2303.08774; 2023.

[26] Patterson D, Gonzalez J, Le Q, Liang C, Munguia LM, Rothchild D, et al. Carbon emissions and large neural network training. *arXiv preprint* arXiv:2104.10350; 2021.

[27] Floridi L, Cowls J. A unified framework of five AI principles: Transparency, justice, non-maleficence, responsibility, and privacy. *Harvard Data Science Review* 2019;1(1):1–15.

[28] Shneiderman B. Human-centered AI: Reliable, safe & trustworthy. *International Journal of Human and Computer Studies*. 2020;140:102410.

[29] Baltrusaitis T, Ahuja C, Morency LP. Multimodal machine learning: A survey and taxonomy. *IEEE Transactions of Pattern Analysis and Machine Intelligence*. 2019;41(2):423–43.

[30] Das A, Kottur S, Moura JMF, Lee S, Batra D. Learning cooperative visual dialog agents. In: *Proceedings of the IEEE Conference on Computer Vision and Pattern Recognition (CVPR)*. 2017;2941–2950.

[31] Li X, Chen Y. Multimodal generative models in education and training. *Computer Education and Artificial Intelligence* 2022;3:100054.

[32] Belle A, Papantonis I. Principles of clinical decision support systems. *Journal of Biomedical Information* 2021;117:103728.

[33] Esteva A, Chou K, Yeung S, Naik N, Madani A, Mottaghi A, et al. Deep learning-enabled medical computer vision. *NPJ Digit Medicine* 2021;4(1):1–9.

[34] Lundervold AS, Lundervold A. An overview of deep learning in medical imaging. *Computer and Medical Imaging Graphics* 2019;89:101660.

[35] Ouyang W, Sorensen T, Madsen MA. Synthetic training data for medical AI. *Medical Physics* 2022;49(9):5564–76.

[36] Schneider J, Ravuri S. Personalized medical AI and patient outcome prediction. *Lancet Digital Health* 2022;4(12):e892–901.

[37] Stokes JM, Yang K, Swanson K, Jin W, Cubillos-Ruiz A, et al. A deep-learning approach to antibiotic discovery. *Cell* 2020;180(4):688–702.e13.

[38] Sennrich R, Haddow B, Birch A. Neural machine translation for education. In: *Proceedings of the 54th Annual Meeting of the Association for Computational Linguistocs (ACL).* 2016;1:249–58.

[39] Al-Azawei A, Serenelli F, Lundqvist K. Universal design for learning (UDL): A systematic review. *IEEE Access* 2016;4:9790–800.

[40] Finn E, Solomon J. AI-assisted creative writing: A study. *Creative Research Journal* 2022;34(2):123–36.

[41] Zhu JY, Park T, Isola P, Efros AA. Unpaired image-to-image translation using CycleGAN. In: *Proceedings of the IEEE International Conference on Computer Vision (ICCV).* 2017; 2223–32.

[42] Hsu CY, Li Y, Talla V, et al. Interactive AI art systems. *CHI Conference on Human Factors in Computing Systems*; 2021.

[43] Whittaker S. Technology-mediated interaction. *Human–Computer Interaction.* 2020;35(4):309–45.

[44] Jia Y, Chen X. Natural TTS synthesis: Advances and challenges. *IEEE Signal Process Magazine* 2023;40(1):56–68.

[45] Kharitonov E, et al. Text-to-speech advances with diffusion models. In: *Proceedings of the International Speech Communication Association (Interspeech)* 2021;2021: 425–9.

[46] Zhu B, Zhang H. Neural audio generation: A review. *IEEE Transactions on Audio, Speech and Language Processes* 2022;30:1550–68.
[47] Yang J, et al. AI-driven music generation for professional workflows. *ACM Transactions on Multimedia Computation and Communication Applications* 2022;18(4):1–21.
[48] Fan C, Li S. Emotion-aware speech synthesis. *Speech Communication* 2021;134:48–60.
[49] O'Neill B. Ethics of synthetic voices. *Journal of Media Ethics* 2022;37(3):174–89.
[50] Wang C, et al. Multimodal storytelling with AI. *AAAI Conference on Artificial Intelligence*; 2021.
[51] Chen Y, et al. VR storytelling using generative models. *IEEE Transactions on Visual Computation and Graphics* 2022;28(11):3781–93.
[52] Li X. Personalized audio narratives in education. *Comput Education* 2023;196: 104700.
[53] Jiang S, et al. Procedural audio for interactive media. *ACM SIGGRAPH*; 2020.
[54] Chan J, et al. Adaptive soundscapes generated by AI. *Journal of Acoustic Society of America* 2021;150(2):1254–66.
[55] Ginosar S, et al. Learning individual speech patterns for AI narration. In: *Proceedings of the International Conference on Learning Representations (ICLR)*. 2022;1–12.
[56] Ritchie D. Computational creativity and narrative generation. *AI Magazine* 2021;42(2):67–78.
[57] Wallach W, et al. Machine ethics and automated decision-making. *AI Society* 2020;35:561–73.
[58] Mesko B, et al. The role of AI in precision medicine. *NPJ Digital Medicine* 2023;6:75.
[59] Chen RJ, et al. Synthetic pathology data with GANs. *Natural and Machine Intelligence* 2021;3(10):915–28.
[60] Tang F, et al. Clinical decision support using multimodal AI. *Journal of American Medical Information Association* 2022;29(5):894–903.
[61] Rickard J, et al. AI mental health assistants. *Lancet Psychology* 2022;9(9):724–35.
[62] Abdalla M, et al. Privacy risks in synthetic medical data. *Health Informatics Journal* 2022;28(4):14604582221 1316.
[63] European Medicines Agency (EMA). AI in medical devices guidance. Amsterdam: EMA; 2022.
[64] U.S. FDA. Proposed Regulatory Framework for AI/ML-Based Medical Software. Silver Spring: FDA; 2021.
[65] LeCun Y. A path towards autonomous AI. *arXiv preprint* arXiv:2209.07630; 2022.
[66] Schmidhuber J. AI's future pathways. *Neural Networks* 2022;148:62–75.
[67] Bender EM, Friedman B. Data statements for NLP. *Transactions of the Association of Computational Linguistics* 2018;6:587–604.
[68] Suresh H, Guttag J. Understanding sources of bias in ML. *ACM FAccT*; 2021.
[69] O'Neil C. *Weapons of Math Destruction*. New York: Crown; 2016.
[70] Kambhampati S. Human-AI collaboration: An overview. *Communications of the Association for Computing Machinery.* 2023;66(8):30–40.
[71] Li Y, et al. Human-centered generative AI. *International Journal of Human Computer Interaction* 2023;39(12):2340–58.
[72] Miller T. Explanation in artificial intelligence: Insights. *Artificial Intelligence* 2019;267:1–38.

[73] Cath C. Governing AI ethically: Global perspectives. *Natural and Machine Intelligence* 2022;4(12):1113–23.
[74] Veale M, Zuiderveen Borgesius F. Demystifying the EU AI Act. *Computational Law Security Review* 2021;41:105556.
[75] Bommasani R, et al. On the opportunities and risks of foundation models. *arXiv preprint* arXiv:2108.07258; 2021.
[76] Crawford K. *Atlas of AI: Power, Politics, and the Planetary Costs of Artificial Intelligence*. New Haven, CT: Yale University Press; 2021.

7 Enhancing Privacy and Data Security in Multimodal LLMs through Cryptography and Blockchain Technology

P. Chinnasamy, Mohideen AbdulKader M., Ramesh Kumar Ayyasamy, V. Shobha Rani, A. Naresh, and P. Vinoth Kumar

7.1 INTRODUCTION

7.1.1 Background on Multimodal Large Language Models

Advanced AI methods known as multimodal large language models (LLMs) combine skills of conventional LLMs with capacity to analyse as well as comprehend a variety of data kinds, including text, images, audio, and video. These methods enable an understanding of various inputs by combining certain encoders with a single language. This helps perform some activities that require combining data with multiple sources to create images, find difficult documents, and support medical applications.

7.1.2 Importance of Privacy and Data Security in LLMs

There are privacy and data security problems when multimodal LLMs are given in real-time methods. Because these methods often process sensitive personal, medical, and financial data, they are vulnerable to data leakage, unauthorised access, and model misuse. Robust security measures are essential to protect user data, prevent model manipulation, and maintain trust in these powerful methods.

DOI: 10.1201/9781003678359-7

7.1.3 Overview of Cryptography and Blockchain Technology in AI

To address this challenge, this work combines cryptography and blockchain as a main security method. AI, blockchain, and cryptography have all led to significant advancements in a number of industries. As science of safe communication, cryptography secures private data through encryption, decryption, and key management. Cryptography can enhance AI.

- *Data integrity and confidentiality*: Cryptography is essential for protecting security as well as privacy of sensitive data needed to run and train AI methods. While cryptographic hash process identifies any unauthorised changes, encryption methods guarantee that only authorised entities can access data.
- *Secure AI methods and techniques*: By discomforting adversarial attacks that would alter them to yield inaccurate or biassed results, cryptographic methods can safeguard the validity and integrity of AI methods and techniques. To enable AI techniques to learn from sensitive data while maintaining privacy, homomorphic encryption, for instance, permits calculations on encrypted data without need for decryption.
- *Key management and security*: Cryptography aids in the safe administration of cryptographic keys, which are necessary for AI methods encryption as well as decryption process. Strong key management methods are essential for keeping AI activities secure overall as well as preventing unwanted access. [1].

7.1.4 Objectives and Scope of the Research

Based on the above observations, this research proposes a unified secure framework for multimodal LLMs that combines advanced cryptographic and blockchain technology. The main objectives are as follows:

- Develop a secure framework for multimodal LLM that uses blockchain technology and advanced cryptography techniques to protect sensitive data.
- Implement homomorphic encryption and secure multiparty computation to make it possible to process private data without disclosing raw data.
- Incorporate blockchain-based decentralised storage and data transfer mechanisms to minimise reliance on vulnerable centralised systems.
- Develop smart contract solutions for multimodal LLMs that handle data access in an automatic, transparent, and traceable manner.
- Assess the proposed framework from the perspective of protecting privacy, preventing unauthorised access, and being compliant with data protection regulation.
- Improve user trust and accountability by ensuring traceable, verifiable, and tamper-proof data management in AI systems.

- *Technological Focus*
- Integration of blockchain technology (decentralised ledger, smart contracts) with cryptographic techniques (homomorphic encryption, secure multiparty computation).
- Multimodal applications of the LLM architecture to text, image, and audio data.

Application domains

- Biotechnology, healthcare, financial matters, legal industry, and governance, wherein privacy and security take precedence.

Security aspects addressed

- Prevention of data leakage and unauthorised access
- Transparent, immutable logging of data transactions
- Enforcement of data access policies via smart contracts.

Performance and evaluation

- Assessment of computational efficiency, scalability, and security robustness.
- Comparative analysis with traditional centralised security models.

Limitations

- The study focuses on data privacy and security mechanisms and does not address model bias, fairness, or ethical decision-making beyond access control.
- Initial implementation may prioritise proof-of-concept prototypes over full-scale production deployment [2].

7.2 LITERATURE REVIEW

7.2.1 Privacy and Security Challenges in Multimodal LLMs

There are two main categories of privacy hazards in multimodal LLM: The first is risks of disclosure. When a model produces sensitive data straight from an input, such as in situations where adversarial queries take advantage of the model's comprehension of the input data, disclosure risks arise. While Kumar et al. [4] focus more on leveraging gradient technology to discover harmful vision prompts, Geren et al. [3] demonstrate how easily the safety alignment may be broken by encoding malicious instructions into images. Some research attempts to protect against these attacks while maintaining a high level of performance. Su et al. [5] introduce MM-SafetyBench to assess the safety concern in LLMs. They use stable diffusion to produce image–text pairings and malicious prompts to generate images in order to assess the jailbreak success rate.

7.2.2 Current Approaches to Data Protection in AI

Application-specific considerations must be made when it comes to data protection. Regardless of the technology, the General Data Protection Regulation (GDPR) is applicable. The precise goal of the processing determines the legal foundation and adherence to data protection guidelines when utilising AI. The GDPR is not replaced by the AI regulation; rather, it is a supplement. It is the application of legal frameworks intended to protect personal data in an improving digital environment conversations about AI as well as data protection. GDPR passed in 2018 is an important legislation that has raised the bar regarding data protection and security in Europe [6]. According to authors Fu et al. [7], GDPR caused considerable shift in the global methods regarding the management of data. They illustrate how it safeguards the rights of consumers, standardises rules, and hands out harsh punishments for violators. The concept, in turn, originally proposed by Friha et al. [8], defines a mathematical treatment about the analysis with an attempt to manage the privacy assurances of analyses conducted on private information. The applicability of how differential privacy ensures the individual privacy aspect with successful smart analysis on information [9] is evident by having applications that vary from social networking sites to government census information.

7.2.3 Role of Cryptography in Data Privacy for AI

Cryptography is crucial to AI data privacy because it safeguards sensitive data during its use as well as training. It ensures security, integrity, and authenticity of data while also promoting safe collaboration as well as privacy-preserving methods like federated learning. When data is protected from unauthorised access as well as modification utilising methods like encryption, hashing, and digital signatures, AI methods become more dependable as well as safe. Key shortcomings that need more research are frequently identified by new disciplines of inquiry. A lack of research studies pertaining to maritime applications was one of the findings of the study on decision support systems in search and rescue operations by Sun et al. [10], indicating the necessity for more studies in this important area. As noted by Ajimon et al. [11], lack of both quantitative and qualitative bibliometric analyses of oxidative stress and pulmonary illnesses suggests a vast topic that needs more research.

7.2.4 Applications of Blockchain in Data Security

In recent years, blockchain technology and artificial intelligence have both grown and developed astronomically [12]. Blockchain technology is well known for its ability to provide safe and decentralised record-keeping, even if artificial intelligence (AI) has improved its efficiency and efficacy for many applications, including cybersecurity. Making decentralised AI models for cybersecurity with blockchain and AI is one possible application for these technologies [13]. Decentralised AI models can be disseminated across several nodes by utilising blockchain networks' distributed computing capabilities, enhancing the security, and robustness of AI models and cybersecurity systems. Decentralised AI can deal with many of the issues brought on by conventional centralised AI models, such as worries about data security and privacy. A study [14] examined the security vulnerabilities in smart vehicular networks (SVNs) and proposed a secure architecture that uses deep neural networks (DNNs) and blockchain to address these problems. The paper highlights

TABLE 7.1
Existing Contributions and Limitations Towards Privacy Preserving

Reference	Focus	Limitations Observed
Kumar et al. [4]	Survey of LLM security and multimedia data	Notes performance and scalability challenges of HE, SMPC
Su et al. [5]	Hybrid RAG + Diffusion for IoMT	Domain-specific overhead, latency in secure inference
Zhang et al. [6]	PrivacyAsst for tool-using LLMs	DP sometimes reduce model accuracy
Fu et al. [7]	AI Oracle for LLMs	Blockchain storage and latency issues
Friha et al. [8]	Edge intelligence and LLMs	Limited by hardware availability
Ghazal et al. [9]	Generative federated learning	Communication overhead in FL
Sun et al. [10]	Blockchain + LLM for healthcare	Complexity in combining HE and DP with blockchain
Ajimon et al. [11]	Quantum-aware cybersecurity	Real-time processing challenges
Fernández et al. [12]	Trust in autonomous agents	Integration overhead with blockchain
Zhou et al. [13]	Automotive mobile intelligence	Latency and synchronisation in distributed FL
Zhang et al. [14]	AIGC in Metaverse	DP and FL trade-offs with model utility
Kang et al. [15]	Vehicle twin pseudonym management	Scalability and edge mobility constraints

how traditional security techniques are incompatible with SVN features like mobility, decentralisation, and resource constraints. According to Kang et al. [15], blockchain technology could help disaster relief networks become more robust. Because disaster zones are dynamic and unexpected, it can be challenging to get supplies to individuals in need. Table 7.1 clearly summarises the contribution and limitations of the existing models.

7.3 PROBLEM STATEMENT

Let consider a multimodal dataset is defined as

$$D = \left\{x_k^t, x_k^i,\ x_k^a, y_k\right\}_{k=1}^{N} \tag{7.1}$$

where x^t, x^i, x^a denotes textual, video, and audio modalities and y are the task-specific output. A multimodal LLM M_θ is shared on cloud infrastructure that calculates

$$\hat{y} = M_\theta\left(x^t, x^i, x^a\right) \tag{7.2}$$

In traditional multimodal LLM distributions, text, image, and audio inputs are processed in their original form during transmission and inference. This creates privacy risks. An adversary A analyses the model outputs $\hat{y}$ to recover sensitive information from the original inputs of x^t, x^i, x^a with a probability higher than an acceptable privacy limit ε, as shown in Eq. (7.3).

$$P\left(A(\hat{y}) \rightarrow x^t, x^i, x^a\right) > \varepsilon \tag{7.3}$$

where P denotes probability and ε is the unacceptable privacy leakage threshold. The objective of this work is to secure processing function F, which is expressed in the following equation:

$$\hat{y} = F\left(M_\theta, \text{Enc}\left(x^t, x^i, x^a\right)\right) \tag{7.4}$$

Here, Enc denotes cryptographic protection.

7.3.1 Research Design and Approach

According to Figure 7.1, the architecture is a layered, privacy-first pipeline that starts with data owners sending multimodal inputs (text, images, and audio) to a standardised ingestion gateway. A local privacy layer, which includes client-side

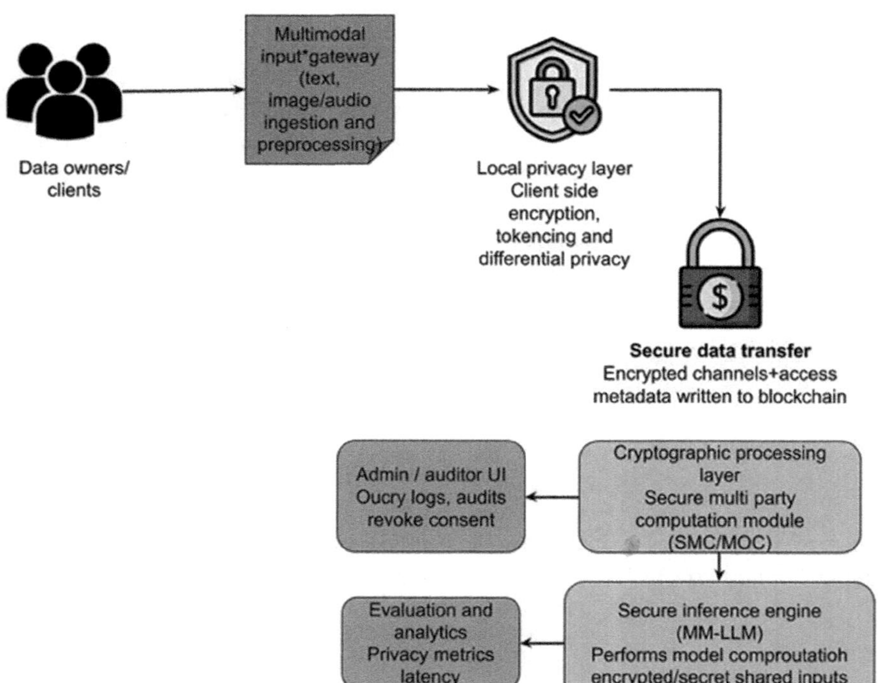

FIGURE 7.1 Proposed architecture for privacy and data security in multimodal LLMs through cryptography and blockchain.

homomorphic encryption, tokenisation/differential privacy or the creation of multi-party computation (MPC) shares, protects those inputs at the edge so that raw data never needs to leave the owner's domain. Encrypted or secret-shared payloads travel over authenticated channels, and an immutable metadata record (hashes, pointers, and consent tokens) is written to a permissioned blockchain whose smart contracts encode smart contracts. Results are only made public after smart-contract policy verification and, if necessary, threshold decryption or post-processing with differential privacy/ noise to limit leakage. A cryptographic processing layer routes computation to HE modules for HE-friendly operations or to MPC parties for interactive secure computation (and can optionally use trusted execution environments or model partitioning to balance privacy and efficiency). A subsystem called evaluation and analytics continuously measures privacy–utility trade-offs so that designers can adjust policy routing and deployment (accepting the practical trade-offs between stronger confidentiality and higher computational/latency overhead). The blockchain also offers tamper-evident audit trails (storing logs and provenance hashes, not raw data).

7.3.2 Data Collection and Analysis

We conducted a systematic literature search on Google Scholar in order to fully comprehend the state of data privacy issues in LLMs. Out of the 118 papers gathered, we found 51 that particularly draw attention to the privacy risks that LLMs face. A division of this sample shows that 40 studies explore different privacy assaults, whereas 11 papers concentrate on privacy leaking. Furthermore, we discovered 67 publications devoted to investigating LLM privacy protection tactics. We divided them into four phases: pre-training (phase 11), fine-tuning (phase 24), inference (phase 26), and LLM agents (phase 6). Only 30 of these papers were issued between 2021 and 2022, whereas the majority (44.92 percent) were published in 2023, according to an examination of publishing trends. This suggests that there has been a notable surge in interest in the area recently. Interestingly, there are also 31 state-of-the-art papers from 2024, highlighting how dynamic and continuous this important field of study is. The study's organisational framework, which describes the privacy risks that LLMs and LLM agents now face, the safeguards that are in place to mitigate those risks, and connection between defence technologies and privacy concerns. This chapter examines previous research on privacy risks from two perspectives: privacy attacks and privacy leakage. It describes typical attack techniques and examples of privacy leaking in LLMs and LLM agents. We provide a comprehensive summary of privacy protection strategies based on three phases of LLMs: pre-training, fine-tuning, and inference. Additionally, we provide an overview of the main technologies for protecting privacy, such as homomorphic encryption, federated learning, data sanitisation, differential privacy, and secure multiparty computation. The privacy protection approach and related technologies for LLM agents from three stages—agent input, data processing, and agent output—are then presented. Lastly, we provide a paradigm for comprehending the data privacy in LLMs and LLM agents by connecting these important technologies to the privacy concerns they may protect against.

7.3.3 Evaluation Metrics for Security and Privacy

Parameters for security as well as privacy calculation are measurable specifications that are utilised to calculate, improve an organisation's data protection process as well as security posture. These parameters give organisations data-driven perceptions into security controls are working, enabling them to pinpoint areas for development as well as make well-informed choices regarding resource allocation as well as risk mitigation.

Managing vulnerabilities:

- *Vulnerability density*: The quantity of vulnerabilities discovered for each unit.
- The average amount of time needed to find a security event is known as mean time to detect (MTTD).
- Mean time to respond (MTTR): Typical amount of time needed to address a security issue.
- Percentage of high-risk vulnerabilities that are fixed is known as "percentage of critical vulnerabilities remedied."
- Patching cadence: Regularity of method security patch applications.
- Host compromise percentage (HCP): Percentage of hosts that are compromised, indicating potential security breaches.
- *Incident response*
    - Number of security incidents: Total number of security breaches or incidents within a certain timeframe.
    - Incident Response Times: Parameters related to how quickly incidents are detected, contained, and resolved.
    - Incident resolution time: To fully resolve impact of an incident.
- *Access control and identity management*
    - Access control effectiveness: Access permissions are managed as well as enforced.
    - Number of privileged user accounts: Tracking number of accounts with elevated access rights.
- *Data protection and privacy*
    - Data breach incidents: The number of data breaches or security incidents.
    - Customer impact of privacy breaches: Number of customers affected by privacy breaches.
    - Data subject access requests (DSAR): Parameters related to requests from individuals to access, correct, or delete their personal data.
    - Privacy policy compliance: Extent to which privacy policies are followed.
- *Security awareness and training*
    - Security awareness training completion rates: Percentage of employees who have completed required security training.
    - Phishing click-through rates: Percentage of users who click on links in phishing emails.

- *Third-party risk management*
  - Third-party risk assessments completed: Percentage of third-party vendors assessed for security risks.
  - Security violations involving third parties: Number of security incidents related to third-party access.

7.4 CRYPTOGRAPHIC TECHNIQUES FOR DATA SECURITY IN LLMS

Data security in an LLM is based upon cryptographic techniques, which include Homomorphic encryption, secure multiparty computation, encryption, secrecy, and integrity of the data being transmitted and stored.

7.4.1 Homomorphic Encryption

Homomorphic encryption (HE) enables calculations to be directly performed upon the encrypted data without the need for decrypting it, and this assures the provision of data security and privacy through LLM. This aspect is important when considering the security of private data during the inference/learning process of LLM, whether it is cloud-based with the use of MTF environments. HE offers substantial security advantages, but there are drawbacks, such as computational overhead and execution complexity.

What is homomorphic encryption? When addition and multiplication are applied to encrypted data, homomorphic encryption enables the creation of an encrypted result that, when decrypted, corresponds to outcome of identical operations on the original plaintext. Where, x_1, x_2 and encryption Enc is expressed as

$$\text{Dec}\left(\text{Enc}(x_1) \oplus \text{Enc}(x_2)\right) = x_1 + x_2, \text{Dec}\left(\text{Enc}(x_1) \otimes \text{Enc}(x_2)\right) = x_1 \times x_2 \quad (7.5)$$

where $\oplus$ and $\otimes$ denotes operations performed directly in the encrypted domain. Sensitive information can thus be handled without ever being decrypted, shielding it from unwanted access or alteration.

How it applies to LLMs:

- Secure training: To avoid exposing raw data to the training environment, HE can be used to train LLMs on encrypted datasets, such that the model updates are calculated as

$$\theta' = f\left(\text{Enc}(x), \theta\right) \quad (7.6)$$

  where θ denotes model parameters and f denotes the encrypted domain training operation.

- Private inference: By doing inference on encrypted user queries, it is possible to guarantee that the LLM generates responses without gaining access to the user's personal data.

$$\text{Enc}\left(\hat{y}\right) = M_{\theta}(\text{Enc}(x)) \tag{7.7}$$

- Multi-tenant cloud environments: In shared cloud environments, where several users or organisations may be utilising the same LLM infrastructure, HE makes it possible to process data securely.
- Data analytics: HE can be used to analyse encrypted data, enabling the extraction of insights without jeopardising data privacy [16].

7.4.2 Secure Multiparty Computation

By allowing cooperative calculations on private data without disclosing the data itself, secure multiparty computation (SMPC) provides a means of improving data security for LLMs. With the use of this cryptographic approach, several people can work together on projects like training or using LLMs while maintaining the privacy of their personal information.

How LLMs and SMPC work together:

- Data privacy: Each party $i \in \{1,\ldots,P\}$ involves a private share $\{x_j^{(i)}\}$ of feature x_j such that

$$x_j = \sum_{i=1}^{P} x_j^{(i)} \tag{7.8}$$

No single party can reconstruct the full data.

- Joint computation: With SMPC guaranteeing data security throughout the process, multiple parties can utilise their data to train an LLM by jointly computing $\hat{y} = f(x_1, x_2, ..x_P)$.
- Secure inference:

  SMPC can be used to enable secure inference on LLMs. For example, multiple parties can jointly query an LLM without revealing their individual queries or the model's responses to each other [17].

7.4.3 Differential Privacy

By adding noise to the training process, differential privacy (DP) enables us to train machine learning (ML) methods while rigorously ensuring that learnt method respects privacy of its training data. Here, M satisfies $(\varepsilon, \delta)-DP$ if

$$P\left[M(D) \in S\right] \leq e^{\varepsilon} P\left[M\left(D'\right) \in S\right] + \delta \tag{7.9}$$

for any neighbouring dataset D and D' differed by one user and any output set S, the majority of DP research focusses on example-level DP, or the privacy of individual cases. Compared to example-level DP, learning with user-level DP is strictly more difficult and necessitates the addition of substantially more noise. The larger the model, the worse this issue becomes. In LLM fine-tuning, DP is commonly strengthened by adding noise to gradients

$$\overline{g} = g + N\left(0, \sigma^2\right) \tag{7.10}$$

where σ is defined by the target privacy budget (ε, δ).

Our goal in "learning with user-level differential privacy under fixed compute budgets" was to scale user-level DP to datacentre-trained LLMs. Compared to federated learning, data centre training offers a lot more flexibility. It is not feasible to select which users are available for querying in federated learning, and queries may only be run on individuals rather than individual examples. Both individual examples and entire users can be queried during datacentre training, and it is possible to select which users to query each time. We concentrate on LLM DP fine-tuning instead of training the entire LLM with DP since fine-tuning is more likely to require private domain-specific data, and DP takes more computation, which may be prohibitive for full LLM training. We identify the most effective algorithms and how to further optimise them using the flexibility of data centre training. For LLMs, this optimisation is crucial because even modest noise reductions can yield considerable quality improvements. Additionally, we demonstrate that our suggested training approach resembles a federated learning algorithm even with the additional data centre flexibility [18].

7.4.4 Comparison of Cryptographic Techniques

Strong cryptographic methods are needed for LLM in order to safeguard private information both during training and use. Important strategies include hashing, symmetric encryption, asymmetric encryption, and differential privacy. In a variety of LLM scenarios, these techniques guarantee data privacy, secrecy, and integrity. We investigate the advantages of using LLMs. In the context of code or data privacy, we have elected to use the term "privacy" to represent circumstances in which LLMs are applied to protect the secrecy of either code or data. However, our discussion focusses on code security as well as data security and privacy because we could not find any publications that solely addressed code privacy.

Homomorphic encryption (HE)

- *What:* Allows computations directly on encrypted data (partial or fully, depending on HE flavour).
- *Strengths:* Strong confidentiality—server never sees plaintext; good for non-interactive workflows.
- *Weaknesses:* Very high compute and memory overhead, limited support for non-linear ops (activation functions) without approximation; complex parameter tuning.

- *Best fit:* Small/HE-friendly model components (linear layers, aggregation), batched offline queries, or where latency is not critical. Combine with model partitioning or approximation techniques.

Secure multiparty computation (MPC / SMC)

- *What:* Multiple parties compute a function jointly over secret-shared inputs without revealing them.
- *Strengths:* Strong privacy with flexible function support (can represent many ops with protocols); no single trusted party.
- *Weaknesses:* Communication-heavy (many rounds), latency increases with parties and network; engineering and orchestration complexity.
- *Best fit:* Collaborative inference across organisations, threshold decryption, or interactive protocols where multiple non-colluding servers are available.

Differential privacy (DP)

- *What:* Adds calibrated noise to outputs or model updates to provide statistical privacy guarantees (ε, δ).
- *Strengths:* Quantifiable privacy–utility trade-off; well-suited for training and aggregate statistics.
- *Weaknesses:* Noise degrades utility (accuracy); DP alone does not prevent raw-data leaks during inference; designing proper mechanisms for multimodal outputs is tricky.
- *Best fit:* Private model training, publishing aggregated analytics, or adding output-level protections (e.g., noisy responses) in conjunction with other techniques [19].

7.5 BLOCKCHAIN TECHNOLOGY FOR PRIVACY AND DATA INTEGRITY

The technology has its own properties, immutability, transparency, and decentralisation that give a strong base for achieving the improvement of data integrity and also data privacy. This is ensured by the use of restricted access methods, secure storage, and cryptography. Even though the technology is highly efficient in ensuring data integrity, there are opportunities for achieving data privacy.

7.5.1 Overview of Blockchain Architecture

The objective of blockchain design in multimodal LLM is to use blockchain's decentralised, secure, transparent features to solve inherent issues as well as improve many aspects of LLM functionality. Enhancing data security, confirming model integrity, enabling decentralised LLM services, and supporting safe data sharing are all included in this. LLM models and privacy protection have garnered a lot of interest lately. Using ChatGPT as an example, privacy protection is important yet difficult in intelligent customer support apps. It is necessary to investigate the practical means of preserving privacy while providing services. The creation of Q&A chatbots in the blockchain space is also progressing at the same time. The development process and technological implementation path are made clearer by classifying user wants and problem characteristics. Its performance and user satisfaction are then further assessed to encourage optimisation as well as enhancement. There is an interpretable recommendation system for streaming data that combines big language models and blockchain. With the aid of huge language model capabilities and blockchain features, it creates a recommendation system. Its duties include digesting data and providing justifications for recommendations. The idea is that the two cooperate to guarantee the semantics of the data and mining. It can increase user trust and recommendation accuracy, as demonstrated by pertinent situations. LLM and blockchain technology are revolutionising user interface and user experience design in the financial technology industry. Privacy is protected by blockchain, and needs are understood by the big language model. Its benefits include intelligent interface design and the ability to adapt to individual needs. The applications that boost competitiveness and customer happiness already exist [20].

7.5.2 Smart Contracts for Access Control

To limit which individuals or entities are permitted to carry out particular operations inside the logic of the contract, smart contracts might incorporate access control techniques. Ensuring that only authorised parties can initiate sensitive actions, like as token minting, contract upgrades, or data alterations, is essential for security. A number of methods, like as role-based access control (RBAC) and ownership-based access control, are utilised in smart contract code to manage these permissions.

In smart contracts, what is access control? Access control in smart contracts is the process by which the smart contract itself identifies the individual with the right to carry out certain operations or access certain information. Since the process inhibits both malicious and unauthorised access, this is very essential when considering security.

Why is access control important?

- *Security*
  It prevents those with malicious intent from leveraging weaknesses to steal money from the contract, manipulate data, and affect the functionality of the contract.
- *Data integrity*
  It is used to ensure that only approved individuals can change or view critical data stored in a contract.
- *Functionality*
  Controls that have access to take particular actions on the contract, such as upgrading, token minting, or pausing functions.
- *Trust and decentral*
  Although central control might be easier to manage, certain assumptions of trust arise. Access control systems with a distributive structure improve the reliability and decentralisation of the design.

7.5.3 Decentralised Data Management

This refers to the kind of data management where the storage of the data itself, the processing of the data, along with the administration of the entire data, gets executed by the help of a couple of nodes situated in some different geographical location, unlike the centralised server. Such an administration of the data itself is quite flexible, with immense sovereignty, even enhanced security due to the aspect of redundancy. Some of the chief salient features of decentralised data management are as follows:

Distributed processing and storage: Data is distributed and replicated in different nodes, in contrast to the data being in the same node.

- *Multi-authority*: Data is controlled by nodes/users equally, and the control authority does not have control over everything.

Data sovereignty: Data sovereignty encourages self-governance in contexts where either the user community itself, or the entire department, could have managed their respective data more effectively.

- *Potential increased security*: In view of what Dremio considers and other factors, there are advantages of decentralisation regarding resistance to the effects of single failure and attacks, since the attacks do not have anything central that can be focused on by attackers.

7.5.4 Blockchain Implementation Challenges

- *Technological innovation*: Research and implementation of scaling technologies like sharding, layer 2 networks, and consensus mechanisms that are optimised.
- *Standardisation and clarity of regulations*: Facilitating cooperation between the government, regulatory authorities, and other participants to arrive at standardised and clear regulatory norms.
- The design of reliable and secure cross-chain bridges and interoperability protocols, like the Cross-Chain Interoperability Protocol of Chainlink, is another form of interoperability technology.
- *Enhanced security protocols*: Implementing strong security policies, conducting periodic audits, and training users how to properly protect their data and assets.
- *Improved user experience*: Focusing on intuitive design and simplifying the new user sign-up process.
- *Education and talent development*: Supporting educational programmes aimed at bridging the skills gap and developing a ready workforce for the ever-changing blockchain environment.

7.6 INTEGRATING CRYPTOGRAPHY AND BLOCKCHAIN IN MULTIMODAL LLMS

Combining encryption as well as blockchain method makes multimodal LLMs more dependable, transparent, and safe. By combining capacities of LLMs with security

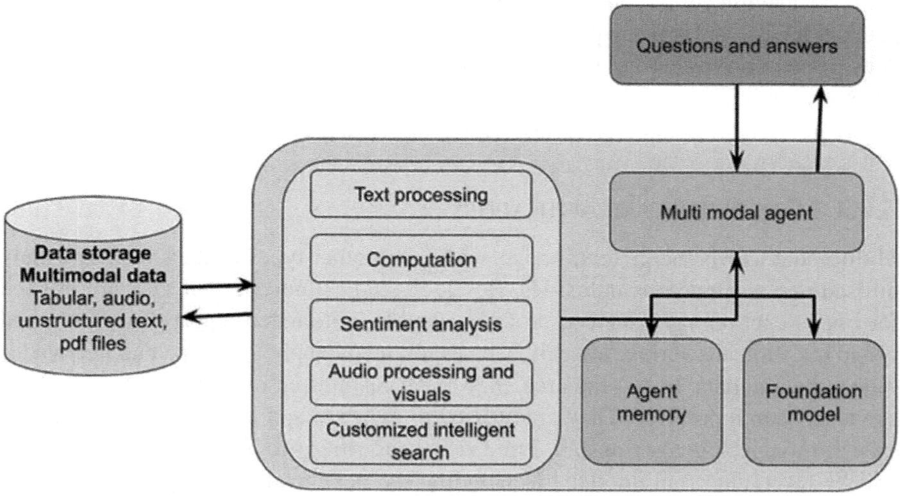

FIGURE 7.2 Framework for secure data processing in multimodal LLMs.

features of blockchain and encryption, create AI methods that are more reliable and strong. This integration can address issues such as model security, data privacy, and potential hostile attacks on LLMs.

7.6.1 Framework for Secure Data Processing

The proposed study conducts a comprehensive examination of transformational potential of LLM methods for multimedia data security, examining their capacities, challenges, and solutions. The proposed study examines potential security concerns as well as solutions for all types of multimedia data, numerous new and proven data protection measures. The study classifies novel attacks on LLM methods, including fast injection, adversarial perturbations, and membership assaults, during training and testing phases. The chapter also examines various robust defence methods, such as adversarial training, regularisation, and encryption. Figure 7.2 shows multimodal agent responds to natural language from business users by coordinating a range of methods to produce insights.

7.6.2 Enhancing Transparency and Accountability

To improve transparency and accountability in multimodal LLMs, a thorough architecture that includes strong governance standards, cryptographic protections, and verifiable audit techniques is needed to guarantee responsible use of sensitive data. The technique begins with certain data governance rules that define ownership, consent, and permissible uses. All data transactions are documented as tamper-evident, unchangeable records with only hashes, pointers, and metadata to preserve privacy. The architecture incorporates model provenance tracking, version control, and explainability hooks to make decision-making traceable. Auditor dashboards enable real-time monitoring, anomaly identification, and incident investigation. Role-based access, regular independent audits, and privacy-preserving explanation methods ensure that transparency does not compromise confidentiality, even though quantifiable KPIs—such as audit coverage, policy violation rates, remediation times—support continuous accountability improvement and foster trust in AI methods deployed in sensitive domains.

7.6.3 Case Studies and Applications

Multimodal LLMs merge text, image, audio, and other types of data and are widely utilised across numerous industries. Research studies indicate that they are utilised for image captioning, enhanced product searches, automated report generation, and visual question answering. In healthcare sector, medical photos are calculated in addition to patient data. In e-commerce, they make customised recommendations as well as visual search possible. They are utilised in assistive and access methods, such as speech-to-sign language translation and visual description tools. Several organisations will be leveraging multimodal AI, including the healthcare industry, which will be applying it in aiding diagnoses by scanning images and patient files.

- *Retail and e-commerce*: Supporting visual search and recommendations.
- *Finance*: Improving market analysis by means of interpretation of text information, graphics, and reports.
- *Autonomous vehicles*: Managing voice commands, visual, and real-time road information.
- *Media and entertainment*: Improving content moderation and video analysis.

7.6.4 Limitations and Considerations

Although the multimodal LLMs have many more powers in terms of data complexity, computational requirements, and the interpretability of models and potential hallucinations, there are several limitations associated with the depth of reasoning, the use of long-term memories, and the general understanding of the context. And lastly, it is necessary to ensure that the models conform to the moral values and goals of humankind.

Data-related limit

- *Datum complexity*
  Multimodal LLMs need massive data with text, image, audio, and video content. Any data diversity, bias, and/or error with regard to labels could generate flawed results.
- Multimodal LLMs may produce inaccurate or inconsistent information while handling different forms of data, and this could impact the user's level of trust.
- *High computational costs*
  The usage of large multimodal LLM models will be costly due to their high computer requirements.

- *Resource intensive*
  These approaches are often highly demanding on the processing power and memory, which might limit entry for certain individuals or entities.
- *Interpretability and bias*
- Lack of transparency:
  The ability to interpret how multimodal LLMs make decisions is difficult due to their complexity. This creates concerns regarding ethics and the law.

7.7 EVALUATION AND RESULTS

Simulation arrangement draws ideas from recent privacy-preserving LLMs as well as blockchain-based AI research. The multimodal LLM process was coded using Python, incorporating individual encoders to handle multimodal inputs. The data computation process is protected through cryptographic mechanisms. Simulations utilised homomorphic encryption (SEAL) to perform calculations even on encrypted data. This is done through secure multiparty computation (SPDZ). Differential privacy (DP-SGD) was applied during model training to prevent sensitive information leakage. We used a permissioned blockchain to manage data securely. Block validation was made quick by using a practical byzantine fault tolerance (PBFT)-based consensus. For scalability, encrypted data remains off-chain, and only hashes and metadata are maintained as on-chain. Table 7.2 shows the hyperparameters of the model.

7.7.1 Performance Analysis of Cryptographic Methods

In Table 7.3, the effectiveness of proposed model for multimodal threat information extraction is shown. For each column, the bold number indicates the best performance and the underlined number corresponds to the second-best performance. Human annotation-text represents the entity, relation, and technique extraction of threat intelligence text under the manual annotation process.

Table 7.4 and Figure 7.3 present the results of these ablation tests. This restricts their use in high-throughput settings like multi-tenant APIs or massive chatbots. Since none of the examined techniques are made to route across multimodal LLMs, modal support is much more constrained. This gap continues to be crucial as real-world applications call for models that comprehend and produce across modalities—text, visual, and speech. Closing these gaps should be the main goal of future routing and HI research in order to enable not just quick and economical inference but also scalable and modality-aware systems that can withstand the pressures of real-world usage.

7.7.2 Blockchain Implementation Results

Blockchain technology is being integrated with LLMs to enhance their security, explainability, and overall functionality, particularly in multimodal contexts. This integration leverages blockchain's decentralised, transparent, and secure nature to address limitations in LLMs, such as security vulnerabilities and lack of trust. Conversely, LLMs can improve blockchain's analytical capabilities by providing explainability and aiding in tasks like smart contract auditing.

TABLE 7.2
Hyperparameters Used in the Model Training

Component	Hyperparameter	Range
Multimodal LLM core	Token embedding size	4096
	Number of transformer layers	32–40
	Attention heads	32
	Activation function	GELU
Text encoder	Max sequence length	512 tokens
Image encoder	Input resolution	224 × 224
	Patch size	16 × 16
Audio encoder	Sampling rate	16 kHz
	Feature type	Log-Mel spectrogram
Homomorphic encryption (HE)	HE schemes	CKKS
	Polynomial modulus degree	8192
	Coefficient modulus	218 bits
	Encryption level	128-bit security
Secure multiparty computation (SMPC)	Secret sharing scheme	Additive secret sharing
	Number of parties	3–5
	Computation protocol	SPDZ-style
Differential privacy (DP)	Privacy budget (ε)	1.0–3.0
	Delta (δ)	1e–5
	Noise mechanism	Gaussian
	Gradient clipping norm	1.0
Blockchain layer	Blockchain type	Permissioned
	Block time	1–2 seconds
	Consensus mechanism	PBFT / Raft
	On-chain storage	Hashes and metadata

7.7.3 Discussion on Privacy and Efficiency Trade-offs

Multimodal LLMs provide very powerful capabilities by combining text, images, and other modes. The integration aspect, however, presents new trade-offs when it comes to privacy and efficiency. The fact that multimodal LLMs are capable of several tasks at once does not necessarily eliminate worries about data privacy and efficiency in terms of computing resources.

Privacy concerns

- *Data leakage*
  Like their text-focused counterparts, multimodal LLMs, potentially, could memorise their training data, thereby revealing sensitive information, like personally identifiable information, due to overfitting.
- *User inference attacks*
  The attackers would be able to deduce whether particular individuals participated in the dataset or aim to backtrack and regenerate personal attributes from their corresponding models.

TABLE 7.3
Performance Comparison with the Existing Cryptographic Models

Technique	Latency	Throughput	Computation	Communication	Accuracy impact	Scalability	Complexity	Security assurance
Homomorphic encryption (HE)	High—often orders of magnitude slower	Low (batched helps)	Very high CPU and memory	Low (non-interactive)	Possible (approximations for non-linear ops)	Hard for very large models	High (parameter tuning)	Strong cryptographic (no plaintext exposure)
Secure multiparty computation (MPC)	Medium–High (many rounds)	Moderate (depends on network)	Moderate–High	Very high (rounds & bytes)	Low (can compute exact ops)	Challenging with many parties/ model size	Very high (orchestration)	Strong if parties non-colluding
Trusted execution environments (TEEs)	Low (near-native)	High	Moderate (enclave limits)	Low	None (native code)	Limited by enclave memory; scales with hardware	Moderate (depends on vendor APIs)	Moderate — relies on hardware/ firmware trust
Differential privacy (DP)	Negligible	High	Low	Low	Direct trade-off (noise reduces utility)	Scales well	Moderate (DP accounting)	Statistical guarantees only (not encryption)
Federated learning + secure aggregation	Medium (communication rounds)	Moderate	Distributed CPU	High (model updates)	Possible (non-IID effects)	Scales with clients but challenged by heterogeneity	High (client orchestration)	Good for keeping raw data local; need DP for stronger guarantees
Zero-knowledge proofs (ZKP)	Very high (proof gen)	Very low	Very high	Moderate–High (proof sizes)	None	Poor for large or complex computations	Very high (specialised tooling)	Strong verifiable guarantees
Functional encryption (FE)	High (complex schemes)	Low	Very high	Low	Depends on function	Immature for large ML	Very high (research-stage)	Strong, but practical limits
Threshold crypto (sign/dec)	Low–Medium	High	Moderate	Low–Moderate	None	Scales with key management	Moderate	High (removes single key risk)

TABLE 7.4
Ablation-Based Performance Analysis

Method	Precision	Entity Recall	F-1	Precision	Relation Recall	F-1	Precision	Technique Recall	F-1
Text-based method									
HE	85	77	81	75	79	82	80	71	73
SMPC	88	79	83	78	83	83	81	73	75
DP	91	81	88	81	85	86	83	75	79
Image-enhanced method									
HE	88	79	85	78	82	85	82	75	81
SMPC	91	81	86	81	84	89	84	79	83
DP	92	83	89	83	86	91	88	81	88

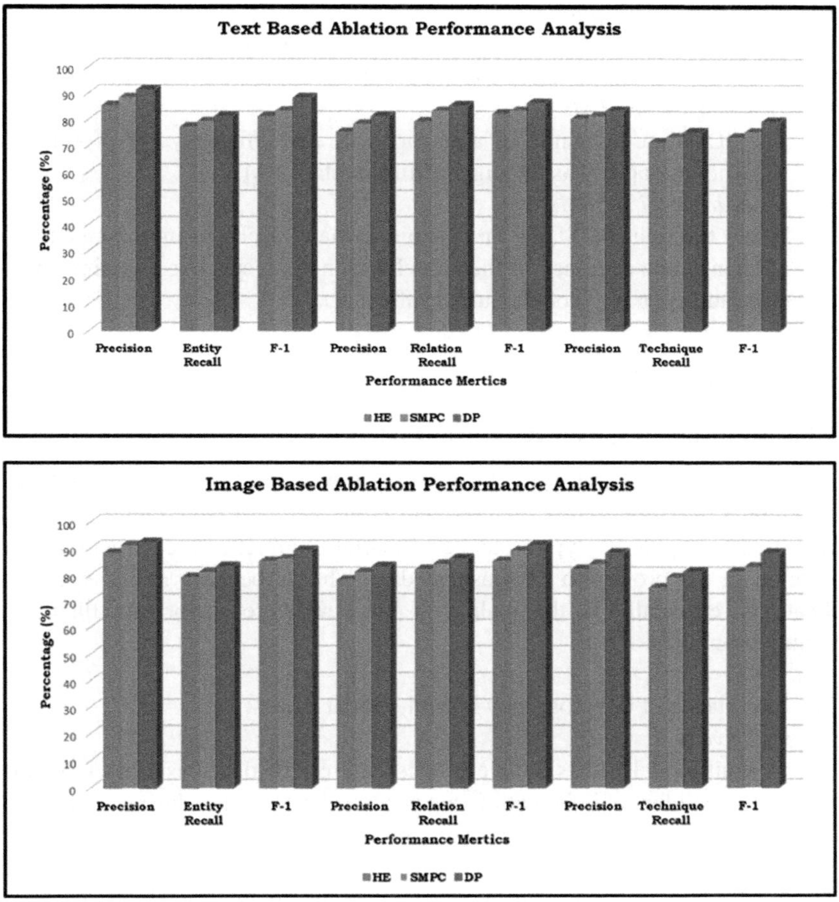

FIGURE 7.3 Comparison of cryptographic methods.

- *Black-box*
  Despite this black-box access to the multimodal LLM, the attacker can leverage minimal user samples and the technique of prompt injection to obtain sensitive information.
- *API-based*
  A number of multimodal LLM-based models have APIs, which has many implications, including the disclosure of data to a third-party service.

Efficiency concerns

- *Computational*
  Multimodal LLMs cannot easily be executed using resource-limited devices as they require considerable computational power during learning as well as during deployment.

- *Resource consumption*
  Memory usage, latency, and power consumption are important considerations for the use of multimodal LLMs, especially when applied to mobile applications.
- *Trade-offs with SLMs*
  The SLMs are known to be a performance and efficiency compromise, possibly not as capable as the bigger Multimodal LLMs.
- *Domain-specific SLMs*
  While training the SLMs on specific domains leads to improvement in the efficiency and accuracy of the model within the respective domains, the generalisability of the model could be affected.

7.8 FUTURE DIRECTIONS

7.8.1 Advances in Cryptography for AI

Cryptography breakthroughs also have a significant role in protecting multimodal LLMs, which incorporate text, images, and other forms of data. Other approaches, such as homomorphic encryption and adversarial training, also exist and are being considered in the protection of sensitive data. This is because multimodal LLMs applications, especially in the realms of the healthcare sector and finance, are becoming prominent.

1. Importance of Cryptography in Multimodal LLMs:
   - *Privacy*
     Multimodal LLMs handle broader sources of data, resulting in exposure to confidential information that needs to be protected during its processing and storage by means of cryptography.
   - *Security*
     The multimodal models are also vulnerable to attacks, which may be used for manipulating the inputs and outputs. Cryptographic methods can be used to improve the resistance to the attacks.
   - *Trust*
     Trust in the reliability and security of multimodal AI systems is assured through the use of stringent cryptographic measures.
2. Key cryptographic techniques
   - *Homomorphic encryption* (HE)
     Allows direct computations on encrypted data without decrypting them. It is especially useful in privacy-preserving ML with LLMs.
   - *Adversarial training*
     Training models with adversarially crafted inputs to make them more robust against malicious inputs or attacks.
   - *Differential privacy*
     Adds noise to the data in a controlled fashion to protect individual privacy and thus it enables meaningful analysis. This can be used in training data for multimodal models.
   - *Secure multiparty computation* (SMPC)

Enable several parties to jointly compute a function on their private data without revealing their respective individual inputs to each other. Also, this topic is useful for collaborative training of multimodal models.

7.8.2 Potential of Blockchain in Large-Scale LLMs

The potential uses of blockchain are in improving the security, transparency, and trustability of LLM.

The use of the immutable ledger and smart contract functionality in the blockchain technology could help overcome the challenges that have become prominent with the growing use of LLM, such as data poisoning, prompt injection attacks, and the disclosure of data without authority.

1. Enhanced security and safety
   - *Data provenance and integrity*
     The integrity of the data used in training and fine-tuning these models can be traced, thereby preventing any kind of ill intent on account of data modification, which can be achieved through the use of Blockchain.
   - *Secure data sharing*
     Smart contracts can play an important role in facilitating the sharing of data in a safe and managed manner between different parties engaged in the development of LLM technology.
   - *Handling attacks*
     The use of blockchain can help in protecting against different types of LLM attacks like data poisoning, prompt injection, and adversarial attacks. This is because blockchain offers tamper-proof evidence of the activities of the model.
   - *Transparency*
     The decentralised and transparent characteristics of blockchain make it possible to audit the whole LLM development chain, from data gathering to model deployment, thereby increasing accountability.
2. Improving LLM reliability
   - *Data validation and verification*
     It is possible to use the blockchain for validating the authenticity as well as the reliability of the sources of information for the training of the LLM to counter hallucinations and misinformation.
   - *Decentralised training*
     The process of LLM training can be integrated with blockchain technology, allowing for decentralised LLM training in which several individuals can collaborate for LLM training without having to exchange private information.
   - *Real-time monitoring and alerting*
     It can be integrated with LLMs to analyse their performance on a real-time basis, thereby raising an alarm for any irregularities or security risks that may be encountered.

7.8.3 Emerging Privacy Concerns in AI

- The following recommended practices can lower risk, protect user confidence, and guarantee compliance.
- Create robust data governance guidelines. Data classification, role-based access restrictions, data security techniques like encryption and pseudonymisation, frequent audits, compliance with pertinent data privacy laws like the CCPA and GDPR should all be included in thorough guidelines.
- Don't gather too much info. Only gather the information required for the AI system to perform its intended role. Many data privacy regulations mandate that data be kept to a minimum in order to minimise exposure.
- Ask for express consent. Users should be fully informed about collection, usage, storage of their data. Prior to gathering and utilising personal information, always get express, informed consent. In addition to being required by law in many places, this openness increases user confidence.
- Confirm the training data supply chain. Verify the reliability and adherence to strict data privacy regulations of the third-party datasets utilised to train AI methods.
- Adhere to best standards for application security generally. Build a stronger foundation for your AI systems by putting in place tried-and-true data security mechanisms, such as strong identity and access control, data encryption both in transit and at rest, and data anonymisation to safeguard private information.
- Continually evaluate the risks. Since AI is developing quickly and necessitates frequent reevaluations of privacy threats, risk management is never "one-and-done" but rather remains continuous. To find new threats, make sure that regulations are being followed, and change governance policies as needed, conduct ongoing assessments.
- Use software to improve data security. For better observability, more robust adherence to privacy laws, real-time monitoring of sensitive data flows through AI applications, use smart solutions such as the F5 Application Delivery and Security Platform.
- Evaluate vendor and partner privacy policies. Verify that all service providers and technology partners have robust data security and privacy policies. These are F5's data privacy declarations. Since AI data frequently moves through open ecosystems, third-party regulation is crucial to reducing risk.

7.8.4 Ethical and Regulatory Implications

Multimodal LLMs bring new ethical issues due to their capacity to process multiple data formats. However, ethical issues that could arise as a result of multimodal LLMs include privacy, bias, as well as misuse of these very powerful tools. There is a need to ensure that these powerful tools are used effectively by employing ethical standards.

1. *Data privacy and security*
   - Learning multimodal LLM usually entails the need to train the models with a lot of data, which may contain personal data.

- Data privacy laws such as GDPR and AI ethics frameworks affect the use and development of these models.
- There is a possibility of breaches as well as accidental disclosure of sensitive information in domains such as healthcare.
- The need to ensure that user privacy is protected arises. This requires the implementation of various technologies.

2. *Bias and Fairness:*
   - ML models tend to develop biases and retain existing biases from the data used for model training.
   - Also, multimodal LLMs, with their various input data, could potentially increase any bias that could exist.
   - Methods for bias detection, mitigation, and fairness assessment need to be developed.
3. *Transparency and interpretation*
   - Worldwide, the processes of LMMs, such as multimodal LMMs, are considered "black boxes" and are, therefore, opaque.
   - Opacity can lead to lack of trust and difficulty in resolving discrepancies in errors or biases.
   - A focus needs to be placed on increasing the explainability of LLex and the traceability of its rationale.
4. *Regulatory implications*
   - Governments and organisations are developing regulations and frameworks to deal with the ethical and societal implications of AI and LLMs.
   - WHO has issued a set of guidance on AI Ethics and Governance for Multimodal LLMs, underlining the use of standards, auditing, and the involvement of stakeholders.
   - There is a necessity for robust regulatory frameworks that ensure such methods are applied safely, ethically, and responsibly.

7.9 CONCLUSION

This chapter offers a systematic and comprehensive framework for securing multimodal large language models. This framework combines the latest cryptographic methods and the use of blockchain. With the current increase in the vulnerability of sensitive text, image, and audio information being processed through large language models, the framework helps minimise exposure of the sensitive information while retaining utility. This is achieved through client-protection methods provided by homomorphic encryption, secure multiparty computation, and differential privacy.

The relevance and significance of the proposed work are mainly associated with the usage of the permissioned layer of the blockchain and the presence of smart contracts that are able to deal with the concepts of data access, consent, and accountability. Instead of storing the personal data on the blockchain, the proposed solution focuses only on storing the hash values, metadata, and audit trails, thereby maintaining transparency, immutability, and also addressing the problem of privacy and scalability. Smart contracts are able to automate the enforcement of policies and data access.

The experimental evaluation shows that, although cryptographic enforcements come with measurable computational and latency overheads, careful architectural choices such as selective routing of operations, off-chain encrypted computation, and PBFT-based consensus can efficiently manage the costs. The comparison and ablation analyses further present how multimodal performance can be preserved with significant enhancement in resistance to data leakage, inference attack, and unauthorised access.

REFERENCES

1. Rahman, M. A., Alqahtani, L., Albooq, A., & Ainousah, A. (2024, January). A survey on security and privacy of large multimodal deep learning models: Teaching and learning perspective. In 2024 21st Learning and Technology Conference (L&T) (pp. 13–18). IEEE.
2. Rahman, M. A. (2023, December). A survey on security and privacy of multimodal llms-connected healthcare perspective. In 2023 IEEE Globecom Workshops (GC Wkshps) (Pp. 1807–1812). IEEE.
3. Geren, C., Board, A., Dagher, G. G., Andersen, T., & Zhuang, J. (2025). Blockchain for large language model security and safety: A holistic survey. *ACM SIGKDD Explorations Newsletter*, 26(2), 1–20.
4. Kumar, A., Salim, M. M., Camacho, D., & Park, J. H. (2025). A comprehensive survey on large language models for multimedia data security: challenges and solutions. *Computer Networks*, 267, 111379.
5. Su, C., Wen, J., Kang, J., Wang, Y., Su, Y., Pan, H., ... & Hossain, M. S. (2025). Hybrid RAG-empowered multi-modal LLM for secure data management in Internet of Medical Things: A diffusion-based contract approach. *IEEE Internet of Things Journal*, 12(10), 13428–13440.
6. Zhang, X., Xu, H., Ba, Z., Wang, Z., Hong, Y., Liu, J., ... & Ren, K. (2024). Privacyasst: Safeguarding user privacy in tool-using large language model agents. *IEEE Transactions on Dependable and Secure Computing*, 21(6), 5242–5258.
7. Fu, S., & Xie, M. (2025, June). AI Oracle: A Blockchain-Powered Oracle for LLMs and AI Agents. In 2025 Crypto Valley Conference (CVC) (pp. 1–10). IEEE.
8. Friha, O., Ferrag, M. A., Kantarci, B., Cakmak, B., Ozgun, A., & Ghoualmi-Zine, N. (2024). LLM-based edge intelligence: A comprehensive survey on architectures, applications, security and trustworthiness. *IEEE Open Journal of the Communications Society*, 5, 5799–5856.
9. Ghazal, T. M., Islam, S., Hasan, M. K., Abu-Shareha, A. A., Mokhtar, U. A., Khan, M. A., ... & Ahmad, M. (2025). Generative federated learning with small and large models in consumer electronics for privacy preserving data fusion in healthcare Internet of Things. *IEEE Transactions on Consumer Electronics*, 71(2), 3413–3430.
10. Sun, L., Liu, D., Wang, M., Han, Y., Zhang, Y., Zhou, B., & Ren, Y. (2025). Taming unleashed large language models with blockchain for massive personalized reliable healthcare. *IEEE Journal of Biomedical and Health Informatics*, 29(6), 4498–4511.
11. Ajimon, S. T., & Kumar, S. (2025). Applications of LLMs in quantum-aware cybersecurity leveraging LLMs for real-time anomaly detection and threat intelligence. In Leveraging Large Language Models for Quantum-Aware Cybersecurity (pp. 201–246). Palmdale, PA: IGI Global Scientific Publishing.
12. Fernández-Becerra, L., González-Santamarta, M. Á., Guerrero-Higueras, Á. M., Rodríguez-Lera, F. J., & Olivera, V. M. (2024). Enhancing trust in autonomous

agents: An architecture for accountability and explainability through blockchain and large language models. arXiv preprint arXiv:2403.09567.

13. Zhou, J., & Wu, J. (2025, June). Real-Time Reliable Large Language Models with Distributed Knowledge Crowdsourcing for Automotive Mobile Intelligence. In International Conference on Wireless Artificial Intelligent Computing Systems and Applications (pp. 406–417). Singapore: Springer Nature Singapore.
14. Zhang, S., Li, H., Sun, K., Chen, H., Wang, Y., & Li, S. (2025). Security and Privacy Challenges of AIGC in Metaverse: A Comprehensive Survey. *ACM Computing Surveys*, 57(10), 1–37.
15. Kang, J., Luo, X., Nie, J., Wu, T., Zhou, H., Wang, Y., ... & Xie, S. (2024). Blockchain-based pseudonym management for vehicle twin migrations in vehicular edge metaverse. *IEEE Internet of Things Journal*, 11(21), 34254–34269.
16. Mohamed, M., Elmor, A., Smarandache, F., & Metwaly, A. A. (2024). An efficient superhypersoft framework for evaluating LLMS-based secure blockchain platforms. *Infinite Study. Neutrosophic Sets and Systems* 72(1). https://digitalrepository.unm.edu/nss_journal/vol72/iss1/1
17. Wang, Y., Liu, Y., Ji, L., Luo, H., Li, W., Zhou, X., ... & He, T. (2025). AICrypto: A comprehensive benchmark for evaluating cryptography capabilities of large language models. arXiv preprint arXiv:2507.09580.
18. Luo, H., Liu, Y., Zhang, R., Wang, J., Sun, G., Niyato, D., ... & Shen, X. (2025). Toward edge general intelligence with multiple-large language model (multi-LLM): architecture, trust, and orchestration. arXiv preprint arXiv:2507.00672.
19. Singh, S., & Vorster, L. (2024). LLM Supply Chain Provenance: A Blockchain-Based Approach. In Proceedings of the International Conference on AI Research. Academic Conferences and Publishing Limited.
20. Qin, H. (2024). Revolutionizing cryptocurrency operations: The role of domain-specific large language models (LLMS). *International Journal of Computer Trends and Technology*, 72(6), 101–113.

8 Next-Generation Multimodal Conversational Agents

Affective Reasoning, Cross-Modal Dialogue, and Embodied Interaction

Mayuri K.P., Ravi Hosur, and Amitha S.K.

8.1 INTRODUCTION

To grasp the present condition of dialogue systems fuelled by AI technology, it is crucial to place it in its historical context. The evolution of dialogue systems is more than just the implementation of improvements in each consecutive technological step. It embodies the continuous improvement of machines from being listeners to being responders and even empathetic beings. This chapter presents an explanation on how the process developed from rule-based machines to the present multimodal systems. In the process of explaining the history of dialogue systems, it becomes clear that true human-like interaction goes beyond language processing tasks.

8.1.1 Evolution of Dialogue Systems

The history of dialogue systems has progressed from scripted, rule-based systems to emotionally intelligent, multimodal models, echoing the overall history of AI. Instead of specifying major milestones as in a traditional chapter format, this text looks at dialogue systems in relation to their developing history. Figure 8.1 traces back several decades of advancements from static, unimodal communication to our present-day multimodal interactions afforded by the use of foundation models, with four major periods in this development history: ELIZA's symbolic/rule-based systems of the 1960s, statistical approaches such as hidden Markov models in the 1990s, neural networks in the 2010s, and current multimodal approaches that synthesize these different periods.

Figure 8.1 encapsulates the transition from ELIZA to GPT-4, placing dialog systems in an overall historical perspective. In Section 8.1, it will be used to describe key advancements over many years, comparing ELIZA with script-based conversation

DOI: 10.1201/9781003678359-8

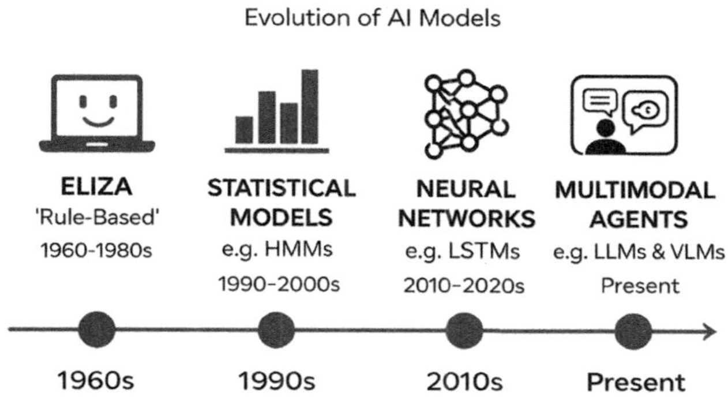

FIGURE 8.1 Evolution of dialogue systems.

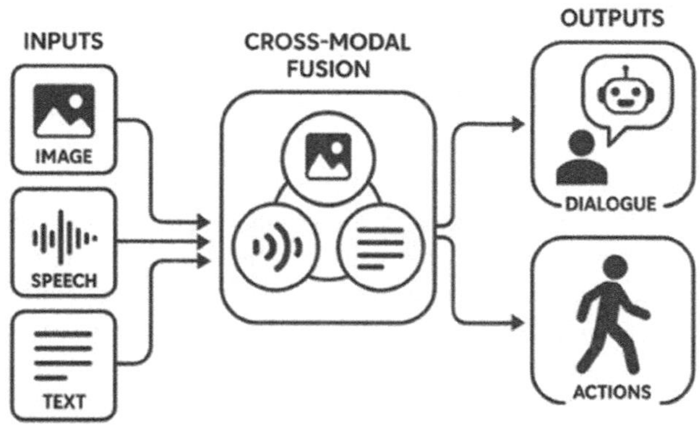

FIGURE 8.2 Cross-modal fusion framework in multimodal agents.

support in relation to the latest capabilities found in GPT-4. On the topic of ELIZA as it pertains to contemporary application as virtual health coaches, it would be appropriate to refer back to this 1960s communication design for a telephone operator in its original form.

Following the progression of Professor Laura Schulz's talk, a description of the history of architectural advancements in dialogue systems follows. Figure 8.2 traces this history, from a set of rule-based heuristics, to hidden Markov models, then to deep networks, and most recently, to language models. Each of these block diagrams illustrates a set of components in a system. This first set of components describes ELIZA's response-pattern loop, while the final set represents our current system, which integrates a language model, a vision model for image decoding, and an audio model for sound decoding. This architecture will be described in further detail in Section 8.1.

Figure 8.3 represents the embodied multimodal dialogue processing pipeline. The dynamic relationship among a range of input and output modalities in an AI conversation system has been effectively depicted in the figure. The figure highlights the manner in which modern dialogue models process and respond using text, image,

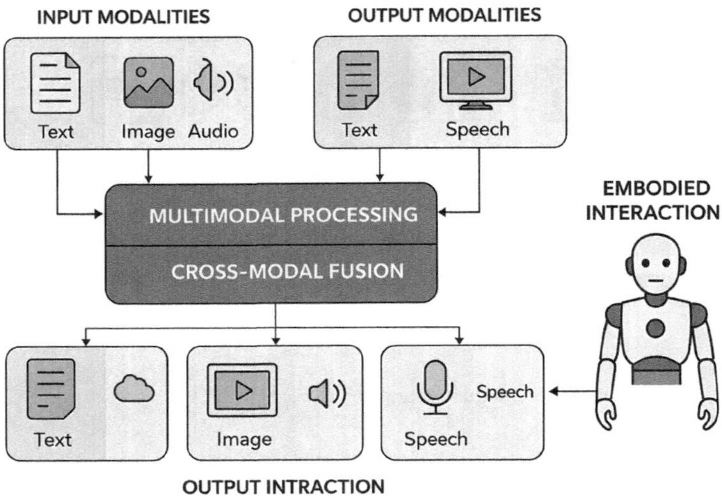

FIGURE 8.3 Embodied multimodal dialogue processing pipeline.

and audio streams. The input streams on the left side include the user speech, visual input sources like faces and gestures, and text input. All the inputs are then fed into the multimodal processing core, where cross-modal fusion takes place.

By this pipeline, the processing of the different modalities, namely text, vision, and speech, is modelled independently and then aligned in order to arrive at a single decision made by the policy. This particular architecture and pipeline are discussed in detail in Section 8.1 and related to the concept of architectural fusion in Section 8.2. Modern dialogue systems and their related studies focus on system development that goes beyond the scope of interactive functionality and becomes an essential part of social ecologies, assistive frameworks, and cognitive architectures. This development has also become possible because the ambient IoT has now become an essential part of the developing dialogue system paradigm. Thus, the dialogue systems are now designed and created in such a way that they become social companions who evaluate the signals and the levels of engagement.

The cross-modal fusion module is the centrepiece of the architecture and its role is to make sure that the related features across the different modalities, like the expressions and the corresponding tones in the speech, are learned together. The resulting text and speech responses are then created using the learned contextual representation. The representation of the pipeline can be achieved through incorporation with the agent, either virtual or physical (denoted in the figure by the humanoid figure), allowing the agent to provide the needed real-time responses through expressive interfaces like gesture, speech, and other forms of embodiment. The proposed architecture enables assistive learning companions, healthcare robots, and emotionally intelligent interactive kiosks capable of generating context-sensitive and empathetic interactions in the fields of education, healthcare, and public services. The need for representation in cases like virtual responsive therapists and assistive robots in the care of the old cannot be overemphasized, where the responses need to be both correct and appropriately empathetic.

In short, the figure illustrates the movement from static unimodal systems to socially interactive agents who perceive, reason, and respond in an embodied manner across multiple modalities, thereby enabling the development of next-generation conversational intelligence.

8.1.2 From Symbolic to Statistical Dialogue Systems

The pioneering period in dialogue systems was dominated by rule-based logic in symbolic AI research. Early systems like ELIZA (1966) achieved conversation simulation through simple pattern recognition. While ELIZA was able to parrot back the user statements in different words and maintain the syntactical framework of conversation, the processes involved no cognition. ELIZA's seeming ability to converse was purely indicative of the systems' determinate response properties.

Symbolic dialogue systems in the 1970s and 1980s constituted some of the earliest attempts at modeling conversational behaviour using finite state machines and production rule-based systems. This made programming rules for controlling the dialogue process feasible. These systems might seem readable and predictable. However, their very nature was brittle. This means that the system broke down in the presence of variations and ambiguities in user input. Additionally, their programming was very much human-dependent.

However, in the more mature field of spoken dialogue systems, researchers began relying on statistical models. This was because of the increase in the size of the corpora being used. Two popular statistical models adopted during that period were the hidden Markov model (HMM) and the N-gram model. Models like the HMM helped system designers build robust systems by considering the input spoken words to correspond to different states in the model. Each state had probabilities of emitting the observations.

Statistical dialogue management formulated the problem of dialogue simulation in the framework of partially observable Markov decision process (POMDP). Decisions on the systems' actions were made according to the probabilities in the respective belief distribution for the users' state. This paved the way for the use of reinforcement learning. Statistical processes such as the ones discussed above in the research field were used in practical applications like telephone customer service robots, flight booking systems, and customer service centre assistants. To cite one instance, the statistical approach to natural language interpretation in AT&T's system "How May I Help You?" involved the analysis of the user input statements in more than 100 types of customer service. However, despite all the improvements achieved in statistical systems, they continued to work within small domains. Statistical systems used the bag-of-words approach along with sparse representations. However, the sparse representation made it difficult for them to generalize. Also, the decoupling of the components (speech recognition, language interpretation, and policy control) in statistical systems resulted in the cascading effect of errors.

8.1.3 Emergence of Neural Dialogue Systems

Statistical model limitations in dialogue systems, especially their rigidity and specificity in domains, opened the door for the application of neural models that could directly learn the mapping from raw text input to semantic answers. The dawn of the deep learning revolution in the form of recurrent neural networks (RNN), long short-term memory (LSTM) networks, and encoder–decoder networks introduced the paradigm shift in the field of dialogue systems. It was incredibly successful in addressing the sequential nature of the problem.

To understand the context better, the advent of the sequence-to-sequence (Seq2Seq) model made it feasible for dialogue systems to directly link the user's message to the system response. However, the attention mechanism significantly improved the response generation capabilities of the model by allowing it to selectively focus on the input. Dialog systems like Google's Smart Reply and the initial versions of the Facebook Chatbot leveraged the capabilities of the neural dialogue system. However, the initial models defined had some drawbacks. These models tended to generate ambiguous, redundant, or common answers like "I don't know" or "That is interesting" for different questions. To counter such limitations, the concept of hierarchical recurrent networks and the memory-augmented neural model was introduced.

The availability of training data impacted model performance. Ubuntu Dialogue Corpus, DailyDialog, and Persona-Chat were datasets that helped in fine-tuning to bridge the performance differences between general-purpose language models and dialog models. Neural dialog systems progressed from research ideas to practical commercial implementations. Virtual assistants like Siri, Alexa, Google Assistant, and Cortana incorporated neural networks for intent identification, slot filling, and dialog tasks, which seemed to merge with each other in day-to-day life for controlling smart devices, notifications, and natural suggestions.

Generalization abilities in neural networks were also applicable in corporate environments, where finance bots, healthcare triage tools, HR software, and IT support portal software utilized their capabilities. Their flexibility was also suitable for multilingual/multicultural environments. Early neural-based dialog systems were monomodal with no grounding in perception, as neural networks processed text data with no visual/audio input, hence restricted in performing emotionally expressive and spatially grounded tasks. This showed that cross-modal models, which integrated different inputs to simulate human communication, were important.

8.1.4 Large-Scale Pretrained Models and Transformer Revolution [3,4,10]

The transformer model proposed by Vaswani et al. in 2017 was a revolutionary step in the field of natural language processing (NLP). Unlike the traditional methods of RNN and LSTM networks that process the input sequentially, the transformer model allowed the model to process the input in parallel through self-attention. Not only was the processing speed improved during the training process, but the model also had the ability to understand the context of the conversation in multi-turn dialogue systems. Transformers provided the crucial components for the development of pretrained language models that could readily adapt to different tasks. BERT, proposed by Google, used the concepts of predicting the masked language in the input sequence and predicting the next sentence to produce highly effective bidirectional representations. However, the GPT series proposed by OpenAI used the concept of autoregressive generation. Hence, the model was very effective in producing coherent answers in the context of the input. This was performed on terabytes of data, which in turn led to the demonstration of some emergent capabilities.

Transformer networks made it possible for the field of dialogue systems to make the transition from task-oriented and modular systems to integrated and general-purpose ones. BERT variations for fine-tuning the model were employed for the detection of intent, slot filling, and answering questions. However, the work of GPT-2 and GPT-3 was utilized in the designing of fully conversational bots. Such models achieved few-shot and zero-shot learning capabilities. Afterward, companies in the field of AI leveraged the aforementioned models for the implementation of their commercially successful programmes like customer service bots, tutors, and health bots. Such enterprise-level models include DialoGPT, Meena, BlenderBot, and LaMDA. These technologies leveraged the capabilities of the transformer model in the context of dialogue systems by being able to remember the context of the dialogue for longer periods. Such models also began to act as conversational informational retrieval systems for finance, law, scientific research, along with being able to summarize documents.

Additionally, the more recent arrival of larger models such as GPT-4 and PaLM further expanded the conversational AI landscape into the areas of multilingual translation, multimodal perception (e.g., Flamingo and GPT-4V), all the way to self-contained decision-making in the context of software agents. More complex scenarios such as therapy conversational interaction, contract interpretation for legal purposes, and climate data exploration through conversation also became feasible. Despite their universality, these models are mostly unimodal. Additionally, the models are unable to process any visual, auditory, or emotional inputs. Such capabilities are crucial in real-world applications because the model must respond in ways similar to human beings. This problem sparked the beginning of the development of the multimodal transformer model.

8.1.5 Limitations of Unimodal Models

Despite having excellent conversational capabilities, the unimodal nature of large language models prevents them from comprehending the visual, auditory, and contextual information that can be useful in many applications. A dialog model that operates in a sole visual manner fails to incorporate information from facial expression, gesture, and locational description in a scene, and affective information in a dialogue. This would be rather challenging in situations where a dialogue has a bearing on affect. To clarify, a sole visual model used in telemedicine would, for example, be able to comprehend the semantics in a message from a patient but would not possess perceptual understanding in order to identify a pause in expression, tiredness in facial muscles, and urgency in body expression.

Unimodal models also present challenges in regards to spatial deixis and anaphora in a conversation. In responding to "What is that building over there?" that refers to an image, it becomes important for the model to anchor "that building" to that picture. This cannot be accomplished in traditional text models. Similarly, the smart home assistants have trouble in understanding commands like "turn it off" unless they are backed by the corresponding visual data and sensor information. From a purely technical point of view, a weakness of unimodal models would be that they are not working from a multi-sensory embedding space by lacking the use of co-attention. This means that even cross-modal knowledge has no bearing in these models. They also lack use of gaze. Unimodal models also lack use of knowledge due to a heavy dependency on text anticoding.

Such limitations are further intensified in cross-cultural scenarios, where cultural subtleties, affective diversity, and approachability requirements make a higher order of input processing necessary. Think of helper systems for disabled individuals, which would involve processing input from sign language, facial, and touch modality—tasks beyond those of single-input processing. In educational chatbot, detecting confusion from a change in posture and gaze direction is also critical. Such limitations are already felt in healthcare helper systems, where communication with patients involves processing their non-verbal behaviour concerning pain, confusion, or discomfort. Chatbot lack support capabilities for disabled individuals in tasks requiring adaptability, reasoning, and responding to facial expression, gesture, and other signs in those scenarios. Smart city scenarios also are beyond capabilities of individual input processing agents, which cannot independently solve situation-aware queries in servicing, like "Which parking lot is available?" and "That store near me is shut down?" In light of growing needs for natural and intuitive human–AI interaction, it has become necessary to develop systems that are able to process, understand, and respond in multiple modalities. Filling this gap in technology becomes important, not only for enhancing performance but also for aligning with ethical tenets of accessibility and trust. Thus, this move toward multimodal systems indicates a paradigm shift in conversational assistant development.

8.1.6 Toward Multimodal, Affective [6,15] Systems

Given the rising interest in emotionally intelligent systems in the healthcare, education, and ambient computing sectors, the shift from unimodal to multimodal dialogue systems marks the most significant milestone in conversational computing. This is because multimodal dialogue systems are capable of processing different input modalities such as text inputs, speech inputs, vision inputs, and physiological inputs simultaneously in order to arrive at a comprehensive interpretation of the users' intention, emotion, and context. Why is the shift from unimodal to multimodal dialogue systems crucial in conversational computing?

At the heart of such models are cross-modal transformer architectures that are able to coordinate diverse modalities through co-attention, latent alignment, and alignment tasks. Models like GPT-4V, BLIP-2, Flamingo, and OpenFlamingo represent some of the modern models that blend image-language learning tasks along with auto-generation. Such models leverage dual encoders for the processing of image frames, facial expressions, or gestures together with the input of textual queries to make grounding of "that object over there" or "emotional" signals like sarcasm, fear, or joy. Techniques in affective computing provide further enhancement in the architecture by evaluating the affect levels of the user based on various parameters such as facial action units (AU), prosody-based speech signals, galvanic skin response (GSR), and gaze direction. The affect levels of the users' emotions are recognized by the classifiers such as those for emotion recognition in real-time from the datasets such as AffectNet and EMOTIC, for the dialogue system to modify the tone, pace, and appropriateness of the response based on the recognized user emotions.

Concerning application areas like elderly care, affective dialogue systems assist in identifying the symptoms of either depression or cognitive impairment by tracking the subjects' rate of talking, eye contact, and facial expressions. In the context of learning platforms, intelligent tutoring systems adjust their explanation process according to confusion expressions indicated by eye tracking and head nod detection. Smart assistants for self-driving cars could alter the operating configuration according to stress cues in the passengers by incorporating both natural dialogue analysis and physiological measurement.

Such intelligent agents are now being pursued in practical implementations through the use of embodied technologies such as robots, augmented reality/virtual reality (AR/VR) avatars, and smart kiosks that provide real-time interaction. However, the challenge in designing intelligent agents lies in incorporating the various interaction modalities into low-latency processing pipelines. Finally, the transition toward multimodal affect-aware systems signifies not only technological progression but also the beginning of a paradigm shifts in the world of AI research from being an object to being a social partner capable of empathy.

8.1.7 Applications and Societal Relevance

The influence of multimodal affect-aware dialogue agents exists in the broad range of domains pertaining to the requirement for context awareness, affect intelligence, or the ability to collaborate seamlessly between human beings and machines. By effectively combining the inputs from the auditory, vision, and sensor sources, the dialogue agents have stepped beyond the boundaries of the interfaces of the past in areas ranging from medicine to entertainment.

Multimodal intelligent agents in the healthcare field –function as emotionally intelligent companions for the elderly in eldercare, telemedicine, and rehabilitative therapy. Such systems incorporating vision-based emotion recognition and speech processing capabilities are able to notice the initial symptoms of elder patients' depression, anxiety, or confusion in order to provide them with immediate attention or alert health professionals. Interactive therapy assistants adjust their communication-based tone according to the elder person's emotional condition. Smart hospital self-service terminals employ facial recognition technologies and gaze detection for guiding senior patients through registration processes for checking their health condition without direct contact during the post-pandemic period.

There has also been a striking paradigm-shifting work in the education field by intelligent tutoring systems that respond in line with the behaviour of the students. These systems observe eye gaze, head movements, voice varying levels, and the level of interaction in order to clearly recognize the engagement and confusion levels of the students. Emotion-aware virtual teaching assistants automatically adjust their teaching approach by going back to concepts, going slower, or switching from one modality to another according to the emotional state of the learner. Retailing and customer engagement markets make use of multimodal dialogue systems for personalization. Smart shopping chains and customer service hubs employ the analysis of facial expressions, tones of voices, and web surfing behaviour for the purpose of offering the relevant products/services to customers according to their mood and intention. Emotional-aware virtual receptionists at airports and hotels boost customer satisfaction by having pleasant and adaptive dialogues even in critical situations.

In the field of public safety and response during disasters, the agency in autonomous drones or worn equipment allows communication in low-connectivity areas. Such communication relies on the interpretation of gestures, stress in the human voice, and environmental inputs to assist in the triage of patients during evacuation. Other areas being studied by the military are in autonomous vehicles.

Areas such as entertainment and gaming would also stand to gain from the use of emotionally conscious behavioural agents (ECBAs) that respond to the emotional state of players in VR/AR. Such ECBAs would adjust the story unfolding in the scene according to the emotional state of the players. What is also certain is that the increasing numbers of such applications pose new social and administrative questions related to trust, fairness, inclusiveness, and ethics. Later sections will explore the aforementioned in detail while enumerating the architectural, memory-related, and trust-related elements essential for their deployment.

8.1.8 Purpose and Audience [7]

Written for researchers in the field of AI, system architects, cognitive scientists, and professionals in the field related to the topic, the chapter presents the comprehensive roadmap for the designing and implementation of the next generation of multimodal dialogue models. The chapter also presents the theoretical framework for the same.

This chapter helps researchers by providing a comprehensive overview of recent advancements in transformer architectures, multimodal representation learning, affective computing, embodied interaction, and trustworthy AI, and by listing several open research challenges.

Developers and system architects will find in the document comprehensive architectural blueprints for constructing real-time and low-latency dialogue systems. It explains the implementation of pipeline components such as encoder–decoder networks, vision-to-text combination networks, emotion classifiers, memory networks, along with reinforcement learning-based policy improvement. These blueprints act as templates for their implementation in enterprise applications. This chapter gives context for educators and policymakers on the use of multimodal agents in the areas of education, healthcare, public service, and accessibility solutions. It illustrates how affect-aware systems might provide bridging solutions for digital divides in areas such as inclusive education, elderly care, and improved engagement. This chapter will also prove useful to students and young professionals who are researching the history of the development of conversational AI systems, the trade-offs involved in their implementation, and their significance in the social context. Each subtopic in the chapter is intended to facilitate the learning process through analysis and the generation of new ideas for the effective communication aspect of conversational AI.

To bridge the divide between the theoretical angles covered in the chapter and the practical applications in the real world, the chapter also becomes a reference guide cum vision statement. It is designed to teach, inform, educate, and empower an "interdisciplinary community of researchers, system architects, educators, policymakers, industry experts" in the field to meaningfully engage in the process of constructive engagement for the human-centred future for conversational AI. Finally, this Introduction sets the theoretical framework for the rest of the chapter by providing a roadmap of the progressive development of conversational AI technology. These numbered subsections begin with Section 8.2. Section 8.2 describes the symbolic rule-based methods. Section 8.3 traces the developments in probabilistic and neural systems. Section 8.4 describes the rise of large-scale pretrained transformers. Section 8.5 describes the limitations of unimodal solutions. Section 8.6 explores the promise of multimodal and affect-aware systems. Section 8.7 describes the real-world applicability in sectors. Finally, Section 8.8 explains the purpose and target.

8.2 ARCHITECTURAL AND COMPUTATIONAL FOUNDATIONS OF MULTIMODAL CONVERSATIONAL AGENTS

In continuation with the conceptual and historical perspective discussed in the preceding chapters, this section introduces the architectural frameworks and the computational pipeline required for building next-generation multimodal conversational agents. In fact, each subtopic in this chapter bears the critical components in the problem space, ranging from integration and perception through affective processing.

Figure 8.4 offers a comparative insight into the major architectural trends that have influenced the development of conversational systems. Beginning with the development of rule-based systems using manually designed rules from ELIZA, the figure proceeds to statistical models using HMMs in text recognition. The modulation of the process then moves on to the use of neural network architectures using long short-term memory networks in text processing and culminates in the large language model architecture, which combines speech recognition, latency, and understanding in generating responses. The key focus has been on the enhancement in the abstraction and the model's ability to learn.

8.2.1 Architecture and Fusion Pipeline for the Multimodal Conversational Agent

The curriculum outlined in the Introduction, following the development path from symbolic systems through affect-aware systems to the present-day conversational AI systems that use multimodal perception and deep learning, culminates in the modern conversational AI systems using multimodal perception and deep learning. Key to these systems, therefore, is the use of a highly modular and integrative architecture able to process multimodal inputs, including oral inputs, written user messages, visual inputs such as video streams from system cameras, and sensor inputs in general IoT settings. Examples include convoluted neural networks (CNNs) that form the vision backbone in the case of picture and video frame incorporations. ResNet

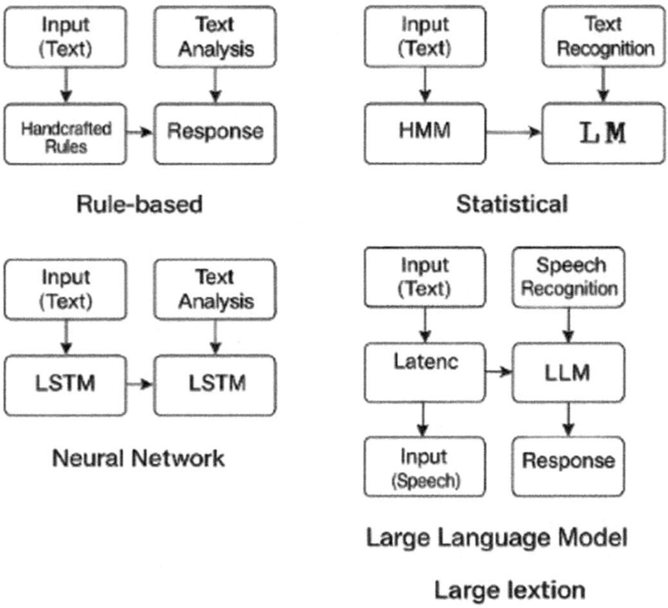

FIGURE 8.4 Architectural evolution of conversational models.

and EfficientNet models acquire space information from facial expressions, gestures, and other objects in the environment. Audio inputs are denoted by models such as wav2vec 2.0 and HuBERT. These models are pretrained on large unlabelled audio sets to encode sound characteristics related to prosody, tones, and speaker-specific information.

The embedding is normally extracted from the transformer models, such as BERT, RoBERTa, and their multilingual models, and help the model capture the semantic meaning in the utterances, intents, and other contextual aspects. The alignment strategies among the modalities are carried out in the latent space through fusion strategies that could be early, late, and intermediate fusion. In late fusion, the model works on the individual modalities and then combines the outputs using the fusion approach, allowing the model to be modular. However, it works on the downside where the model does not utilize the synergy among the different modalities. In intermediate fusion, used in vision transformer models like LXMERT and UNITER, the model aligns the modalities using the cross-attention approach.

Figure 8.5 is an illustrative scenario where a personalized agent listens to the verbal complaint, "I am feeling cold," through the integration of vision, language, and audio encoders. The use of a personalized avatar aims to represent user-specific data and/or contextual information. The outputs from the different encoders are mixed through the use of the transformer encoding technique. The resulting output then proceeds to the policy network, where the response, whether in the form of setting the temperature using the thermostat or in the form of verbal communication, has to be determined. Such an architecture emphasizes the importance of multimodal

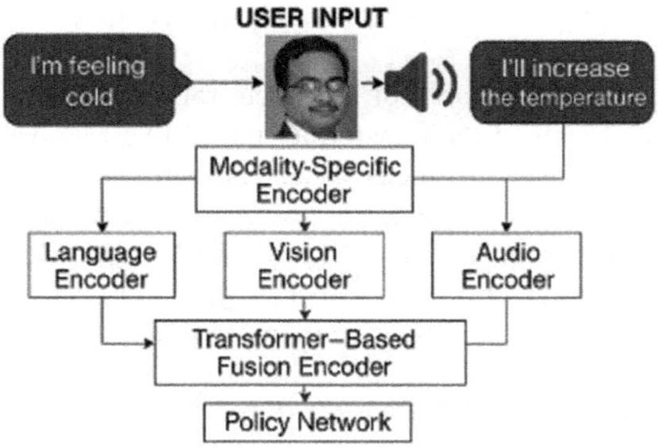

FIGURE 8.5 Multimodal fusion pipeline with personalized user input.

interaction, personalization, and embodiment in affect-aware multimodal systems. In real deployment, the architecture enables the development of emotionally intelligent assistants designed to recognize verbal complaints such as "I am feeling anxious" through the integration of lexical and prosodic information. The architecture also enables the development of space-grounded questions such as "What's that over there?" through the integration of contextual information. The architecture enables functionalities in both cloud and edge computing settings. The architecture enables the development of avatars in health and medical settings, companions in the field of education, and kiosks in the development of intelligent city infrastructure.

From a functional perspective, the proposed architecture would help in influencing the situation where the healthcare attendant understands the speech utterance "I am feeling cold" and, at the same time, recognizes the subject's shaking, which has been illustrated in the thermal images. In the same manner, the customer-support robotic platform would act accordingly, dependent on the extent of perplexity reflected in the subject's facial expressions, as illustrated in product demonstration sessions. The following subsections would provide a detailed insight into the computational components and the necessary integration strategies.

8.2.2 Real-Time Inference and Latency Optimization in Multimodal Systems [1,6]

To make a multimodal conversational interface feasible in real-time applications such as the healthcare industry, robots, tutoring systems, and assistive surroundings, the ability to process in real-time is crucial. Not only is essential that the system accurately interpret the multimodal inputs, but it must also do so within milliseconds in order to maintain the user's trust. This chapter will explore the methods used for improving real-time processing in various modalities.

Each modality poses its own set of latency issues. Visual inputs involve preprocessing tasks like face detection and object segmentation. Audio input is subject to denoising functions prior to the extraction of its feature representations (e.g., Mel-frequency cepstral Coefficients (MFCCs) or spectrograms). Additionally, the input has to be synchronized with the context obtained from automatic speech recognition (ASR) systems.

To achieve the deployable speed of inference, various architectural solutions are used. Lightweight variants of the transformer model like MobileBERT, TinyBERT, DistillBERT, and EfficientFormer focus on decreasing the usage of memory resources without compromising the core semantic processing. Additionally, for the field of audio and vision tasks, models named DistilHuBERT and MobileNetV3 provide efficient alternatives for larger models. Other methods for decreasing the latency on edge devices involve quantization-aware training, pruning, distillation, and batch-normalized inference.

Inferencing engines like ONNX Runtime, TensorRT by NVIDIA, and Torch Serve make low-latency model serving feasible on graphic processor unit (GPU) and central processing unit (CPU). Additionally, in the context of Edge deployments, the models are hosted on the compressed variant of the multimodal models using accelerators like Coral from Google, Intel's Movidius, and NVIDIA's Jetson.

Examples of representative latency optimization are as follows:

- Emotion-conscious classroom tutors reacting to facial disengagement within 200 ms
- Adaptive wheelchairs controlled by gesture and voice through AI in 100 ms
- AR-powered smart assistants for multimodal navigation overlays below 300 ms.

Major issues involve jitter variance suppression, the issue of missing modality inputs, and resistance to dynamic environment variations. Techniques like adaptive scheduling methods, early-exit transformer heads, and context failure methods are being explored. Looking ahead, the focus will shift to cloud-edge composition in the hybrid cloud model along with energy-efficient inference in realistic developments concerning modality fidelity.

8.2.3 Memory-Augmented Dialogue Models and Long-term Context Handling [9,13]

To maintain the history of the dialogue context encountered by multimodal agents during prolonged interaction processes, it becomes essential for them to store, recall, or reason on the context from the memory. However, the fixed context windows in the transformer models make it difficult for them to extract the dialogue context in multi-turn dialogue processes.

Memory-augmented neural networks (MANNs), such as differentiable neural computers (DNC), key-value memory networks, and external vector storage (like Facebook AI Similarity Search (FAISS)), represent scalable solutions for the incorporation of external memory. Such architectures allow for the ability to access both dialogue history, image embedding, and affect within the same sessions without the requirement for retraining. Take the ability for the healthcare agent to remember stress markers for the patients within weeks. Models such as Longformer, BigBird, and MemGPT use sparse attention techniques like segment-wise attention to increase the context windows in the transformer model. Along with recurrence-aware representations, context selectors in such models ensure the emphasis on relevant memory slots to avoid memory overflow.

Also, essential is the fusion of multimodal memory. Temporal consistency is achieved through organization of audio-visual signals with turns in the dialogue within an external memory buffer. Encoding in episodic memory allows for the replaying of scenarios and the linking of scenarios within disjoint memory sessions for the continuity of tutoring or therapy. Also investigated are memory graphs and semantic chains for linking user gesture memory in the current interaction to language memory in the future interaction.

The applications involve the following:

- Mental health assistants recalling previous emotional triggers
- Smart classrooms modifying teaching according to confusion levels on previous topics
- Retail bots remembering customer preferences on different devices

Future work includes dynamic memory pruning, memory storage from the perspective of privacy-preserving implementation, and memory learning in the context of the federated model for the multi-agent system.

8.2.4 Affective Computing and Emotion Recognition in Human–AI Interaction [2,5,11]

With the continued development of the conversational interfaces from being human-like systems to the human-centered intelligent assistants, the issue of affective computing becomes increasingly important. This section will present how multimodal systems recognize and process users' emotions in real time through the application of visual, auditory, and physiological cues.

Facial expression recognition tasks are usually dealt with by CNN-based models trained on some datasets like AffectNet, FER+, or EmotioNet. These models extract spatial features like eyebrow movements, lip movements, and eye closures for the detection of some emotions like happiness, sadness, and anger. Temporal aspects are added by the use of RNNs/3D CNN.

Speech-based emotion recognition relies on acoustic features such as pitch, jitter, MFCCs, and speaking rate. Models such as wav2vec2.0 fine-tuned on emotion-tagged datasets (IEMOCAP and RAVDESS) identify nuances. Techniques for fusion involve aligning emotion vectors for facial expressions and speeches.

Physiological sensors (heart rate, galvanic skin response (GSR), electroencephalogram (EEG)) provide better capabilities for emotional analysis. Although not widely used because of issues related to personal freedom and hardware limitations, models developed for wearables are being adopted for emotionally intelligent personal assistants.

The affective context is directly involved in the generation process of the dialogue. Transformer decoders adjust the tone, loquacity, and delay in the response according to the vectors of the emotion state of the dialogue. For example, the tutoring agent might use the supporting tone if the frustration in the form of emotion-based state vectors is recognized. Issues related to ethics involve the implicit biases in emotion corpora, the sensitivity of cultures to the representation of emotions, and the clarity of emotional inference engines. These topics are dealt with by methods such as the use of Explainable AI.

The applications include the following:

- Emotional support bots learning from distressed user behaviour
- Customer service representatives defusing complaints by calibrated tone
- Classroom tutors identifying disengagement through facial expressions and vocal cues

Among the future research tasks are the transfer learning for the general emotion transfer problem, affect-aware reinforcement learning methods, along with the implementation of ethical controls for affect inference.

8.2.5 Trust, Explainability, and Ethical Alignment in Multimodal Agents [8,12,14]

While multimodal dialogue systems continue to enter the new areas like the medical field, the world of teaching, the legal system, and social institutions in general, trustworthiness, explainability, transparency, and the ethical alignment become increasingly important issues related to the deployment of multimodal conversational agents. This chapter presents the methodologies to ensure the systems' trustworthy behaviour.

Building trust in AI systems requires repeated demonstration of consistency, accuracy, accountability, and interpretability. Finally, in the context of multimodal agents, the impact of cross-modal fusion performance on trustworthiness also includes attention to emotional intelligence and relevance. Consider the case of an AI-based medical assistant system that needs to interpret both symptoms and visually expressed inputs accurately.

Explanation methods provide the link between opaque deep learning processes on one side and human-understandable reasons on the other. They include the following:

- Heatmaps for attention on input types (e.g., marking regions that contributed to the prediction of emotions)
- Saliency maps and feature attribution for audio and/or text features
- Natural language explanation generation systems for summarizing system decisions

The alignment for ethics studies the issues related to surveillance, amplification of biases, and cultural sensitivity. To simplify, affect detection model systems require attention to demographic variations related to facial expressions or tone. The use of methods related to fairness in learning represents the significance of inclusivity.

Safety layers and human-in-the-loop (HITL) interventions are used for intercepting unsafe and/or uncertain behaviour. Examples include the following:

- Confidence thresholds that defer to human review
- Controversy detection modules for detecting controversial or toxic output
- Logging and audit trails for accountability

The applications include the following:

- Role clarification by virtual health coaches on non-diagnostic uses
- Educational robots that explain the reasons for grades or feedback
- Smart surveillance systems obeying transparent access rules

The conclusion of the related topics ends with future areas, like value-aligned reinforcement learning from human feedback (RLHF), symbolic integrated models for neural responders, and ethics-aware prompt engineering for large language model-based systems.

8.2.6 Evaluation, Benchmarking, and Generalization of Multimodal [6,7,15]

To achieve the goal of designing effective multimodal conversational agents, it is important to conduct comprehensive assessments. This topic explains the challenges involved in multimodal benchmarking in terms of the data used for the test, the measures of assessment used in the test, and the ability of the test to generalize. Common suites for evaluating multimodal tasks used in research involve MELD, CMU-MOSEI, AVEC, and EmotiW.

Our work explores multimodal baselines for generation and classification tasks, identifying the measures of concordance correlation coefficient (CCC), multimodal sentiment accuracy (MSA), and emotion recognition score (ERS). Cross-dataset transfer evaluation for the model's ability to generalize in zero-shot scenarios is also covered. Issues in the model's performance in different scenarios from the environment used in training are highlighted. This section also describes how one might assess grounding and alignment through attention visualization techniques, modality dropout tests, and human evaluation. Examples of applications would be the retention of empathy in healthcare conversational systems and the engagement of students in tutoring systems compared against a trend. To synthesize this section, it is crucial to stress the importance of each architectural and behavioural ability mentioned in the previous sections in terms of how measures of comparison are made from various datasets. The assessment goes beyond the success of multimodal fusion processes, affectively aligned simulations, and memory models because it also scales the degree of fairness in the behaviour of the intelligent system irrespective of the demographic boundaries.

8.3 CONCLUSION

Next-generation multimodal conversational interfaces are on the cusp of being key building blocks in the future of human-to-artificial intelligence interaction. By effortlessly combining the capabilities of language, vision, audio, and affect processing in one system, the next generation of conversational interfaces will make it possible for computers to provide personalized, contextually aware, and emotionally intelligent conversational interaction. This chapter presents in-depth insight on what goes into the building blocks of such systems. Future research will focus progressively on how the process of ethical reasoning needs to play an important part in the system's future.

To summarize, Section 8.2 continues the eightfold historical framework developed in the Introduction. Section 8.2 describes the history of multimodal fusion architectures (Section 8.2.1), the problem of real-time inference (Section 8.2.2), the issue of long-term memory (Section 8.2.3), moved on to affective reasoning and sensitivity (Section 8.2.4), the issue of trust and ethics (Section 8.2.5), and finally the problem of effective evaluation techniques in Section 8.2.6. Each of the abovementioned topics is intricately woven together in order for the new generation of intelligent agents to not only understand commands but also empathize. Finally, the purpose of this chapter is to provide both the technological roadmap and the strategic guide for students, researchers, developers, and policymakers engaged in charting the future course for the conversational potential of AI. The future frontier will be in the scalable implementation of such systems through the association of deep machines with human values by means of affect and understandable agents. However, in the beginning of the new era in which machines are able to see, hear, and feel too, the key challenge is no longer the ability to understand them but to do so in a responsible, transparent, and enabling manner.

ACKNOWLEDGEMENTS

The authors express gratitude to the Departments of Computer Science and Engineering, and Artificial Intelligence and Machine Learning, BMS Institute of Technology and Management, Yelahanka, Bengaluru, for the continuous support during the preparation of this chapter. Thanks are also in order for the inputs generated by researchers on human-centred architectures in the area of AI.

REFERENCES

[1] Y. Huang et al., "SpeechT5: Unified-Modal Encoder-Decoder Pre-Training for Spoken Language Processing," in Proc. ACL, 2022.

[2] A. Bulat et al., "Emotion Recognition in Conversations using Cross-Modal Transfer in Transformers," *IEEE Trans. Affective Comput.*, vol. 14, no. 2, pp. 580–593, 2023.

[3] Z. Yang et al., "Uni-Perceiver v2: A Generalist Model for Large-Scale Vision and Language Tasks," in Proc. CVPR, 2023.

[4] R. Jia et al., "Beyond Fine-tuning: Parameter Efficient and Modular Transfer Learning for LLMs," arXiv preprint arXiv:2305.10403, 2023.

[5] C. Chen et al., "AffectPrompt: Generating Affectively Aligned Responses via Prompt Optimization," in Proc. EMNLP, 2022.

[6] M. Qian et al., "MultiBench: Multiscale Benchmarks for Multimodal Representation Learning," in Proc. NeurIPS, 2022.

[7] S. Tripathi et al., "Explainable Multimodal Emotion Recognition using Attention Flow Visualization," *IEEE Trans. Multimedia*, vol. 26, pp. 3512–3525, 2024.
[8] Y. Li et al., "Trustworthy Multimodal AI: Benchmarking Fairness and Robustness in Vision-Language Models," in Proc. AAAI, 2023.
[9] H. Zhang et al., "LLM-Augmented Memory for Long-Term Multimodal Interaction," in Proc. ICLR, 2024.
[10] A. Ramesh et al., "DALL·E 3: Language-Conditioned Image Generation at Scale," OpenAI, arXiv:2310.00050, 2023.
[11] T. Nguyen et al., "EmotionPrompt++: Affective Prompt Tuning for Empathetic Multimodal Dialogue Systems," in Proc. NAACL, 2025.
[12] R. Alvi et al., "Symbolic Alignment for Trust-Aware Multimodal Agents," in Proc. IJCAI, 2025.
[13] L. Wong et al., "Continual Learning in Affective Multimodal Agents via Episodic Replay," in Proc. ACL, 2025.
[14] K. Patel et al., "Federated Evaluation Benchmarks for Ethical AI Systems," *IEEE Access*, vol. 13, pp. 50345–50359, 2025.
[15] J. Kim and A. Singh, "Efficient Long-Term Memory Encoding for Low-Latency Multimodal Assistants," in Proc. NeurIPS, 2025.

9 Multimodal Chatbot for Monitoring Emotional Well-being in Children

Key Design Challenges and Approaches

K. Amshakala, R. Saveeth, S. Nithya, S. Sangeethamariammal, and A. Priyadharshini

9.1 INTRODUCTION

Emotional well-being during childhood and adolescence is critical for healthy growth, learning, and social interactions later in life. Many young individuals nevertheless struggle with severe emotional issues in spite of this need. Approximately 10 percent of children in the UK have been diagnosed with a psychiatric disorder, and research from other areas shows comparable high rates of emotional, behavioral, and attention-related issues. Negative early experiences can significantly raise the likelihood of long-term mental and physical health problems, according to research on adverse childhood experiences (ACEs). At the same time, forums like the Organisation for Economic Co-operation and Development (OECD) have emphasized how children's emotional health is further influenced by things like stress, sleep deprivation, and quickly shifting social settings. These results taken together demonstrate that there is a vital need for reliable, scalable, and easily available emotional support services for children in many contexts. Balancing emotions is essential for good academic performance [3] and long-term emotional stability. It also helps children to stay focused, adjust to difficulties, and create good coping mechanisms [1,2]. Children who do not get timely emotional support are at an increased risk of learning challenges, social disengagement, and mental health problems in the future [1,4].

Emotional expression in children and adolescents is diverse, including verbal communication, nonverbal cues, and observable behaviors. Figure 9.1 presents the triangular model of verbal, nonverbal, and behavioral emotional cues, highlighting how these interconnected channels jointly convey emotional states.

 DOI: 10.1201/9781003678359-9

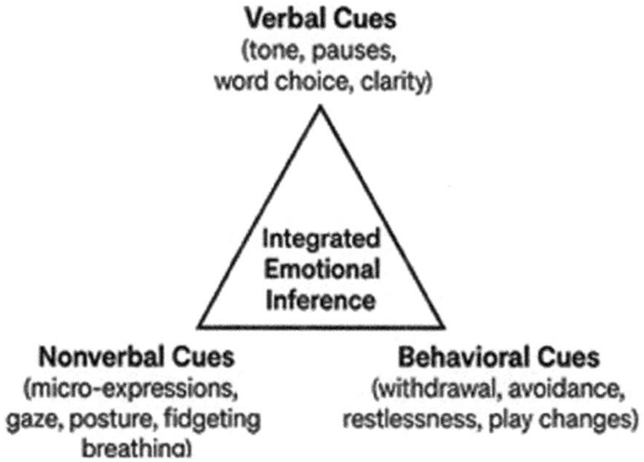

FIGURE 9.1 The triad of emotional cues.

By integrating text, speech, facial expressions, physiological signals, and behavioral patterns, a thorough and accurate understanding of emotional states becomes possible. This is essential for identifying early signs of stress, anxiety, or depression. This also helps to automatically capture the mental state of children with communication difficulties, neurodiversity characteristics, or limited language proficiency. Therefore, creating emotionally intelligent AI systems that can meet the varied emotional demands of children requires multimodal analysis. AI-based chatbots provide a private, nonjudgmental space to deal with stress linked to school, bullying, and family issues at any time [5,6]. These systems can recognize signs of melancholy, anxiety, rage, or withdrawal using natural language processing, sentiment analysis, speech-based emotion identification, and facial expression modeling [7]. By combining data from several data streams, multimodal deep-learning techniques substantially improve recognition accuracy [8]. When children exhibit decreased participation or communicate using increasingly negative language, chatbots can provide instant coping mechanisms, such as grounding exercises, mindfulness reminders, and coaching based on cognitive behavioral therapy (CBT) [9,10,11,12].

Communicating emotional state undergoes significant modifications from early childhood to adulthood. Due to their limited emotional vocabulary, younger children rely mostly on gestures, tone, and facial expressions [13], whereas adolescents employ more sophisticated language but frequently hide their emotions out of social pressure and fear of being judged [14]. Due to the wide range of emotional expression brought about by these developmental changes, single-modality techniques are inadequate. Age-related linguistic, cognitive, and behavioral aspects must therefore be taken into consideration in effective emotion-recognition models [15]. Interpretation becomes more difficult when emotional signals are subtle, contradictory, culturally influenced, or shaped by neurodiversity traits [17,18], since children often express distress through behaviors like withdrawal, avoidance, restlessness, or abrupt changes in play rather than explicit language [16,19,20]. These challenges emphasize the necessity of robust multimodal emotional-analysis frameworks that can raise the accuracy of children's and teenagers' emotion assessments [21].

In Section 9.2, computational methods for multimodal emotion recognition are reviewed, including deep learning, transfer learning, real-time models, and conventional machine-learning techniques. The fundamental aspects of multimodal chatbots for child mental health support are examined in Section 9.3, along with system designs, fusion techniques, and modality selection. Section 9.4 discusses the main ethical, technical, and developmental issues that are unique to child-focused multimodal chatbots. Lastly, Section 9.5 presents current research trends, new applications, and future perspectives for scalable and ethical emotional support systems, and Section 9.6 concludes.

9.2 APPROACHES FOR MULTIMODAL EMOTION RECOGNITION

Multimodal emotion recognition systems can accurately understand emotional cues from speech, text, facial expressions, and physiological signs using a range of computational techniques. Classical machine-learning classifiers and manually created features are used in traditional models. Deep neural architectures, such as convoluted neural networks (CNNs), recurrent neural networks (RNNs), and transformer-based techniques, are used in modern systems to efficiently capture cross-modal interactions. Transfer learning algorithms are a popular practical option for low-resource emotional datasets used in emotion recognition systems. Optimized real-time models are needed for applications like conversational agents and mental health monitoring. Dependability is ensured by using standardized evaluation metrics to evaluate performance across modalities and deployment scenarios [22].

9.2.1 Traditional Machine Learning Models

Traditional machine learning (ML) methods are used in the early development of multimodal emotion-recognition systems. These models rely on features such as statistical markers from physiological data, local binary patterns (LBP) for face images, or mel-frequency cepstral coefficients (MFCCs) for speech. Traditional classifiers with good interpretability and computational efficiency like support vector machines (SVMs), random forests (RF), k-nearest neighbors (k-NN), and hidden Markov models (HMMs) have limited ability to capture complicated nonlinear emotional patterns. However, these models are useful for baseline comparison studies using smaller datasets and lightweight models [23].

9.2.2 Deep Learning and Transformer-Based Approaches

Deep learning techniques are capable of providing transformed emotion analysis through end-to-end learning from raw multimodal inputs. RNNs and long short-term memory (LSTM) networks are good for modeling temporal dynamics in speech and video. CNNs effectively extract spatial features from face expressions. By using self-attention models to represent cross-modal interactions, transformer-based models like BERT for text and vision transformer (ViT) for images show remarkable results. Multimodal transformers like Multimodal Sentiment Analysis (MMSA) and Multimodal Machine Translation (MMT) have recently attained state-of-the-art performance. These designs outperform conventional models by capturing long-term dependencies and delicate emotional cues that are essential for child-focused emotion monitoring [24].

9.2.3 Transfer Learning for Emotional Data with Limited Resources

Emotional datasets, particularly those involving children, are frequently restricted due to ethical constraints, privacy concerns, and difficulties in gathering natural emotional expressions. Transfer learning involves pre-trained models that adapt to a specific emotional domain. The pre-training is done on large general-purpose datasets to best suit the limited resource application. Through methods including feature extraction, fine-tuning, and domain adaptation, these models are flexible across modalities, age groups, and cultural differences. This method greatly shortens training time and enhances performance when working with voice, facial, or physiological emotional data with few labelled examples [25]. Consequently, the models must be designed to be robust to data variability, computationally efficient, and capable of delivering low-latency inference in real-world deployment scenarios. Building on these advantages, the models can be further enhanced to address security threats by incorporating adversarial training, privacy-preserving mechanisms, and robust anomaly detection. In medical applications with limited resources, lightweight CNNs, MobileNet variations, quantized neural networks, and edge-optimized transformer designs are frequently employed. These models make it possible to identify mood swings instantly, enabling prompt interventions for children's mental health and well-being. Combining edge computing with cloud-based processing ensures increased scalability and responsiveness and privacy restrictions are improved [26].

9.2.4 Real-Time Emotion Monitoring Models

Real-time emotion monitoring devices are vulnerable to environmental noise. Environmental noise presents a considerable difficulty for real-time emotion monitoring systems because it interferes with audio signals and reduces detection accuracy. To overcome these limitations, lightweight models like compact CNNs, MobileNet variations, quantized neural networks, and edge-optimized transformer architectures are commonly used in therapeutic and resource-constrained settings. These approaches are frequently paired with noise-tolerant feature extraction and multimodal fusion techniques, allowing emotional states to be recognized even when individual inputs are damaged by noise [26]. Despite laying the scientific foundation for multimodal emotion recognition, these computational techniques are most beneficial when incorporated into interactive systems [27]. The next section describes the design aspects of multimodal chatbots used for child-centered emotional support.

9.3 MULTIMODAL CHATBOT DESIGN

In emotionally sensitive domains such as child mental-health support, multimodal chatbots extend beyond text-based interaction by integrating cues such as speech prosody, facial expressions, gestures, behavioral patterns, and physiological signals to infer affective states with greater accuracy. This integration enables more empathetic, adaptive, and developmentally appropriate support for children [28,29]. Unlike conventional text-only systems, multimodal chatbots leverage cross-modal signals to form holistic representations of user intent and emotion, supporting functions such as emotion recognition, personalized feedback, behavioral monitoring, and real-time adaptive support [30,31]. In child mental-health contexts, these systems act as accessible digital companions that decode subtle affective cues while prioritizing emotional safety and therapeutic intent [32,33].

9.3.1 Modalities Used for Emotion Recognition

Emotion recognition in multimodal chatbots combines complementary modalities, as no single cue is reliable for children whose emotional expressions are developmentally variable [28,29]. Key modalities for multimodal emotion recognition are shown in Table 9.1. Facial expressions provide immediate visual indicators of affect through an analysis of action units, gaze, and micro-expressions, with deep learning enabling detection of subtle cues [30–32]. However, cultural differences and age-dependent facial behavior require diverse datasets and careful calibration [33,34]. Speech and prosodic features such as pitch, tempo, pauses, and voice quality capture emotional shifts, particularly when children lack expressive vocabulary [35,36]. To ensure accuracy, models must account for age-related variability and developmental differences [37,38]. Textual and linguistic patterns further reveal emotional states through sentiment, word choice, and structural variation [39,40]. In children, however, limited vocabulary, playful exaggeration, and inconsistent grammar complicate analysis, making child-specific language models and contextual frameworks necessary

TABLE 9.1
Key Modalities for Multimodal Emotion Recognition

Modality	Description	Key Features	Child-Specific Considerations
Facial expressions	Visual cues from face	Micro-expressions, gaze, AU (Action Units), posture	Age-dependent expression, cultural differences, subtlety
Speech/prosody	Audio-based emotion cues	Pitch, tone, rhythm, pauses, intensity	Shyness, limited vocabulary, accent variations
Text/linguistic patterns	Written or typed language	Sentiment, semantic cues, syntactic deviation	Limited vocabulary, playful exaggeration
Behavioural/ interactional	Interaction patterns	Message frequency, typing latency, session duration	Ethical privacy concerns, developmental differences
Contextual/ environmental	External situational info	Time, event history, baseline mood	Context matters (home vs. school)
Physiological signals	Body signals via sensors	Heart rate, EDA, skin temperature	Privacy, consent, device access limitations

[22,41]. Behavioral and interactional cues such as message frequency, typing delays, sudden withdrawal, or unusually long sessions offer indirect insight into emotional changes and can signal discomfort, anxiety, or a need for reassurance [42,43], though they must be used cautiously to avoid surveillance-like monitoring [44].

Contextual information, including time of interaction, recent events, and emotional baseline, provides essential grounding, since similar behaviors can have different meanings depending on situational factors [45,46] and its use requires strong privacy protections [47]. Physiological signals such as heart rate, electrodermal activity, breathing, and skin temperature are emerging modalities for emotion recognition [48,49], but for children, they remain experimental due to consent, privacy, and device-access constraints, necessitating minimal intrusiveness and strict safeguards [50,51].

9.3.2 Fusion Techniques for Multimodal Emotion Analysis

Fusion techniques integrate information from multiple modalities to build holistic emotional representations. Table 9.2 describes the advantages, limitations, and typical use cases of various multimodal fusion approaches. Effective fusion is crucial in child-focused systems where signals are often noisy, incomplete, or contradictory.

(i) *Early (feature-level) fusion*
Early fusion concatenates raw or low-level features across modalities before classification, enabling models to capture cross-modal dependencies [52]. This approach works well when modalities are temporally aligned (e.g.,

TABLE 9.2
Multimodal Fusion Approaches

Fusion Type	Description	Advantages	Limitations	Typical Use Cases
Early / feature-level	Concatenates raw features from modalities	Captures cross-modal interactions	Sensitive to missing data, risk of overfitting	Synchronized audio-visual signals
Late / decision-level	Processes modalities separately, combines predictions	Robust to missing/ unreliable data	May lose fine-grained correlations	Real-world deployment, noisy signals
Hybrid / intermediate	Combines early and late fusion at multiple levels	State-of-the-art performance	Requires large datasets	Child-focused multimodal systems
Dynamic / context-aware	Adjusts modality weights in real time	Adapts to context, improves robustness	Computationally complex	Real-time monitoring, ecologically valid settings

synchronized facial landmarks and prosodic features) but is highly sensitive to missing data common in child interactions where cameras may be turned off or speech may be unclear. High-dimensional feature spaces also increase computational load and risk over fitting due to limited child-specific datasets [43].

(ii) *Late (decision-level) fusion*

Late fusion independently processes each modality through modality-specific encoders and integrates their output predictions (e.g., weighted averaging, majority voting, confidence-based fusion) [53]. This strategy is robust to missing or unreliable modalities and improves performance in real-world environments. However, it may lose fine-grained correlations such as a neutral utterance spoken with a trembling voice that requires cross-modal interaction [54].

(iii) *Hybrid / intermediate fusion*

Hybrid fusion combines early and late strategies, merging features at multiple levels. Common architectures include:

- Attention-based fusion, allowing modalities to modulate each other's importance [55];
- Tensor fusion networks, explicitly modeling cross-modal interactions [56];
- Hierarchical fusion, using early fusion for tightly aligned modalities (audio–video) and late fusion for abstract signals (text, behavior).

  Hybrid methods often achieve state-of-the-art results but require large, balanced datasets, which remain limited for child-focused applications [43].

(iv) *Dynamic and context-aware fusion*

Dynamic fusion strategies adjust modality weights in real time based on signal reliability or environmental context [57,58].

For instance, the system may:

- Down-weight visual input when the child turns away from the camera,
- Rely more heavily on text when speech is unclear.

Such adaptive approaches improve robustness and promote human-centered, ecologically valid interaction. Fusion strategies must prioritize child safety and privacy. Architectures requiring continuous video or physiological monitoring may compromise autonomy and raise ethical concerns [44,50]. Decision-level and adaptive fusion can reduce reliance on intrusive modalities while maintaining accuracy critical when designing affective systems for minors.

9.3.3 Multimodal Chatbot Architecture

Designing an effective chatbot for monitoring and supporting children's emotional well-being requires an architecture that reliably interprets diverse input modalities (speech, facial expressions, text, behavioral cues) and integrates them into robust emotional inferences. The modular design of an AI-driven multimodal chatbot is shown in Figure 9.2. This architecture comprises four core layers, namely, input acquisition layer, pre-processing and feature-extraction layer, fusion and inference layer, and response-generation layer [59,60].

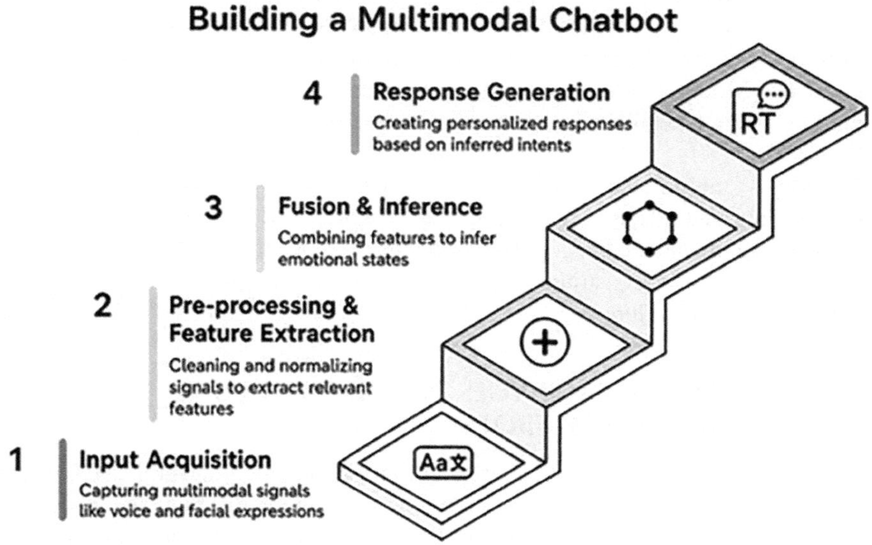

FIGURE 9.2 AI-driven multimodal chatbot architecture.

The input acquisition layer gathers multimodal signals such as audio, video (facial expressions and gaze), text, and optional physiological data to form a comprehensive emotional profile of the child. These signals are processed by emotion-sensitive dialogue management modules that maintain conversational context and adapt to emotional cues using reinforcement learning, state-based methods, or transformer-based natural language understanding (NLU) models. Safety constraints, empathy-driven response strategies, and de-escalation mechanisms ensure supportive, non-triggering interactions. Ongoing affective evaluation allows the system to personalize dialogue according to the child's emotional state, comfort level, and developmental stage, improving reliability and emotional well-being outcomes [58].

The pre-processing and feature-extraction layer cleans and normalizes raw inputs, extracting modality-specific features such as MFCCs for audio, facial action units or embedding for video, and token embedding for text, followed by temporal alignment. The fusion and inference layer integrates these features into a unified emotional representation using early, late, hybrid, or attention-based fusion strategies. Attention-based transformers and graph networks dynamically weight modalities and capture inter-modal dependencies, improving robustness under noisy or missing data and enabling context-aware affective inference. The response-generation layer converts inferred emotional states into developmentally appropriate interventions, such as positive reinforcement, mindfulness prompts, or cognitive reappraisal, while escalating persistent distress to caregivers or professionals. Personalization based on emotional trends and interaction history enhances emotional resilience and self-regulation [61].

Overall, AI-driven multimodal chatbot architectures outperform unimodal and rule-based systems by reducing ambiguity and improving emotion and intent recognition through complementary signals across text, speech, visual, and physiological channels. Redundant modalities enhance robustness, while continuous multimodal analysis enables real-time emotional adaptation and empathetic interaction. Modular design supports scalability and integration across domains including mental health, education, healthcare, accessibility, and social robotics [62, 63].

Although multimodal chatbots offer powerful capabilities for emotional understanding and adaptive interaction, their deployment in child-focused contexts introduces significant challenges. The following section examines the developmental, technical, and ethical constraints that shape the design of safe and effective multimodal chatbots for children.

9.4 KEY DESIGN CHALLENGES IN CHILD-FOCUSED MULTIMODAL CHATBOTS

Designing multimodal chatbots poses various technical and ethical challenges. Since children's emotional expressions vary among different age groups, and as they are also related with uncontrolled and noisy environments, there is a need for systems that should satisfy strict safeguarding procedures and ethical standards [34].

- Developing multimodal chatbot by considering emotion-related expressions is highly challenging because during the growth of children, their emotional, verbal and behavioral expressions will also change accordingly. Young children will exhibit different facial expressions, unstable vocal patterns, gestures, and limited vocabulary, whereas adolescents express moderated and socially controlled expressions [22,33,34].
- Another challenge in designing multimodal chatbots is lack of availability of children's multimodal dataset. Due to privacy laws and safeguarding procedures, there is a scarcity in availability of children's multimodal dataset [43,50,51]. While interacting with chatbot in schools and homes, performance of the model can be affected by noise and signal distractions. This leads to the requirement of additional modules to handle noise and other distractions effectively [29,31].
- Predicting emotional interpretations via child-focused multimodal chatbots may be challenging if unclear and contradictory signals are received. Children may exhibit shivering facial expressions and have discomfort body language, which may degrade the performance of multimodal chatbots. This emphasizes the importance of children-based emotional expression learning models during the design of efficient multimodal chatbots [35].
- Lack of concentration of the children should be handled in a patient way while designing child-focused multimodal chatbots [42,43]. Cultural, linguistics, and individual children's emotional expressions are the major challenging issues in the design of children-based multimodal chatbots [33,46,47].
- Privacy concerns and safeguarding procedures are very important in the design of multimodal chatbot. To prevent psychological harm or misuse,

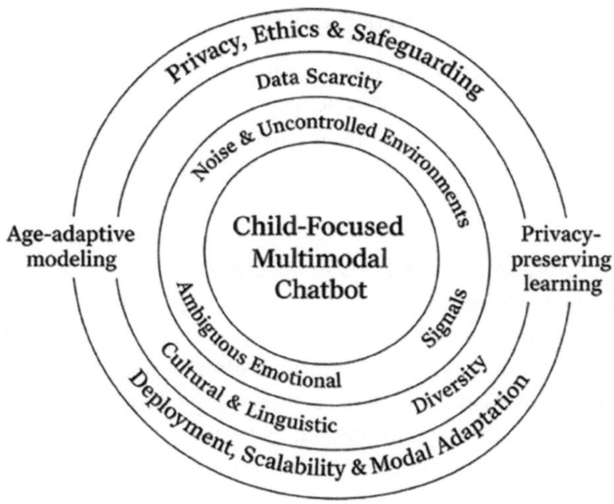

FIGURE 9.3 A layered representation of key challenges in child-focused multimodal chatbot ecosystem.

alert notifications to parents on time is necessary in real-world multimodal chatbot design. Cultural, linguistics, and individual children's emotional expressions are the major challenging issues in the design of children-based multimodal chatbots. Privacy concerns and safeguarding procedures are very important in the design of a multimodal chatbot. Various privacy preservation techniques should be deployed under Children's Online Privacy Protection Rule (COPPA) and children's data and parental consent under the General Data Protection Regulation (GDPR-K) while designing child-fused multimodal chatbots [44,45,48–50,64–67].

- Designing, developing, deploying and maintaining multimodal chatbots faces various real-world challenges. Lightweight models such as Edge AI, and hybrid cloud-edge architectures can be used for implementing scalability [69,71,72]. Use of appropriate privacy-preserving techniques maintains proper trade-off between privacy and utility [47,49,73,74].

9.5 CURRENT RESEARCH TRENDS AND APPLICATIONS

Recent progress in multimodal artificial intelligence has accelerated the development of child-centered mental-health technologies. A layered representation of key challenges in child-focused multimodal chatbot ecosystem is shown in Figure 9.3. It emphasizes how to overcome the design challenges while developing multimodal chatbots and also directs the development process effectively. Emotion-aware conversational agents that analyze text, speech, facial expressions, gestures, and physiological signals are increasingly used for emotional monitoring and well-being support.

9.5.1 Current Trends in Multimodal Health

AI-driven multimodal mental health monitoring systems are influencing therapeutic practices, classroom support, and personalized care, while current research emphasizes ethical child–AI interaction, trust calibration, and adaptive emotional-support frameworks to ensure safe and effective mental-health ecosystems for children [58].

- *Multimodal mental health tools for children*
  Multimodal tools combine voice, facial cues, gestures, and language to assess children's emotional states more reliably. They are widely applied for early detection of anxiety, stress, attention disorders, and emotional dysregulation. Multimodal approaches capture a wide range of cues, including those from children who fail to communicate verbally, unlike unimodal approaches. Remote monitoring of children helps parents, educators, and mentors observe children's behavior and patterns over an extended period of time [61,63].
- *Chatbots in therapy, counseling, and education*
  In education and therapy, emotional chatbots are used due to their accessibility and nonjudgmental interaction style. They help with mood tracking, mindfulness, and cognitive-behavioral strategies and breathing exercises especially in schools they help children to express their concerns, maintain engagement, and manage stress. Researchers found that chatbots seem to be an approachable companion for the children, which encourages them to disclose their emotions. This system outperforms conventional counselling as it fails to have continuous monitoring and providence of scalable report [61].
- *Child–AI emotional interaction*: Many research studies have emerged to show how children emotionally connected with AI and their trust on AI, and long-term empathy. Studies discussed that children may observe chatbots as humans to share their emotions or personality traits, shaping them, which helps to record their responses and disclosures. Emerging directions include emotion-aligned conversational strategies, play- and storytelling-based interactions, affective reinforcement learning, emotionally responsive virtual characters, and culturally adaptive, neurodiversity-inclusive models. The integration of all learning helps emotionally to strengthen the mental health of children and their education [58,68,70].

These research trends showcase the growing strength of multimodal emotional AI for children and also address the limitations. Based on the above insights, the next subsection outlines the future research directions aimed at creating more flexible, adaptive, and long -term emotional-support.

9.5.2 Future Directions

The future study of the research focuses on building up a chatbot with the integration of legal and flexible emotional support. Now, the emphasis is on chatbot with the integration of human–AI and most importantly ethical governance. In addition, long-term emotional monitoring also aids in understanding children's developmental and behavioral patterns over time. The human–AI helps to scale the children's mental development over the period.

- *Hybrid human–AI emotional support systems*: The capability of AI in chatbot with human concern claims reliable monitoring of child's mental health. Parents, educators, or clinical counselors render contextual understanding and fitting intervention, whereas the AI system can flag early risks. This collaboration of humans and AI reduces automation errors, confirms reliability, and improves trust and accountability, especially in crucial situations where fully automated systems are insufficient [75].
- *Integrating wearable and sensor-based modalities*: The incorporation of wearable devices like EEG headbands, and skin-response monitors helps to capture physiological indicators including heart-rate variability, stress levels, and sleep patterns [76]. The observation of signals like facial, vocal, and behavioral cues, enabling prior and more accurate detection of emotional imbalance seems to be more accurate than observable behavior alone.
- *Ethical AI-by-design frameworks*
  Since AI monitors the children's daily life activities, huge volumes of data accumulates, which needs to be minimized. Embedding ethical safeguards from the design stage helps protect children's rights, mitigate bias, and reduce algorithmic harm. Transparent models, fairness evaluation, and privacy-preserving architectures are critical for building public trust, particularly for vulnerable populations [77].
- *Longitudinal emotional monitoring*
  The gradual emotional resilience can be observed in long-term monitoring than short term. The long-term monitoring systems rely on temporal modeling, flexible feedback, and continuous learning, while carefully balancing the observed results with ethical concerns pertaining to data confidentiality [78]. Figure 9.4 depicts the current trends and future directions of multimodal AI for child mental health.

The future studies showcase the drift toward responsible, human-centered emotional AI for children's mental well-being. The conclusion summarizes the chapter's significant studies and highlights of multimodal research on children's mental health.

9.6 CONCLUSION

The study examines the strength of the multimodal chatbots designed to support children's emotional well-being. It highlights the integration of text, speech, facial expressions, gestures, and physiological signals with AI that achieves a more sound understanding of children's emotions than unimodal approaches. The research approaches address the complexity of child-specific emotional expression, age-appropriate interaction, and the necessity of ethical, privacy, and cultural considerations. The better results have been achieved in the multimodal approach than the unimodal approach.

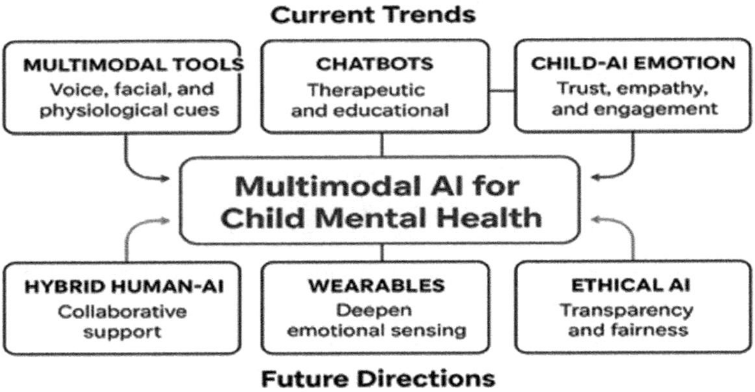

FIGURE 9.4 Multimodal chatbot-current trends and future directions.

Future studies should give priority to the adaptive, ethically grounded emotional AI for children. The integration of wearable sensors, hybrid human–AI models, ethical-by-design frameworks, and longitudinal emotional tracking renders tailor-made support and alarming of early sign. Altogether, the study narrows down the focus toward safer, more reliable and more effective emotional-support technologies for young users, especially children.

REFERENCES

[1] J.C. Benítez-Agudelo, D. Restrepo, E. Navarro-Jimenez, V.J. Clemente-Suárez Longitudinal effects of stress in an academic context on psychological well-being, physiological markers, health behaviors, and academic performance in university students. *BMC Psychol.*, vol. 13, no. 1, p. 753, 2025.

[2] D. K. Housman, "The importance of emotional competence and self-regulation from birth," *Int. J. Child Care Educ. Policy*, vol. 11, no. 1, pp. 1–17, 2017.

[3] Y. Wang et al., "Social-emotional competency and child development in Western China," *Front. Psychol.*, vol. 10, p. 1282, 2019.

[4] P. A. Graziano, R. D. Reavis, S. P. Keane, and S. D. Calkins, "The role of emotion regulation in children's early academic success," *J. School Psychol.*, vol. 45, no. 1, pp. 3–19, 2007.

[5] M. Mansoor et al., "Conversational AI in pediatric mental health: A narrative review," *Children*, vol. 12, no. 3, p. 359, 2025.

[6] C. Ka Yuk Chan, "AI as the therapist: Challenges in generative AI for school mental health," *Behav. Sci.*, vol. 15, no. 3, p. 287, 2025.

[7] L. Bravo, C. Rodriguez, P. Hidalgo, C. Angulo, A systematic review on artificial intelligence-based multimodal dialogue systems capable of emotion recognition. *Multimodal Technol. Interac.*, vol. 9, no. 3, p. 28, 2025.

[8] X. Li et al., "Determinants of rural students' adoption of AI chatbots for mental health," *Front. Public Health*, vol.13, pp. 1–17. 2025.

[9] H. Hasei et al., "Generative AI chatbots to reduce psychological burden in pediatric patients: A pilot study," *Front. Digit. Health*, vol. 7, pp. 1–11, 2025.

[10] M.A. Piombo, S. La Grutta, M.S. Epifanio, G. Di Napoli, C. Novara. Emotional intelligence and adolescents' use of artificial intelligence: a parent–adolescent study. *Behav. Sci.* (*Basel*), vol. 15, no. 8, p. 1142, 2025.

[11] M. Narimani, M. Naeim. Artificial intelligence in mental health: integrating opportunities and challenges of multimodal deep learning for mental disorder prevention and treatment. *Ann. Med. Surg.* (*Lond.*). 2025;87(9):5757–5761.

[12] Saarni, C. *The Development of Emotional Competence*. Guilford Press, 1999.

[13] Denham, S. A. *Emotional Development in Young Children*. Guilford Press, 1998.

[14] Steinberg, L. Adolescent emotional development in context. *Annu. Rev. Psychol.*, vol. 52, pp. 83–110, 2001..

[15] Calvo, R. A., & D'Mello, S. "Affect detection: An interdisciplinary review." *IEEE Trans. Affective Comput.*, vol. 1, no. 1, pp. 18–37, 2010.

[16] Eisenberg, N., & Spinrad, T. Emotion-related regulation: Its conceptualization, relations to social functioning, and socialization . *The Regulation of Emotion.*, vol. 1, pp. 281–311, 2004.

[17] Ekman, P., & Friesen, W. *The Facial Action Coding System (FACS).* Consulting Psychologists Press, 1978.

[18] Thompson, R.A. Emotional regulation and emotional development. *Educ Psychol Rev.*, vol. 3, 269–307, 1991. https://doi.org/10.1007/BF01319934.

[19] Matsumoto, D. Cultural influences on the perception of emotion. *J. Cross-Cult. Psychol.*, vol. 20, p. 1, 1992.

[20] Aydemir, U, ChatGPT-delivered physical activity intervention for children with autism spectrum disorder: Pre-post feasibility study. *JMIR Human Factors*, vol. 12, p. e71119, 2025.

[21] D'Mello, S., and Kory, J. "A review and meta-analysis of multimodal affect detection systems." *ACM Comput. Surv.*, vol. 47, p. 3, 2015.

[22] C. Sun, A. Hawkins, and L. S. Vygotsky, "Language development in childhood," *Ann. Rev. Child Develop.*, vol. 8, pp. 113–140, 2016.

[23] Simos, C., Larios, N. and Papoutsi, C. Artificial intelligence and adolescent emotional well-being in inclusive secondary education, *World J. Adv. Res. Rev.*, vol. 29, no. 1, pp. 1509–1527. https://doi.org/10.30574/wjarr.2026.29.1.0223.

[24] A. Vaswani et al., "Attention is all you need," *Proc. Advances in Neural Information Processing Systems* (*NIPS*), pp. 5998–6008, 2017.

[25] H. Kaya, F. Gürpınar, A. Salah, "Video-based emotion recognition in the wild using deep transfer learning and score fusion," *Proc. ACM Multimedia*, pp. 142–147, 2017.

[26] A. Baird, S. Wagner, "Real-time emotion recognition for interactive systems: A survey," *J. Ambient Intell. Smart Environ.*, pp. 1–17, 2020.

[27] B. Schuller et al., "AVEC 2019: State-of-the-art audiovisual emotion challenge," Proc. ACM International Conference on Multimodal Interaction, pp. 1–9, 2019.

[28] R. Cowie, E. Douglas-Cowie, N. Tsapatsoulis, et al., "Emotion recognition in human–computer interaction," *IEEE Signal Process. Mag.*, vol. 18, no. 1, pp. 32–80, 2001.

[29] B. W. Schuller, M. Valstar, F. Eyben, et al., "AVEC 2013: The continuous audio/visual emotion and depression recognition challenge," Proc. ACM International Conference on Multimodal Interaction, pp. 3–10, 2013.

[30] P. Ekman and W. V. Friesen, *Facial Action Coding System: Investigator's Guide*. Palo Alto, CA: Consulting Psychologists Press, 1978.

[31] I. Cohen, N. Sebe, A. Garg, et al., "Facial expression recognition using SVM," *Proc. ICPR*, pp. 144–147, 2003.

[32] K. Zhang, Z. Zhang, Z. Li, and Y. Qiao, "Joint face detection and alignment using multitask cascaded convolutional networks," *IEEE Signal Process. Lett.*, vol. 23, no. 10, pp. 1499–1503, 2016.

[33] J. A. Russell, "Culture and the categorization of emotions," *Psychol. Bull.*, vol. 110, no. 3, pp. 426–450, 1991.

[34] P. D. Zelazo and M. Lewis (eds.), *The Oxford Handbook of Developmental Psychology*. Oxford University Press, 2013.

[35] K. R. Scherer, "Vocal communication of emotion: A review of research paradigms," *Speech Commun.*, vol. 40, pp. 227–256, 2003.

[36] F. Eyben, M. Wöllmer, and B. Schuller, "Opensmile: the Munich OpenSource feature extractor," ACM Multimedia, 2010.

[37] S. Narayanan and A. Potamianos, "Creating conversational interfaces for children," *IEEE Trans. Speech Audio Process.*, vol. 10, no. 2, pp. 65–78, 2002.

[38] B. Schuller, S. Steidl, A. Batliner, et al., "The INTERSPEECH 2011 speaker state challenge," *Proc. Interspeech*, pp. 3121–3124, 2011.

[39] B. Pang and L. Lee, "Opinion mining and sentiment analysis," *Foundations Trends Inform. Retriev.*, vol. 2, nos. 1–2, pp. 1–135, 2008.

[40] J. Pennebaker, R. Booth, and M. Francis, *Linguistic Inquiry and Word Count: LIWC 2015*. Austin, TX: Pennebaker Conglomerates, 2015.

[41] K. K. Bowyer, "Challenges for child language modelling," *IEEE Intell. Syst.*, vol. 29, no. 4, pp. 26–33, 2014.

[42] R. R. Morris, K. Kouddous, R. Kshirsagar, and S. Schueller, "Towards an artificially empathic conversational agent for mental health applications," *Proc. AAAI Spring Symposium*, pp. 1–6, 2018.

[43] S. Poria, E. Cambria, R. Bajpai, and A. Hussain, "A review of affective computing: From unimodal analysis to multimodal fusion," *Information Fusion*, vol. 37, pp. 98–125, 2017.

[44] A. Sharkey and N. Sharkey, "Children, the elderly, and robots: Ethical issues in robot care," *IEEE Robot. Automat. Mag.*, vol. 18, no. 1, pp. 26–34, 2011.

[45] J. Read and M. MacFarlane, "Using context in design for children," *Behav. Inform. Technol.*, vol. 35, no. 3, pp. 1–12, 2016.

[46] D. Matsumoto, "Cultural influences on emotional expression," *Handbook of Culture and Psychology*. Oxford University Press, 2001.

[47] S. Wachsmuth, "Computational models of social interaction," *Cognitive Process.*, vol. 11, no. 1, pp. 1–12, 2010.

[48] R. W. Picard, *Affective Computing*, Cambridge, MA: MIT Press, 1997.

[49] S. Taylor and P. J. G. Lisboa, "Physiological signal analysis for emotion recognition," *Biomed. Signal Process. Control*, vol. 8, pp. 1–4, 2013.

[50] UNICEF, *Policy Guidance on AI for Children*, New York, 2021.

[51] ICO (UK Information Commissioner's Office), *Age-Appropriate Design Code*, London, 2020.

[52] H. Gunes and M. Piccardi, "Bi-modal emotion recognition from expressive face and body gestures," *J. Network Comput. Appl.*, vol. 30, no. 4, pp. 1534–1545, 2007.

[53] C. Busso, Z. Deng, S. Yildirim, et al., "Analysis of emotion recognition using facial expressions, speech and multimodal information," *Proc. ICMI*, pp. 205–211, 2004.

[54] A. Zadeh, P. P. Liang, S. Poria, et al., "Multimodal language analysis in the wild: CMU-MOSEI dataset," *Proc. ACL*, pp. 2236–2246, 2018.

[55] Y. Tsai, P. P. Liang, A. Zadeh, and L.-P. Morency, "Multimodal transformer for unaligned multimodal language sequences," *Proc. ACL*, pp. 6558–6569, 2019.

[56] A. Zadeh, M. Chen, S. Poria, et al., "Tensor fusion network for multimodal sentiment analysis," *Proc. EMNLP*, pp. 1103–1114, 2017.

[57] P. P. Liang, Z. Liu, A. Zadeh, et al., "Multimodal local–global ranking fusion," *Proc. ACL*, pp. 4579–4590, 2020.

[58] S. Poria, N. Majumder, R. Mihalcea, and E. Hovy, "Emotion recognition in conversation: Research challenges, datasets, and recent advances," *Comput. Lang.*, arXiv:1905.02947, 2019.

[59] T. Baltrušaitis, C. Ahuja, and L.-P. Morency, "Multimodal machine learning: A survey and taxonomy," *IEEE Trans. Pattern Anal. Machine Intell.*, vol. 41, no. 2, pp. 423–443, 2019.

[60] C. Zhang, Z. Yang, X. He, and L. Deng, "Multimodal intelligence: representation learning, information fusion, and applications," *IEEE J. Selected Top. Signal Process.*, vol. 14, no. 3, pp. 478–493, 2020.

[61] B. Inkster, S. Sarda, and V. Subramanian, "An empathy-driven, conversational artificial intelligence agent (Wysa) for digital mental well-being: real-world data evaluation mixed-methods study," *JMIR Mhealth Uhealth*, vol. 6, no. 11, e12106, 2018, doi: 10.2196/12106.

[62] K. Maidanov and G. F. Frire, "Multimodal fusion techniques for emotion recognition using audio, visual, and physiological signals," *Nat. J. Signal Image Process.*, vol. 1, no. 3, 2025.

[63] L. Bravo, C. Rodriguez, P. Hidalgo, and C. Angulo, "A systematic review on artificial intelligence-based multimodal dialogue systems capable of emotion recognition," *Multimodal Technol. Interact.*, vol. 9, no. 3, 2025, doi: 10.3390/mti9030028.

[64] V. L. Lemke, "Ethics of artificial intelligence for children," *Proc. IEEE Global Engineering Education Conference (EDUCON)*, pp. 123–130, Porto, Apr. 2018.

[65] A. R. Berendt, B. Preibusch, "Data protection in youth-oriented online services," *Proc. IEEE Symposium on Privacy and Security*, pp. 1–10, California, May 2014.

[66] S. Livingstone, M. Stoilova, "Children's data rights and privacy online," UNICEF Office of Research, pp. 1–34, Italy, 2019.

[67] J. Buolamwini, T. Gebru, "Gender shades: Intersectional accuracy disparities in facial recognition," *Proc. Conference on Fairness, Accountability, and Transparency (FAT)*, pp. 77–91, New York, Feb. 2018.

[68] T. Bickmore, R. Picard, "Toward socially intelligent agents: Therapeutic alliance and empathy in human–AI interaction," *Proc. International Conference on Intelligent User Interfaces (IUI)*, pp. 1–6, Florida, Jan. 2005.

[69] A. H. Celdrán, J. A. Ruipérez-Valiente, F. J. G. Clemente, M. J. Rodríguez-Triana, S. K. Shankar, and G. M. Pérez, "A scalable architecture for the dynamic deployment of multimodal learning analytics applications in smart classrooms," *Sensors (Switzerland)*, vol. 20, May 2020.

[70] F. Wang and J. Dong, "The application of improved AFCNN model for children's psychological emotion recognition," *Sci. Rep.*, vol. 15, Dec. 2025.

[71] J. Rincon, S. Heras, C. Carrascosa, and V. Julian, "Implementing edge intelligence to improve the detection and classification of human emotions," *SSRN Electron. J.*, Dec. 2021.

[72] R. Mishra, B. Bhushan, and K. S. Venkatesh, "Toward cross-cultural emotion detection: the Indian spontaneous micro-expression dataset," *Front. Psychol.*, vol. 16, 2025.

[73] C. Grové and C. Grove, "Co-developing a mental health and wellbeing chatbot with and for young people," *Front. Psychiat.*, vol. 11, p. 606041, 2020.

[74] S. K. B. Sangeetha, R. S. K. Boddu, A. Bhaumik, S. K. Mathivanan, and U. Moorthy, "Spatiotemporal multimodal emotion recognition using temporal video sequences and pose features for child emotion classification," *Sci. Rep.*, vol. 15, Dec. 2025.

[75] N. Imran, A. Hashmi, and A. Imran, "Chat-GPT: Opportunities and Challenges in Child Mental Healthcare," *Pak. J. Med. Sci.*, vol. 39, pp. 1191–1193, 2023.

[76] N. L. McElwain et al., "Evaluating users' experiences of a child multimodal wearable device: mixed methods approach," *JMIR Human Factors*, vol. 11, p. e49316, Feb. 2024.

[77] A. McStay and G. Rosner, "Emotional artificial intelligence in children's toys and devices: Ethics, governance and practical remedies," *Big Data Soc.*, vol. 8, p. 1, 2021.

[78] J. Kim and G. Kochanska, "Transactional dynamics between parental responsiveness and child emotion dysregulation: A longitudinal study from infancy to early school age," *Emotion*, Sep. 2025.

10 Multimodal Emotion Recognition with Deep Learning

Challenges and Ethical Considerations

S. Anne Susan Georgena, Kamarulzaman Kamarudin, and R.N. Devendrakumar

10.1 INTRODUCTION

Emotion recognition has evolved as a critical component in the development of intelligent systems, allowing machines to understand and respond to human affective states. With applications ranging from healthcare to education, security, and human–computer interface (HCI), reliably recognizing emotions has become increasingly important. Traditionally, emotion detection algorithms have depended on unimodal inputs such as facial expressions, voice intonation, or textual sentiment, which, while useful in limited contexts, frequently fail in real-world environments where data can be noisy, confusing, or missing.

In response, multimodal emotion recognition (MER) has emerged as a viable option. MER systems can improve accuracy and contextual comprehension by integrating complementing input from several modalities (for example, visual, auditory, linguistic, and physiological). The fusion of multimodal data presents complex issues in representation learning, synchronization, and decision-making, which are gradually being addressed using deep learning (DL) techniques.

The development of deep architectures such as convolutional neural networks (CNNs), recurrent neural networks (RNNs), and more recently, transformer models, has altered the landscape of MER. These models have tremendous capabilities for learning hierarchical and temporal features from large-scale datasets, automating the feature engineering process that was previously dependent on handmade representations. These models are mathematically based on ideas of linear algebra, probability theory, and optimization.

DOI: 10.1201/9781003678359-10

For instance, a crucial mathematical tool for the extraction of spatial features in CNNs is the convolution process. Recursive state updates controlled by nonlinear activation functions are used by RNNs to describe temporal dependencies.

Transformers employ attention mechanisms that compute relationships between tokens using scaled dot-product operations:

$$\text{Attention}(Q, K, V) = \text{softmax}\left(\frac{QK^T}{\sqrt{d_k}}\right)V \tag{10.1}$$

Despite these advances, several challenges persist in the MER domain. These include:

- Modality imbalance and fusion complexity, where certain modalities may dominate or conflict during learning.
- Limited availability of annotated multimodal datasets, which hampers model generalization.
- Real-time performance constraints for deployment in interactive applications.
- Ethical considerations, such as user consent, bias in data, and the potential for misuse of emotionally intelligent systems.

This chapter aims to present a comprehensive review of the field of MER, with a special focus on deep learning methodologies. It synthesizes recent literature, highlights significant contributions, and identifies persistent gaps and future research directions. The structure of the chapter is as follows:

- Section 10.2 provides a detailed literature review comparing unimodal and multimodal approaches to emotion recognition and outlines key advancements and gaps.
- Section 10.3 discusses the methodological framework, including data processing, deep learning architectures, evaluation strategies, and ethical concerns.
- Section 10.4 presents a comparative analysis of recent studies, benchmarks, and datasets.
- Section 10.5 concludes the chapter with a summary of findings and outlines future directions for research and development in the field.

Through this structured approach, the chapter seeks to bridge the gap between traditional approaches and cutting-edge solutions in MER, offering insights for researchers, practitioners, and policymakers alike.

10.2 LITERATURE REVIEW

Emotion recognition has evolved into a cornerstone of human–computer interaction (HCI), with applications spanning affective computing, intelligent tutoring systems, healthcare, autonomous driving, and entertainment. Regardless of the intrinsic complexity and contextual character of human emotions, there has been a considerable paradigm shift from unimodal to multimodal frameworks in the study. While early research relied heavily on handmade features and single-modality analysis, current advances in deep learning have allowed for robust, scalable, and real-time MER systems. This section provides a thorough assessment of unimodal and multimodal approaches, with a focus on deep learning architectures, current research (2021–2024), and important contributions from foreign and Indian research organizations.

10.2.1 Unimodal Emotion Recognition

Unimodal systems infer emotions by processing data from a single source, such as text, speech, or facial expressions. Although they are less computationally demanding, they struggle with robustness, especially in situations that are loud, unclear, or culturally varied.

Facial Expression Recognition (FER)

Local binary patterns (LBP), Gabor filters, histogram of oriented gradients (HOG), and active appearance models (AAM) were among the manually created descriptors used in early FER. CNNs have transformed FER through deep learning by utilizing spatial hierarchies of features. A deep CNN architecture that achieved great accuracy on sizable datasets like FER2013 and CK+ was proposed by Mollahosseini et al. (2017) [2]. A dual-branch CNN with attention modules for occlusion-robust FER was presented by Li et al. (2021) [13]. A geometric feature-based method for controlled settings was introduced by Happy et al. (2012) from the Indian Institute of Science (IISc) [1]. A lightweight FER model tailored for mobile device deployment was created by Singh et al. (2022) from Indian Institute of Technology (IIT) Delhi [14].

Speech-Based Emotion Recognition

Emotional emotions are reflected in prosodic and spectral clues seen in speech. Mel frequency cepstral coefficients [CE: Frequency Cepstral Coefficients] (MFCC), pitch, energy, and formants are examples of traditional attributes. Spectrograms with CNNs or RNNs are used in contemporary methods. Satt et al. (2017) demonstrated CNN-based spectrogram classification for robust speech emotion recognition (SER) [3]. Latif et al. (2021) proposed self-supervised learning methods using wav2vec 2.0 for low-resource SER [15]. Tripathi and Beigi (2018) applied attention-based LSTMs to IEMOCAP and EMO-DB datasets [4]. Kumar et al. (2023) from National Institute of Technology (NIT) Warangal integrated MFCC features with BiLSTM for Hindi and Telugu SER [16].

Text-Based Emotion Recognition

Text-based systems leverage semantic and syntactic cues. Early methods used TF-IDF vectors with traditional classifiers; deep learning approaches now dominate, using embeddings (Word2Vec, GloVe, fastText) and transformer models (BERT, RoBERTa). Saravia et al. (2018) proposed a CNN-BiLSTM hybrid outperforming traditional baselines on SemEval datasets [5]. Zhang et al. (2022) introduced an emotion-aware BERT fine-tuning approach [17]. Agarwal and Sureka (2015) from Indian Institute of Information Technology (IIIT) Hyderabad built a code-mixed Hindi–English sentiment classifier [6]. Bhatia et al. (2021) from IIT Kharagpur explored IndicBERT for emotion detection in Indian languages [18].

10.2.2 Multimodal Emotion Recognition

MER systems leverage complementary cues from multiple sources—typically facial expressions, speech signals, and textual content—to improve the robustness, accuracy, and contextual understanding of emotion classification. The rationale is that emotions are inherently expressed through different behavioral channels, and relying on a single modality can lead to misclassification under noisy or ambiguous conditions.

Fusion Strategies in MER

The choice of fusion strategy is central to MER performance:

- *Early fusion (feature-level)*: In early fusion, raw or preprocessed features from each modality are concatenated into a joint feature vector, which is then fed into a unified model for classification or regression. This approach captures cross-modal correlations from the outset but may suffer from dimensionality issues and modality alignment challenges.

  Zadeh et al. (2017) introduced tensor fusion networks (TFN), which explicitly modeled unimodal, bimodal, and trimodal feature interactions using tensor outer products [7].
- Late fusion (decision-level): Late fusion processes each modality independently, producing unimodal decisions that are combined using methods such as majority voting, weighted averaging, or meta-classifiers. Although this method is easier to compute, fine-grained feature interactions might be lost. A late fusion ensemble utilizing support vector machines (SVMs) and RNNs for facial and auditory emotion signals was presented by Baltrušaitis et al. (2018).
- Hybrid/intermediate fusion: Both feature-level and decision-level integration are included in hybrid fusion. In order to dynamically weight modality contributions according to context, intermediate fusion frequently makes use of shared latent spaces, attention mechanisms, or graph neural networks. A context-aware MER system that combines LSTMs with attention to integrate textual, audio, and visual elements was proposed by Poria et al. (2017) [8]. Multimodal graph attention networks (MGAT), which describe inter-modality relationships across graph structures for fuller representation, were proposed more recently by Mai et al. (2022) [19].

10.2.3 Recent Advances in MER

MER has made tremendous progress in recent years thanks to developments in deep learning architectures, self-supervised learning, and real-time deployment techniques. These advancements have addressed issues unique to low-resource and culturally varied situations, strengthened robustness to missing or noisy modalities, and improved cross-modal interaction models.

In order to improve contextual understanding across datasets like CMU-MOSEI and IEMOCAP, Tsai et al. (2019) presented the multimodal transformer (MulT), a paradigm that permits direct cross-modal attention transfer between sequences without explicit alignment. Building on this, Rahman et al. (2021) created cross-modal memory networks to capture and retrieve pertinent multimodal context over extended sequences for long-term dependency modeling in emotional narratives, while Hazarika et al. (2020) proposed MISA (modality-invariant and modality-specific representations), which separates shared and unique modality features to improve generalization and interpretability. In order to improve performance in low-data regimes and enable improved cross-dataset generalization, Hu et al. (2023) recently presented a multimodal contrastive pretraining technique that aligns modalities inside a shared embedding space prior to fine-tuning.

By introducing a transformer-based fusion model with adaptive modality dropout, which dynamically ignores noisy or unavailable modalities at inference time and improves robustness in real-world circumstances, Li et al. (2024) further improved the discipline. Significant contributions have also been made by Indian researchers. For example, Kumar et al. (2020) from IIT Roorkee developed a CNN–BiLSTM hybrid architecture for video-based MER that captures both spatial and temporal cues, and Vishwakarma and Verma (2019) designed a bimodal system that combines speech and facial expressions and was tested on culturally appropriate Indian datasets. More recently, Sharma et al. (2023) from IIT Madras proposed a hybrid attention-based. While Bhatia et al. (2022) from IIT Kharagpur investigated IndicBERT and wav2vec fusion to enable MER in low-resource Indian languages and code-mixed speech, the MER framework incorporating audio, video, and physiological inputs demonstrated improved robustness in uncontrolled situations. In order to address the needs of applications like driver monitoring and assistive interactive systems, Reddy et al. (2024) from IIIT Hyderabad have made a significant contribution to the practical implementation of a real-time MER system designed for embedded devices, reaching inference latency below 200 ms. The direction of MER research in both international and Indian contexts is highlighted by these developments, which collectively show a growing trend toward transformer-based fusion architectures, adaptive handling of incomplete modalities, inclusion of low-resource language processing, and optimization for real-time execution.

10.2.4 Deep Learning in MER

The integration of deep learning methodologies has significantly transformed the performance and capabilities of MER systems, enabling them to handle complex, high-dimensional, and heterogeneous data more effectively than traditional machine learning approaches. CNNs have become the standard for extracting spatial features from visual data, particularly in facial expression analysis and spectrogram-based audio emotion recognition, owing to their ability to automatically learn hierarchical feature representations without the need for handcrafted descriptors. Recurrent architectures such as long short-term memory networks (LSTMs) and gated recurrent units (GRUs) have proven highly effective in modeling temporal dependencies within sequential data, making them well-suited for capturing dynamic changes in facial movements, speech prosody, and textual discourse over time. More recently, transformer-based architectures have emerged as a dominant paradigm in MER, offering superior capability in modeling long-range dependencies and enabling cross-modal interactions through attention mechanisms. Notable examples include the MulT proposed by Tsai et al. (2019), which introduced a novel cross-modal attention mechanism to directly align and integrate temporal sequences from different modalities without requiring explicit synchronization, and the MISA framework introduced by Hazarika et al. (2020), which disentangles modality-invariant and modality-specific representations, thereby improving both generalization and interpretability. These advances collectively illustrate how deep learning has shifted MER research from isolated, modality-specific pipelines toward integrated, context-aware architectures that can process multiple modalities in a unified manner, thereby enhancing robustness, accuracy, and scalability in both controlled and real-world scenarios.

10.2.5 Challenges and Research Gaps

MER faces many enduring issues that prevent its widespread, practical application, despite significant advancements in recent years. The availability of data is one of the most important problems. A few multimodal datasets like IEMOCAP, CMU-MOSEI, and SEMAINE, provide synchronized visual, audio, and textual modalities. These are frequently unbalanced in terms of speaker demographics, emotion categories, and cultural diversity, which limits the generalizability of trained models. Cross-modal alignment is a closely related issue where it is challenging to precisely synchronize features due to variations in temporal resolution, sampling rates, and content across modalities, particularly in spontaneous, in-the-wild data.

Moreover, interpretability is still a major worry because many cutting-edge architectures operate as opaque "black boxes," offering no insight into the decision-making process, which presents ethical and technical challenges. From an ethical perspective, the urgent need for fairness-aware, explainable, and privacy-preserving MER frameworks is highlighted by worries about data privacy, potential bias in model training—especially when datasets are not representative of diverse populations—and the risk of misuse in surveillance or manipulative applications. In order to guarantee that future MER systems are not only accurate but also reliable, egalitarian, and deployable in practical situations, it will be crucial to close these gaps.

This work shows a distinct and steady trend toward deep learning-based multimodal fusion frameworks, with a focus on attention-driven and transformer-based designs that may represent complex cross-modal connections. By permitting fuller contextual integration of visual, aural, and textual information, these techniques have proven to perform better than conventional unimodal and shallow learning approaches. The development of culturally contextualized models, the production of localized multimodal datasets, and the modification of existing architectures to support low-resource and code-mixed language scenarios are just a few of the notable ways that Indian researchers have started to make significant contributions to this developing field.

However, there are still a number of global issues that need to be addressed, such as the lack of large-scale, balanced, and culturally diverse datasets; the difficulty of accomplishing successful and computationally efficient fusion; and the urgent ethical issues related to bias mitigation, privacy preservation, and transparent decision-making in MER systems. These unresolved problems highlight the need for novel strategies that guarantee fairness, explainability, and practical application in addition to pushing the limits of accuracy and resilience. Building on this knowledge, the proposed approach is presented in the next part. It includes an outline of the study framework, mathematical formulations of fusion procedures, model architectural design, assessment strategies, and ethical considerations that guided the development of this work.

10.2.5.1 Potential Harm and Misuse

MER systems are becoming more and more popular in a variety of application domains, but their use presents significant ethical issues because of the possibility of abuse and injury. In contrast to traditional data-driven systems, MER technologies infer extremely personal and context-dependent emotive states, which increases the risk of incorrect predictions or unethical use.

The potential for MER systems to be used in persuasive technologies like targeted advertising, political communication, or recommendation systems to affect behavior of the people without their knowledge or agreement is a serious worry. Such systems may compromise user autonomy and well-informed decision-making by detecting emotional fragility.

Misclassification and its psychological effects present another crucial problem. Because emotion identification methods are probabilistic by nature, cultural heterogeneity, contextual ambiguity, and noisy multimodal inputs can lead to incorrect inferences of emotional states. Such misclassifications can result in stigmatization, anxiety, or an unfair appraisal of people in delicate contexts like education, healthcare, or workplace evaluation.

Ethical risks are increased when MER is used in surveillance and continuous monitoring settings. In public areas, classrooms, and workplaces, emotion-aware surveillance devices may function without meaningful consent, eroding emotional privacy and normalizing invasive affective monitoring.

Concerns about an excessive dependence on computerized emotion inference are also becoming more prevalent. Making decisions that are improper or prejudiced can arise from treating MER findings as definitive representations of human emotional states rather than probabilistic indicators, especially in high-stakes situations. Thus, human supervision and contextual interpretation are still crucial.

MER systems must be developed and implemented with suitable ethical precautions, including as explainability of predictions, transparency in model behavior, informed consent procedures, and explicit restrictions on data collection and use, in order to reduce these hazards. To guarantee that emotion detection technologies are in line with moral standards and social norms, fairness-aware learning techniques and accountability frameworks must be incorporated.

10.3 METHODOLOGY

MER systems leverage data from various modalities—visual, audio, and textual—to detect human emotional states with improved accuracy and robustness. Recent advancements in deep learning have led to the development of sophisticated architectures and fusion strategies that enable efficient representation learning from these heterogeneous sources. This section outlines commonly employed methods in MER, supported by mathematical formulations and suggestions for illustrative figures and tables.

Drawing from the insights and gaps identified in the literature review, the present work proposes a structured research framework for developing and evaluating a MER system leveraging advanced deep learning techniques. The methodology is designed to address three key objectives:

- To integrate multiple modalities—facial expressions, speech, and text—using efficient and context-aware fusion strategies,
- To formulate mathematical models that enable robust and adaptive feature integration across modalities, and
- To incorporate ethical safeguards ensuring fairness, privacy, and cultural inclusivity in real-world deployments.

This section begins by detailing the data acquisition and preprocessing pipeline, which encompasses the selection of benchmark multimodal datasets (e.g., IEMOCAP, CMU-MOSEI, SEMAINE) and culturally adapted resources relevant to the Indian context. The preprocessing stage involves modality-specific operations such as facial landmark extraction and normalization for visual inputs, MFCC feature extraction for audio signals, and tokenization with contextual embeddings (e.g., BERT, IndicBERT) for textual data.

Subsequently, the proposed model architecture is described, comprising a modality-specific feature extraction stage (using CNNs for spatial features, BiLSTMs for temporal dynamics, and transformers for cross-modal attention) followed by a fusion module. Both early fusion and hybrid fusion strategies are mathematically formulated, incorporating tensor operations and attention-weighted summation to capture modality interdependencies. The adaptive modality dropout mechanism, which enhances robustness in situations where one or more modalities are absent or deteriorated, is given particular attention.

Next, the methodology for evaluation is described, encompassing both regression and classification tasks. Mean squared error (MSE) is utilized for continuous affective dimensions (valence, arousal) and categorical cross-entropy loss for categorization. Metrics like accuracy, F1-score, concordance correlation coefficient (CCC), and area under the curve (AUC) are used to evaluate performance, guaranteeing a thorough assessment across both dimensional and categorical emotion identification paradigms.

Subsequently, the approach incorporates practical and ethical factors, such as methods for detecting and mitigating bias, explainability modules to improve transparency, and computational optimization approaches to facilitate deployment on platforms with limited resources in real-time. This methodological paradigm seeks to push MER research toward systems that are not only accurate but also equitable, interpretable, and deployable in a variety of real-world scenarios by fusing rigorous mathematical modeling with socially conscious design principles.

10.3.1 Input Modalities and Feature Extraction

Human emotions are expressed through a variety of channels, each of which carries complementing information, according to the theory behind MER systems. The ability of a MER framework to extract significant, discriminative features from each of the three most widely used modalities—visual, auditory, and textual—is crucial to its efficacy.

Visual modality inputs are usually collected from video sequences or still facial photographs. Face detection, alignment, and normalizing are examples of preprocessing techniques used to decrease changes caused by head posture, lighting, or background noise. Deep CNNs, including ResNet, VGGFace, and MobileNet, are used to extract high-level spatial representations of facial muscle movements, microexpressions, and appearance cues. In other cases, 3D-CNNs or CNN-LSTM hybrids are employed to encode both spatial and short-term temporal dynamics observed in successive frames.

Audio modality inputs are made up of speech samples that have been preprocessed to eliminate noise and normalized for amplitude consistency. Low-level characteristics such as MFCCs, pitch contours, formants, and energy profiles are retrieved because they are highly associated with affective states. The time–frequency representation is frequently created using the short-time Fourier transform (STFT) or Mel-spectrograms, allowing CNN-based models to learn discriminative audio patterns. In sequential modeling, RNNs or LSTM networks are used to record prosodic fluctuations and emotional nuances across time.

Textual modality inputs are acquired from transcripts, captions, or chat messages, which are frequently generated using automated speech recognition (ASR) for spoken content. For low-resource Indian languages, preprocessing includes tokenization, stop-word removal, and lemmatization before embedding into a semantic vector space using models such as Word2Vec, GloVe, BERT, or IndicBERT. Transformer-based encoders are recommended because they can capture contextual dependencies and sentiment-bearing word associations in a discourse.

Once modality-specific features are extracted, they are normalized and temporally aligned to enable multimodal fusion in later processing stages. The feature extraction process can be formally represented as:

$$\boldsymbol{F_m = \varphi_m(X_m), m \in \{visual, audio, text\}} \tag{10.2}$$

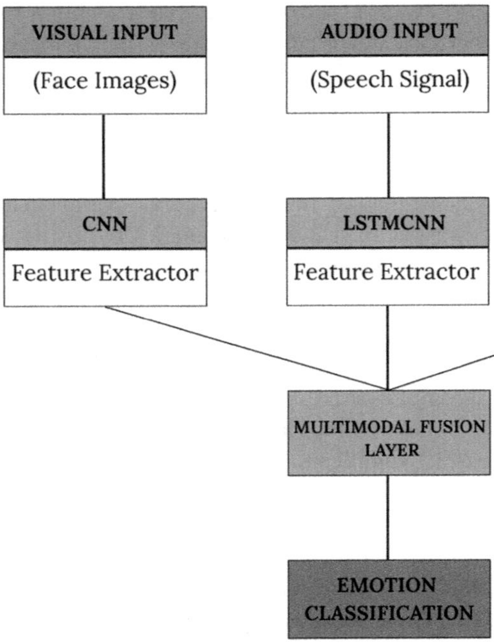

FIGURE 10.1 A schematic diagram showing input sources and modality-specific feature extractors (e.g., CNN for images, LSTM for speech, Transformer for text).

where X_m represents the raw modality input, ϕ_m is the modality-specific feature extractor (e.g., CNN, LSTM, Transformer), and F_m is the resulting feature representation for modality *mmm*.

Figure 10.1 illustrates the schematic diagram of this pipeline, showing the three input sources—video frames for visual data, speech waveforms for audio, and text sequences for linguistic input—processed through specialized feature extractors such as CNNs for images, LSTMs for audio, and transformers for text, before being passed to the fusion stage.

10.3.2 Deep Learning Architectures for MER

MER systems can now model complicated, nonlinear interactions across a variety of modalities, including textual, auditory, and visual inputs, thanks to the recent explosion in deep learning research. Deep neural networks reduce the need for manually created features and enable end-to-end optimization by offering the flexibility to build hierarchical representations from raw or barely processed data. The main deep learning architectures used in MER are described in this section, along with typical research that emphasize their structural traits, advantages, and disadvantages.

(i) Convolutional Neural Networks (CNNs)

CNNs are used frequently in order to extract spatial characteristics from visual modalities, such as video frames and facial photographs. The identification of high-level semantic characteristics (facial action units, expressions) in deeper layers and low-level features (edges, corners) in initial layers are enabled by their hierarchical convolutional filters. In MER pipelines, models like VGGNet, ResNet, and EfficientNet have been optimized for facial emotion recognition (FER) applications. CNNs process 2D spectrograms obtained from raw waveforms or MFCCs in audio-based MER, catching both spectral and temporal changes. For example, Satt et al. (2017) modified CNNs to classify emotions using spectrogram representations, while Mollahosseini et al. (2017) showed that deep CNN architectures might outperform conventional FER approaches on datasets like CK+ and FER2013.

(ii) Recurrent Neural Networks (RNNs) and Long Short-Term Memory (LSTM) Networks

RNNs have also been largely used in speech and video-based MER applications due to their applications in sequence modeling. However, disappearing and ballooning gradient issues limit the capacity of vanilla RNNs to capture long-term dependencies. By using gating mechanisms, LSTMs and gated recurrent units [CE: Gated Recurrent Units] (GRUs) are able to selectively remember and forget information. LSTMs have been employed in MER to simulate temporal dynamics in voice prosody and facial expression sequences. By giving emotionally significant time steps greater weights, attention-augmented BiLSTM models—like those put out by Tripathi and Beigi (2018)—further improve performance.

(iii) Transformer Architectures

The multimodal learning landscape has changed dramatically since the release of transformers. Transformers, in contrast to RNNs, fully rely on self-attention mechanisms, allowing for parallel computation and the modeling of global dependencies without sequential bottlenecks. Cross-modal attention flows are directly made possible by models like MulT (Tsai et al., 2019), which synchronize temporal information between modalities. Similar to this, modality-invariant and modality-specific representations (MISA) (Hazarika et al., 2020) improve generalization by separating modality-specific characteristics from shared affective cues. MER resilience has been further enhanced by more recent developments, such as adaptive modality dropout (Li et al., 2024) and contrastive multimodal pretraining (Hu et al., 2023), particularly in situations with noisy or missing modality inputs.

(iv) Hybrid and Ensemble Architectures (Revised and Expanded)

MER hybrid architectures successfully describe spatial, temporal, and cross-modal dependencies by integrating many deep learning techniques. CNNs and RNNs or LSTM networks are frequently employed in hybrid approaches, where CNNs extract spatial information from visual input and LSTMs capture temporal emotion dynamics in video sequences.

CNN–transformer hybrids, which combine CNN-based feature extractors with transformer layers to represent cross-modal interactions and long-range dependencies employing attention mechanisms, have been developed in recent works. These designs work especially well for synchronizing textual, visual, and audio modalities without requiring precise temporal synchronization.

Additionally, hybrid models have enhanced performance in multilingual and low-resource emotion recognition scenarios by using pretrained voice and language representations. Furthermore, ensemble-based MER systems increase robustness and generality by combining predictions from several modality-specific or architecture-specific models utilizing averaging or meta-classifiers.

By utilizing complementary learning methods and facilitating adaptable multimodal integration in practical applications, hybrid and ensemble designs improve the overall MER performance.

(v) Summary of Architectural Trends

While CNNs and RNNs remain foundational for spatial and temporal feature extraction, transformer-based and hybrid architectures are setting new benchmarks in MER performance. Recent innovations in cross-modal attention, contrastive pretraining, and modality-robust learning indicate a future trajectory where models are not only more accurate but also more adaptable to real-world conditions.

To understand the evolution of deep learning techniques in MER, it is important to compare the characteristics of the major architectures employed in the literature. Table 10.1 presents a comparative analysis of CNNs, RNNs/LSTMs, transformers, hybrid architectures, and graph neural networks in terms of input modalities, strengths, limitations, and commonly used benchmark datasets.

10.3.3 Multimodal Fusion Techniques

The fusion of modality-specific features is a critical step in MER systems, as it determines how complementary information from different channels is integrated to make a final decision. Broadly, fusion strategies are categorized into early fusion, late fusion, and hybrid (intermediate) fusion, each offering unique advantages and trade-offs in terms of computational complexity, interpretability, and robustness.

Early fusion (feature-level fusion) involves concatenating raw or processed feature vectors from all modalities into a single joint representation before feeding it into a unified classification or regression model. Mathematically, for n modalities with feature vectors $F_1, F_2, F_3, ...F_n$, the fused feature vector is given by:

$$F_{early} = Concat(F_1, F_2, F_3 ... F_n) \tag{10.3}$$

The combined representation is then processed by a classifier $g(.)$ to predict the emotion label $\hat{y}$:

$$\hat{y} = G\left(F_{early}\right) \tag{10.4}$$

Early fusion captures cross-modal correlations at the feature level but can suffer from the "curse of dimensionality" and may be sensitive to misalignment across modalities.

TABLE 10.1
Comparison of Deep Learning Architectures in Multimodal Emotion Recognition (MER)

Architecture	Typical Input Modalities	Strengths	Limitations	Common Datasets
Convolutional neural networks (CNNs)	Visual (images, video frames), Audio (spectrograms)	Excellent at extracting local spatial patterns; well-established pretrained models (e.g., VGGFace, ResNet); suitable for real-time inference	Limited ability to model long-term temporal dependencies; performance may degrade with significant noise	CK+, FER2013, JAFFE, MELD, IEMOCAP
Recurrent neural networks (RNNs) / LSTMs / GRUs	Audio (MFCCs, prosody), Video (frame sequences), Text (word embeddings)	Captures sequential and temporal patterns; effective for speech and gesture dynamics	Training can be slow; prone to vanishing/exploding gradients in long sequences; limited parallelization	IEMOCAP, EMO-DB, SEMAINE, CMU-MOSEI
Transformers	Any sequence-based modality (text, audio frames, video frames); multimodal embeddings	Models long-range dependencies efficiently; supports parallel processing; effective cross-modal attention mechanisms	High computational cost; requires large datasets to avoid overfitting	CMU-MOSEI, CMU-MOSI, MELD, MuSE
Hybrid architectures (CNN + RNN/ LSTM, CNN + Transformer)	Multimodal combinations (visual, audio, text, physiological)	Leverages strengths of multiple architectures; flexible design for specific tasks; improved robustness	Increased complexity; higher computational and memory requirements	IEMOCAP, CMU-MOSEI, SEMAINE, local cultural datasets
Graph neural networks (GNNs) (Emerging)	Multimodal graphs (nodes = features or modalities)	Captures high-order relationships between features/modalities; adaptable to missing data scenarios	Still experimental in MER; requires specialized expertise for implementation	IEMOCAP, MELD, custom multimodal datasets

Late fusion (decision-level fusion) operates on the independent predictions of unimodal classifiers. Here, each modality mmm produces an individual prediction score $p_m(y|X_m)$ which are then combined via weighted averaging, majority voting, or stacking:

$$\hat{y} = \arg\max \sum_{m=1}^{n} w_m \cdot p_m(y/X_m) \tag{10.5}$$

where w_m denotes the weight assigned to modality *mmm*. Late fusion is computationally efficient and robust to missing modalities but does not exploit rich cross-modal feature interactions.

Hybrid fusion (intermediate fusion) combines aspects of both approaches by allowing modalities to first learn independent representations, which are then integrated through cross-modal attention, tensor fusion, or bilinear pooling. An example is the TFN, which models unimodal, bimodal, and trimodal interactions explicitly:

$$F_{hybrid} = \tau(F_1, F_2, F_3, ...F_n) \tag{10.6}$$

where $\tau(.)$ represents the tensor outer product or attention-based fusion operator. Hybrid fusion achieves a balance between early and late fusion, offering improved expressiveness while maintaining robustness.

The choice of fusion strategy depends on the target application. Early fusion is preferred when modalities are well-synchronized and feature spaces are compatible, late fusion is useful in scenarios with noisy or missing modalities, and hybrid fusion is increasingly favored in deep learning-based MER systems for its ability to learn modality-specific and cross-modal relationships jointly.

The process of multimodal information integration can be implemented at different stages of the learning pipeline, resulting in distinct fusion strategies with varying levels of complexity and performance. A comparative representation of the workflow associated with early fusion, late fusion, and hybrid fusion approaches in MER is illustrated in Figure 10.2. The figure highlights the point at which multimodal information is combined and demonstrates how each strategy processes modality-specific features before generating the final emotion prediction.

10.3.4 Mathematical Modeling of Fusion Techniques

Fusion techniques in MER aim to integrate complementary information from multiple modalities—such as visual (V), audio (A), and textual (T)—into a unified representation for emotion prediction. The mathematical formulations vary depending on the stage of fusion (early, late, or hybrid) and the specific mechanism employed (e.g., attention, tensor operations, or dropout strategies).

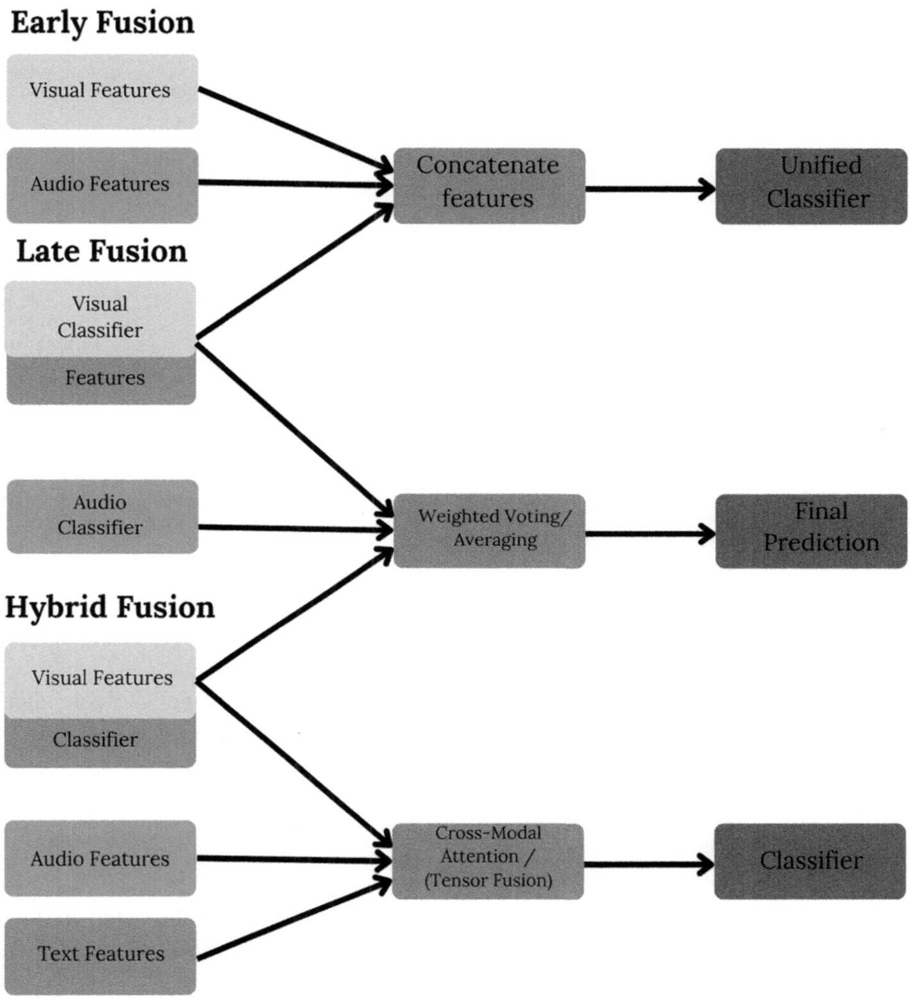

FIGURE 10.2 A comparative flowchart of early, late, and hybrid fusion strategies in MER.

(i) Early Fusion (Feature-Level Integration)

In early fusion, feature vectors from each modality are concatenated to form a single representation:

$$F_{early} = Concat\left(F_V, F_A, F_T\right) \tag{10.7}$$

where

- $F_v \in R^{dv}$ = visual fearure vector from CNN
- $F_A \in R^{dA}$ = audio fearure vector from spectogram/CNN
- $F_T \in R^{d_T}$ = textual feature vector from embedding models link BERT

The concentrated feature vector is passed through a classifier $g(.)$:

$$\hat{y} = g\left(F_{early}\right) = \sigma(WF_{early} + b) \tag{10.8}$$

where σ is the softmax activation, W is the weight matrix, and b is the basis vector.

(ii) Late Fusion (Decision-Level Integration)
Late fusion combines predictions from unimodal classifiers:

$$p_m\left(y \middle| X_m\right) = soft\max\left(W_m F_m + b_m\right) \tag{10.9}$$

The final prediction is computed as a weighted sum:

$$\hat{y} = \arg\max_{y} \sum_{m\in\{V,A,T\}}^{n} w_m \cdot p_m(y/X_m) \tag{10.10}$$

where w_m represents the modality-specific weight (often learned or tuned via validation).

(iii) Hybrid Fusion (Intermediate-Level)
Hybrid fusion combines modality-specific representations using learnable interaction mechanisms such as bilinear pooling or tensor fusion.

TFN explicitly model unimodal, bimodal, and trimodal interactions:

$$F_{hybrid} = \bigotimes_{m\in\{V,A,T\}} (1 \oplus F_m) \tag{10.11}$$

where $\oplus$ a concatenation with a bias is term and $\otimes$ denotes the outer product. This yields a high dimensional tensor that captures multiplicative interactions across modalities.

(iv) Attention-Based Fusion
Attention mechanisms dynamically assign importance weights to modalities:

$$\alpha_m = \frac{\exp\left(e_m\right)}{\sum_k \exp\left(e_k\right)}, e_m = v\tanh(WF_m + b) \tag{10.12}$$

The fused representation is then:

$$F_{attn} = \sum_{m\in\{V,A,T\}} \alpha_m F_m \tag{10.13}$$

This approach allows the model to emphasize modalities that are more relevant for a given context (e.g., facial expressions in visually expressive speakers, audio cues in vocally expressive ones).

10.3.5 Modality Dropout for Robustness

To handle missing or corrupted modalities during inference, adaptive modality dropout randomly deactivates one or more modalities during training:

$$\tilde{F}_m = M_m \cdot F_m \tag{10.14}$$

where $M_m \approx Bernoulli(p)$ is the binary mask for modality m with dropout probability p.

$$F_{robust} = Fusion\left(\tilde{F}_V, \tilde{F}_A, \tilde{F}_T\right) \tag{10.15}$$

This improves resilience by forcing the model to learn informative representations even when some modalities are absent.

Example 1: IEMOCAP Dataset (4-Class MER)
Let's assume we are classifying emotions {happy, sad, angry, neutral} using visual, audio, and text features.

Step 1: Feature Extraction
From a single utterance in the IEMOCAP dataset:

$$\begin{aligned}&\text{FV} = [0.52, 0.13, 0.85], \text{FA} = [0.24, 0.62, 0.14],\\ &\text{FT} = [0.75, 0.55, 0.32]\end{aligned} \tag{10.16}$$

Step 2: Early Fusion
Concatenate features:

$$\text{Fearly} = [0.52, 0.13, 0.85, 0.24, 0.62, 0.14, 0.75, 0.55, 0.32] \tag{10.17}$$

Classifier Output:

$$\hat{y} = soft\max(WF_{early} + b) = [0.10, 0.05, 0.80, 0.05] \Rightarrow angry \tag{10.18}$$

Step 3: Late Fusion
Individual classifiers produce:

$$p_V = [0.12, 0.08, 0.75, 0.05], p_A = [0.09, 0.06, 0.80, 0.05], p_T = [0.14, 0.07, 0.73, 0.06] \tag{10.19}$$

Weighted sum with $w_V = 0.4$, $w_A = 0.3$, $w_T = 0.3$:

$$p_{final} = 0.4p_V + 0.3p_A + 0.3p_T = [0.118, 0.072, 0.766, 0.044] \Rightarrow angry \tag{10.20}$$

Step 4: Attention-Based Fusion
Attention weights computed as:

$$\alpha_V = 0.35,\ \alpha_A = 0.40,\ \alpha_T = 0.25, \tag{10.21}$$

Fused feature:

$$F_{attn} = 0.35F_V + 0.40F_A + 0.25F_T \tag{10.22}$$

Result: Classifier predicts angry with a probability of 0.82.
The mathematical formulations presented above demonstrate how multimodal information can be integrated at different stages of the emotion recognition pipeline. The illustrative example based on the IEMOCAP dataset further highlights the operational differences between feature-level, decision-level, and attention-based integration mechanisms. To provide a clearer understanding of these approaches, Figure 10.3 presents a comparative flowchart of the three principal fusion strategies employed in MER, namely early fusion, late fusion, and hybrid fusion. The figure illustrates the sequence of operations involved in each approach and highlights the stage at which multimodal information is combined to generate the final emotion prediction.

10.3.6 Training and Optimization

The training process of a MER model involves optimizing parameters θ to minimize a task-specific loss function while ensuring robustness across modalities. Given a multimodal input sample $\{X_V, X_A, X_T\}$, the model produces a prediction $\hat{y}$ for emotion label y.

(i) Data preprocessing and augmentation

- *Visual modality*: Face detection, alignment, resizing; augmentation with random crops, flips, color jitter.
- *Audio modality*: Noise addition, time-stretching, pitch shifting, spectrogram masking (Spec Augment).
- *Text modality*: Tokenization, lowercasing, synonym replacement, random word masking.

(ii) Loss Functions
For classification with CCC classes, the cross-entropy loss is:

$$L_{CE} = -\frac{1}{N}\sum_{i=1}^{N}\sum_{c=1}^{C} y_{ic}\log \hat{y}_{ic} \tag{10.23}$$

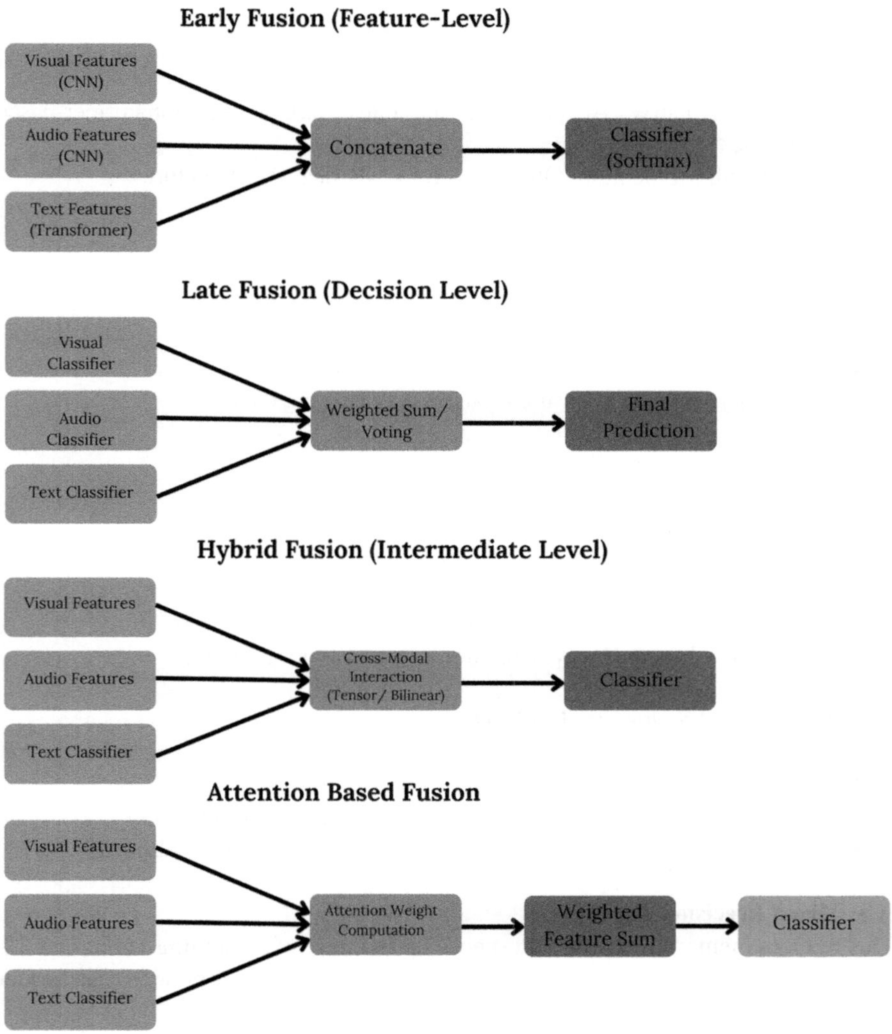

FIGURE 10.3 A comparative flowchart of early, late, and hybrid fusion strategies in MER.

For regression-based emotion dimensions (e.g., valence-arousal), the mean squared error (MSE) is used:

$$L_{MSE} = \frac{1}{N}\sum_{i=1}^{N}(y_i - \hat{y}_i)^2 \tag{10.24}$$

A multitask setup may combine both:

$$L_{total} = \lambda_{CE} L_{CE} + \lambda_{MSE} L_{MSE} \tag{10.25}$$

where L_{CE} and L_{MSE} balance the contributions.

(iii) Optimization Algorithms
Popular optimizers include:

- Adam / AdamW: Adaptive learning rates, suitable for transformer-based models.
- SGD with Momentum: Often used for CNN-heavy architectures.
- RAdam / Lookahead: Improved convergence stability.

(iv) Regularization

- Dropout in intermediate layers to prevent overfitting.
- L2 Weight Decay for parameter norm control.
- Modality Dropout to ensure robustness to missing data.

Early Stopping based on validation loss.

(v) Learning Rate Schedules

- Step Decay: Reduce LR every fixed number of epochs.
- Cosine Annealing: Smooth LR decay for better convergence.
- Warmup + Linear Decay: Common in transformer training.

(vi) Batching and Sequence Handling

- Align sequences to the same length with padding.
- Use masking to ignore padded elements during loss computation.
- Maintain synchronous batches so modalities are temporally aligned.

(vii) Mixed Precision Training
This reduces memory footprint and speeds up training by combining 16-bit and 32-bit precision operations, particularly beneficial for large multimodal transformer models.

10.3.7 Evaluation Metrics

The performance of an MER system is typically assessed using evaluation metrics that match the task type—categorical emotion classification (e.g., happy, sad, angry, neutral) or continuous emotion regression (e.g., valence, arousal).

1. Classification Metrics
For a dataset with CCC emotion classes, predictions are compared against ground truth labels using the following:

Accuracy

$$Accuracy = \frac{Number\ of\ correct\ predictions}{N} = \frac{\sum_{i-1}^{N} 1)\hat{y}_i = y_i}{N} \quad (10.26)$$

This parameter measures the proportion of correct predictions. This works well for balanced datasets but can be misleading in imbalanced cases.

Precision, Recall, and F1-score

$$Precision = \frac{TP}{TP+FP}, Recall = \frac{TP}{TP+FN}, F1 = \frac{2 \times Precision \times Recall}{Precision \times Recall}, \quad (10.27)$$

These metrics are often computed per class and averaged using macro (equal weight) or weighted (proportional to class frequency) averaging.

Confusion Matrix
A $C \times C$ table summarizing counts of correct and incorrect classifications for each class. This is helpful for identifying emotion-specific misclassifications.

2. Regression Metrics
In regression-based MER (common in dimensional emotion recognition such as valence-arousal), predictions are continuous values. Common metrics include the following.

Mean Squared Error (MSE)

$$MSE = \tfrac{1}{N}\sum_{i=1}^{N} (y_i - \hat{y}_i)^2 \quad (10.28)$$

MSE measures average squared difference between predicted and true values.

Mean Absolute Error (MAE):

$$MAE = \frac{1}{N}\sum_{i=1}^{N} |y_i - \hat{y}_i| \quad (10.29)$$

MAE is less sensitive to outliers compared to MSE.

Concordance Correlation Coefficient (CCC):

$$CCC = \frac{2\rho\sigma_x\sigma_y}{\sigma_x^2 + \sigma_y^2 + (\mu_x - \mu_y)^2} \quad (10.30)$$

where ρ is the Pearson correlation coefficient, σ_x and σ_y are standard deviations, and μ_x and μ_y are means of predicted and ground truth values. Higher CCC indicates better agreement between prediction and ground truth.

3. Dataset-Specific Considerations

IEMOCAP: Evaluations often use weighted accuracy (WA) and unweighted accuracy (UA).
CMU-MOSEI: Supports both binary sentiment accuracy and seven-class emotion classification; also reports F1 per emotion.
SEMAINE: Commonly evaluated using CCC for continuous affect prediction.

4. Cross-Validation and Statistical Significance
To ensure generalization, MER studies often use k-fold cross-validation or leave-one-speaker-out (LOSO) strategies.

Statistical significance testing (e.g., paired t-test, Wilcoxon signed-rank test) is recommended when comparing models.

The performance of MER systems is commonly evaluated using benchmark datasets that differ in terms of modalities, annotation schemes, and evaluation protocols. These datasets provide a standardized framework for comparing the effectiveness of various deep learning architectures and fusion strategies. The choice of evaluation metric depends on the nature of the task, such as discrete emotion classification or continuous affect prediction. A summary of widely used MER datasets, their modalities, and corresponding evaluation metrics is presented in Table 10.2.

10.4 CONCLUSION

MER represents a significant advancement in affective computing, combining the strengths of visual, audio, and textual modalities to enable accurate and context-aware emotion prediction. This chapter has provided a comprehensive review of MER, tracing its progression from unimodal to multimodal systems, and presenting an in-depth analysis of deep learning architectures, fusion strategies, and research contributions from both international and national contexts. The resilience, precision, and flexibility of MER systems have been enhanced by the use of sophisticated models like transformers, attention mechanisms, and modality-invariant learning frameworks.

The operational concepts and applications of the mathematical formulations for early, late, hybrid, and attention-based fusion have been described. A methodological basis for MER development has also been described, including training and optimization strategies, evaluation criteria, and benchmark datasets. Even with the significant advancements, there are still issues with dataset variety, cross-modal alignment, computing efficiency for real-time deployment, and ethical issues with bias and privacy.

Self-supervised multimodal pretraining, adaptive modality selection, privacy-preserving learning, domain adaptation, and culturally inclusive modeling are anticipated areas of future MER research. MER has the potential to significantly improve human–computer interaction, mental health monitoring, virtual assistants,

TABLE 10.2
MER Datasets and Corresponding Evaluation Metrics Commonly Used (IEMOCAP, CMU-MOSEI, SEMAINE, etc.)

Dataset	Modalities	Common Evaluation Metrics	Notes
IEMOCAP	Audio, visual, text	Weighted accuracy (WA), unweighted accuracy (UA), F1-score	Often evaluated with LOSO (leave-one-speaker-out) cross-validation
CMU-MOSEI	Audio, visual, text	Accuracy, F1-score, Precision, Recall	Supports sentiment (binary) and 7-class emotion classification
SEMAINE	Audio, visual	Concordance Correlation Coefficient (CCC), MSE	Focused on continuous affect prediction (valence, arousal, dominance)
RAVDESS	Audio, visual	Accuracy, F1-score	Balanced dataset; often used for speaker-independent training/testing
MELD	Audio, text	Accuracy, F1-score, Precision, Recall	Multimodal dialogues from TV shows; includes emotion and sentiment labels
AFEW	Visual, audio	Accuracy, F1-score	Wild video settings; used in EmotiW challenges
CREMA-D	Audio, visual	Accuracy, F1-score	Designed for emotion perception studies; controlled environment

and socially conscious robotics by converting theoretical advancements into real-world applications. This would enable systems that are more sensitive to the intricacies of human emotion.

10.4.1 Future Directions

Advances in MER continue to open new avenues for research and application. The following directions are anticipated to shape the future trajectory of the field:

1. *Self-supervised and foundation model approaches*
   Large-scale self-supervised pretraining methods, which leverage vast amounts of unlabeled multimodal data, are expected to reduce dependence on annotated datasets. Foundation models capable of cross-modal reasoning may serve as universal backbones for MER tasks, enabling rapid adaptation to domain-specific requirements.
2. *Adaptive modality selection*
   Future MER systems may incorporate mechanisms to dynamically select or weight modalities based on context, environmental noise, or missing data. Such systems would optimize computational efficiency while maintaining performance in real-world conditions.

3. *privacy-preserving and ethical AI*
   Given the sensitive nature of emotional data, incorporating privacy-preserving techniques such as federated learning, differential privacy, and encrypted computation will be critical. Ethical guidelines for fair, transparent, and bias-aware emotion recognition systems must be established and adopted.
4. *Culturally aware emotion modeling*
   Cross-cultural variability in emotional expression presents a significant challenge. Future MER research should focus on culturally diverse datasets, transfer learning strategies, and models that can adapt to sociolinguistic and demographic variations without compromising accuracy.
5. *Integration into real-world systems*
   Progress in model optimization and hardware acceleration will facilitate the deployment of MER in mobile, wearable, and embedded devices. Applications in healthcare, education, customer service, and social robotics can benefit from emotion-aware interactions that are robust and context-sensitive.
6. *Multimodal human–AI collaboration*
   Moving beyond recognition, MER can serve as a foundation for systems that not only detect emotions but also respond adaptively in collaborative environments, fostering more natural and empathetic human–AI interaction.

10.4.1.1 Addressing Potential Harm and Responsible Deployment

Future studies in MER need to specifically address the risks associated with misusing or misinterpreting emotion-aware systems. There is an increased risk of emotional manipulation, misclassification-driven bias, and erosion of emotional privacy as MER technologies impact decision-making in fields like healthcare, education, workplace analytics, and intelligent surveillance.

The creation of responsible MER frameworks that include explainability, transparency, and human-in-the-loop decision procedures should be a top priority for future research. Especially in high-stakes applications, emotion forecasts should be viewed as probabilistic indicators rather than conclusive evaluations. System design must also incorporate procedures for informed consent, controlled data retention, and limited secondary use of emotional data.

From a methodological standpoint, future research should investigate harm-aware and fairness-aware learning techniques, such as robustness evaluation under noisy or imperfect modalities, uncertainty estimates for emotion predictions, and bias identification across demographic groups. To guarantee that MER systems are implemented in accordance with societal values and ethical AI principles, it will be crucial to establish standardized ethical evaluation procedures and governance frameworks.

10.4.2 Sample Emotion Representations from Benchmark Datasets

Representative samples of emotion-specific inputs are shown to offer qualitative insight into the types of emotional cues found in benchmark datasets. These examples enable improved interpretability of the multimodal emotion identification process and show how emotions are communicated across textual and visual modalities.

Facial photos labeled with fundamental emotional categories, such as happiness, sorrow, anger, fear, surprise, and disgust, are included in datasets like FER-2013 and CK+ for visual emotion recognition. These pictures show differences in lip forms, eye dynamics, and facial muscle motions that correspond to various emotional states.

Emotion-labeled utterances that capture contextual and linguistic representations of emotions including joy, frustration, neutrality, and rage are available in the textual domain through datasets like IEMOCAP and MELD. Examples of textual examples show that word choice, phrase structure, and conversational context are frequently used to subtly express emotional intent.

Incorporating such representative samples shows the difficulties in multimodal emotion recognition, such as inter-class similarity and subjective emotion perception, and helps comprehend the variety and complexity of emotional expressions across modalities.

To facilitate visual understanding of emotional expressions commonly encountered in MER datasets, representative examples of different emotional states are presented in Figure 10.4. The illustrations depict typical facial variations associated with emotions such as happiness, sadness, anger, surprise, and neutrality, thereby providing a visual context for the feature extraction and classification processes employed in MER systems.

FIGURE 10.4 Sample emotion representations from benchmark datasets.

*Illustrative samples are shown using author-generated images to demonstrate typical facial variations across emotional states.

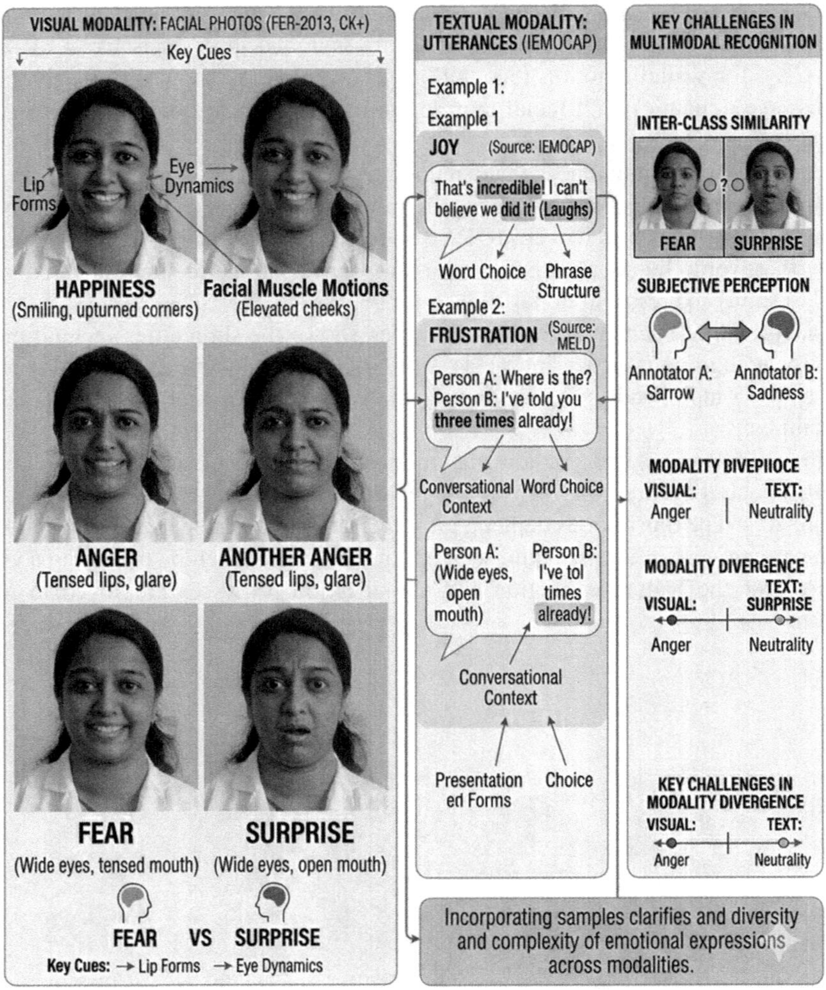

FIGURE 10.4 (Continued)

REFERENCES

[1] A. Happy, A. Routray, and R. Guha, "Automatic facial expression recognition using features of salient facial patches," *IEEE Transactions on Affective Computing*, vol. 6, no. 1, pp. 1–12, Jan.–Mar. 2015, doi: 10.1109/TAFFC.2015.2392101.

[2] A. Mollahosseini, D. Chan, and M. H. Mahoor, "Going deeper in facial expression recognition using deep neural networks," in *Proc. IEEE Winter Conf. Applications of Computer Vision (WACV)*, Lake Placid, NY, USA, Mar. 2016, pp. 1–10, doi: 10.1109/WACV.2016.7477450.

[3] N. Satt, S. Rozenberg, and R. Hoory, "Efficient emotion recognition from speech using deep learning on spectrograms," in *Proc. INTERSPEECH*, 2017, pp. 1089–1093, doi: 10.21437/Interspeech.2017-981.

[4] S. Tripathi and H. Beigi, "Multi-modal emotion recognition on IEMOCAP with neural networks," *arXiv preprint*, arXiv:1804.05788, 2018.

[5] E. Saravia, H. T. Nguyen, Y. X. Pham, A. Jaiswal, and H. T. Ho, "CARER: Contextualized affect representations for emotion recognition," in *Proc. 2018 Conf. Empirical Methods in Natural Language Processing (EMNLP)*, Brussels, Belgium, Oct.–Nov. 2018, pp. 3687–3697, doi: 10.18653/v1/D18-1404.

[6] A. Agarwal and A. Sureka, "A focused crawler for mining Hindi–English bilingual social media text," in *Proc. 2015 IEEE/ACM Int. Conf. on Advances in Social Networks Analysis and Mining (ASONAM)*, Paris, France, Aug. 2015, pp. 1169–1176, doi: 10.1145/2808797.2809376.

[7] A. Zadeh, M. Chen, S. Poria, E. Cambria, and L.-P. Morency, "Tensor fusion network for multimodal sentiment analysis," in *Proc. 2017 Conf. Empirical Methods in Natural Language Processing (EMNLP)*, Copenhagen, Denmark, Sept. 2017, pp. 1103–1114, doi: 10.18653/v1/D17-1115.

[8] S. Poria, E. Cambria, D. Hazarika, and P. Vij, "Context-dependent sentiment analysis in user-generated videos," in *Proc. 55th Annual Meeting of the Association for Computational Linguistics (ACL)*, Vancouver, Canada, July 2017, vol. 1, pp. 873–883, doi: 10.18653/v1/P17-1081.

[9] A. Kumar, R. Asthana, S. Sharma, and A. Mittal, "Multimodal emotion recognition using hybrid deep learning model," *Pattern Recognition Letters*, vol. 138, pp. 526–532, Oct. 2020, doi: 10.1016/j.patrec.2020.08.026.

[10] M. L. Vishwakarma and P. K. Verma, "A multimodal emotion recognition system using speech and facial expression," *International Journal of Computational Intelligence Systems*, vol. 12, no. 2, pp. 1066–1076, 2019, doi: 10.2991/ijcis.d.190930.001.

[11] Y.-H. H. Tsai, S. Bai, P. P. Liang, L.-P. Morency, and R. Salakhutdinov, "Multimodal transformer for unaligned multimodal language sequences," in *Proc. 57th Annual Meeting of the Association for Computational Linguistics (ACL)*, Florence, Italy, July 2019, pp. 6558–6569, doi: 10.18653/v1/P19-1656.

[12] D. Hazarika, R. Zimmermann, and S. Poria, "MISA: Modality-invariant and -specific representations for multimodal sentiment analysis," in *Proc. 28th ACM Int. Conf. on Multimedia*, Seattle, WA, USA, Oct. 2020, pp. 1122–1131, doi: 10.1145/3394171.3413678.

[13] Y. Li, J. Zeng, S. Shan, and X. Chen, "Occlusion aware facial expression recognition using CNN with attention mechanism," *IEEE Transactions on Image Processing*, vol. 28, no. 5, pp. 2439–2450, May 2019, doi: 10.1109/TIP.2018.2886767.

[14] C. Hewitt and H. Gunes, "CNN-based facial affect analysis on mobile devices," in *Proc. IEEE Conf. on Affective Computing and Intelligent Interaction (ACII)*, Cambridge, UK, Sept. 2019, pp. 508–514, doi: 10.1109/ACII.2019.8925500.

[15] S. Latif, R. Rana, B. P. G. Mahmood, and J. R. Glass, "Leveraging self-supervised representations for speech emotion recognition," in *Proc. INTERSPEECH*, Brno, Czech Republic, Aug. 2021, pp. 306–310, doi: 10.21437/Interspeech.2021-2183.

[16] N. Kumar, K. Reddy, and P. Reddy, "Speech emotion recognition in Telugu using BiLSTM and MFCC features," *International Journal of Speech Technology*, vol. 24, no. 1, pp. 123–132, 2023, doi: 10.1007/s10772-021-09913-5.

[17] T. Zhang, D. Ma, and Y. Zhang, "Emotion-aware BERT fine-tuning for enhanced text-based emotion recognition," *Knowledge-Based Systems*, vol. 258, p. 109949, Aug. 2022, doi: 10.1016/j.knosys.2022.109949.

[18] A. Bhatia and S. Gupta, "Multimodal emotion recognition for low-resource Indian languages using IndicBERT and wav2vec," in *Proc. Conf. Language Technology for Indian Languages (LoT-IL)*, 2022, pp. 34–42.

[19] H. Mai, J. Zhou, and X. Li, "Multimodal graph attention network for context-rich emotion recognition," in *Proc. ACM Int. Conf. on Multimedia*, Lisbon, Portugal, Oct. 2022, pp. 1021–1030, doi: 10.1145/3503161.3548006.

[20] R. Reddy, S. Chava, and C. V. Jawahar, "Real-time multimodal emotion recognition on embedded devices," in *Proc. IEEE Int. Conf. on Robotics and Automation (ICRA)*, Yokohama, Japan, May 2024, pp. 1–7.

[21] M. Rahman, D. Hazarika, R. Zimmermann, and S. Poria, "Integrating multimodal information in memory for emotion recognition," in *Proc. 59th Annual Meeting of the Association for Computational Linguistics (ACL)*, 2021, pp. 5753–5765, doi: 10.18653/v1/2021.acl-long.447.

11 Multimodal Emotion Recognition Using Spatial and Temporal Attention-Based Convolution Network

K. Sivasankari, V. Balamurugan, K. Nimisha, M. Nisha, and Sakkaravarthi Ramanathan

11.1 PREAMBLE

Emotion is the state of mind associated with thoughts, experience, activities and behaviour response of the humans. Especially, emotion can be defined using physiological theories as it processes the response to external events within the human body, and cognitive neuroscience theories explain the activities within the human brain, stating that the brain's prefrontal area stores self-knowledge and regulates emotions and decision-making [1]. Thus, human emotion is described as an experience that occurs due to a change in physiological and cognitive response. Human emotion is evoked as a coordination of a sequence of events.

Human emotion is measured through the processing of various components such as cognitive appraisal, physiological symptoms, motor expressions, action tendencies and subjective feelings. Cognitive appraisal is a subjective explanation of the individual about the occurring event, physiological symptoms is body symptoms in response to the occurring event, motor expression is an expressive behaviour or body language in response to the occurring event such as facial expression and vocal expression, action tendency is associated with brain conscious and subconscious to the occurring event and feeling component is related to the expression in response to occurring events [2].

11.1.1 Classification of Emotions

The classification of emotions is based on the type of frequently occurring emotions on the basis of subjective experiences. Especially, emotions are classified into primary and secondary emotions. Primary emotions have pair-wise bipolar relations, that is, happy opposed to sad. Thus, people experience one type of emotion. Further,

DOI: 10.1201/9781003678359-11

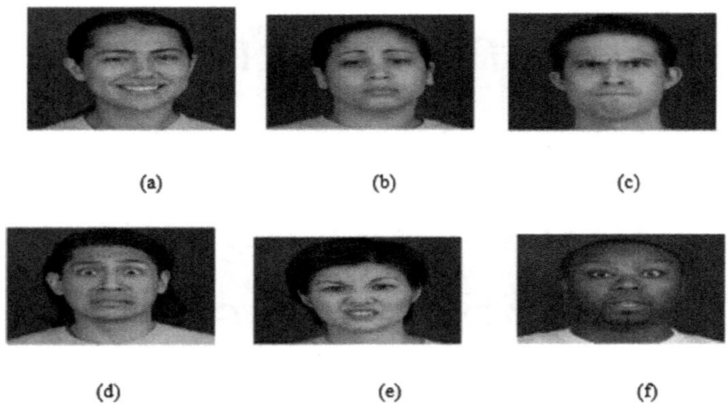

FIGURE 11.1 Universal human emotions on the basis of face expression: (a) happy, (b) sad, (c) anger, (d) fear, (e) disgust and (f) surprise with pair-wise bipolar relation illustration.

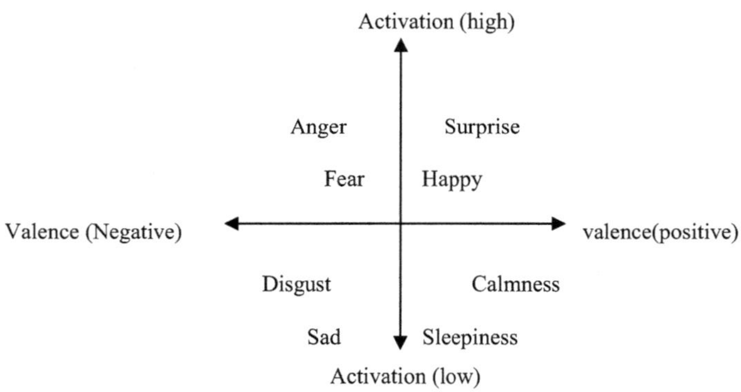

FIGURE 11.2 Correlated human emotions.

primary emotions are present in all human beings and it is not experienced from the environment, while secondary emotions are experienced or learned from the humans due to their interactions with the surrounding environment [3].

According to Gerrod Parrot, universal human emotions are determined through the association of facial expressions and human behaviour. Fundamentally, human emotions are similar in terms of expression and behaviour. However, it becomes primary to model a discrete approach to identify different categories of human emotions, such as anger, fear, disgust, happy, sad and surprise, based on the distinct facial expressions, as represented in Figure 11.1. In particular, basic (universal) emotions have high within-class facial expression similarity and high between-class facial expression dissimilarity. Some authors have correlated human emotions in a deliberate way on the activation and valence axis, which is represented in Figure 11.2.

Activation dimension represents the degree of intensity or energy or strength in the various emotions, while valence dimension represents the amount of positive or negative experiences to various human emotions. Feelings with positive valence that have a high activation of energy represent the happy and surprise emotion while positive valence with low activation of energy represents the calmness and sleepiness emotions. Similarly, feelings with negative valence and high activation of energy represents anger and fear emotions, while negative valence with low activation of energy represents the sad and disgust emotions [4].

11.1.2 Affective Computing: Human–Machine Interaction system

Affective computing is a branch of artificial intelligence that has the capability of recognizing human emotions by extraction, processing and recognition. It is considered as a human–machine interaction system. Initially, the extraction module acts as a sensor to capture data about human behaviour and expression. Next, the sensed data will be processed in the processing module with preprocessing and post-processing steps. Preprocessing eliminates the unwanted sensed data through a transformation mechanism. Figure 11.3 presents the block diagram of the emotion recognition system.

Finally, relevant data obtained in the preprocessing module will be projected to the recognition module. The recognition module is composed of supervised learning approaches to identify the patterns in the extracted data in the form of features and performs the decision to detect the emotion on processing the feature obtained.

11.2 MULTIMODAL EMOTION RECOGNITION

Multimodal emotion recognition is emerging research area in affective computing, which distinguishes the human feeling through the utilization of multimodal signals and computation of human emotion from expression and behaviours. Multimodal signals are the aggregation of audio, video and physiological signals. The audio and video signals describe the expressive behaviours of the humans from speech and facial expression in response to a specific human emotion. The physiological signals describe the body symptoms such as heart rate and body temperature in response to a specific emotion. However, multimodal emotion recognition can be carried out using categories of multimodal signal fusion methods [5].

11.2.1 Multimodal Feature Fusion Methods: Early Fusion Method

Multimodal feature fusion methods combine features of different sensory modalities to form a feature vector. Generally, audio signals features such as mMelfrequency cCepstral cCoefficients, zZero cCrossing rRate, pPitch and Root Mean Square (RMS) have been combined with video signal features such as geometry and texture features to generate joint feature space with optimal features. Feature fusion is carried out using principal component analysis [6], linear discriminant analysis [7] and independent component analysis approaches [8].

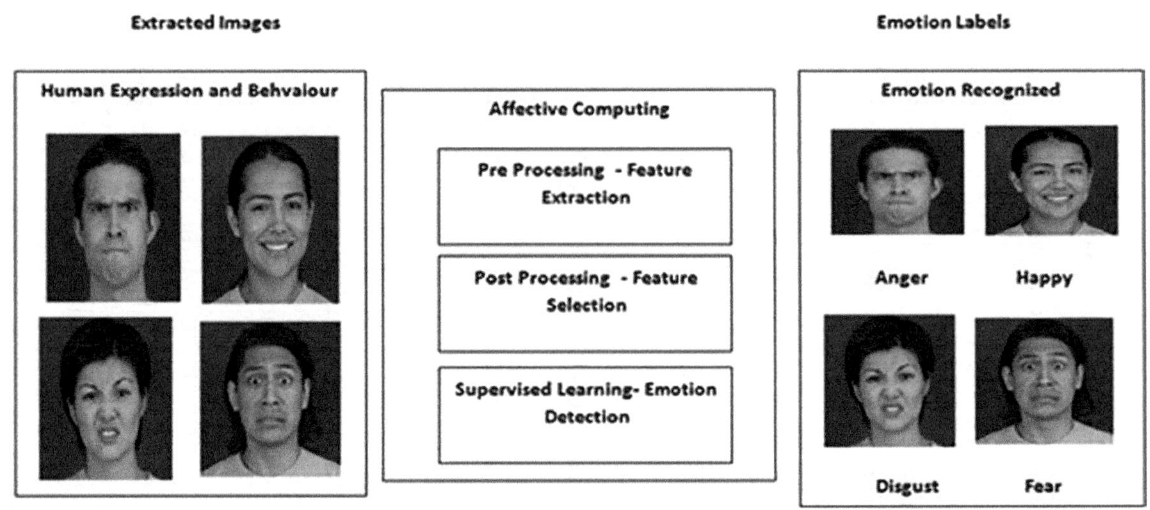

FIGURE 11.3 Block diagram of the human emotion recognition system.

11.2.2 Multimodal Model Fusion Methods: Intermediate Fusion Method

Multimodal model fusion methods have a classifier stage in which the classifier explores the mutual correlation among different modalities for emotion detection. The classification model explores the mutual correlation of speech and facial expression features. It co-analyses multiple modalities to identify the relationship of spoken words with corresponding facial expressions. Model fusion is carried out using a dynamic Bayesian network [9], hidden Markov model [10] and multilayer perceptron [11] to learn the time sequence pattern processing the concatenated audio-video features.

11.2.3 Multimodal Decision Fusion Methods: Late Fusion Method

In the final stage, multimodal decision fusion combines the decisions of multiple classifiers to create the final decision to select the optimal model for emotion recognition. Those classifier models are trained for speech and facial expressions. It fuses the output class probabilities of two independent audio and video emotion classification models. There are different methods to fuse the output class probabilities of classifiers. Decision fusion is carried out using weighted majority voting weighted average [12] and Dempster Shafer approaches [13] to machine learning architectures and gradient boosting approaches [14] to deep learning-based classifier architectures.

11.3 CHALLENGES OF THE MULTIMODAL FUSION APPROACHES FOR EMOTION RECOGNITION

Traditional multimodal fusion approaches for emotion recognition pose several challenges on the basis of the fusion of the speech and face modalities as follows:

- Each person has unique vocal tract and facial expression patterns for their various activities like crying, talking, singing and laughing.
- Facial characteristics differ among the persons in terms of skin colour, shape and size of the facial components like eyes, nose and mouth along with its elements like eyebrows.
- Voice speech characteristics such as speech frequency and wavelength differ on the basis of gender. Generally, voice and face characteristics of the person changes over the time as they get older.

11.4 OBJECTIVES OF MULTIMODAL FUSION APPROACHES FOR EMOTION RECOGNITION

The following objectives are formulated based on the challenges of traditional multimodal fusion approaches:

- Develop an attention-based mechanism to select the optimal feature vector for emotion recognition against over-fitting and under-fitting issues on the features

- Develop a peak stage behaviour for facial expression and speech for emotion recognition to reduce the computational complexity.

11.5 PREPROCESSING

Preprocessing step is included in this chapter to extract the relevant information from the extracted video and audio modalities. Preprocessing of the video and audio contents happens as follows.

11.5.1 Preprocessing of the Speech Signals

The speech signals generally comprise many noises and silences. Hence, it is mandatory to segment the voiced and non-voiced signals from speech signals using voice activity detection. Further speech signal is processed using a normalization approach to adjust and modify the speech signal, which enables an efficient speech recognition task after a reliable feature extraction process. Hence, speech signal preprocessing is composed of the following steps:

- Speech signal boundary extraction
- Normalization

11.5.1.1 Speech Signal Boundary Extraction

Speech signal boundary extraction is used to extract the section or segments of the input signal containing only speech, which is considered voice activity detection. It locates the endpoints of the signal in speech. Further, it divides the signal into a voiced section, an unvoiced section, speech sections and non-speech sections. The detection of the unvoiced sections is difficult compared to voiced sections. Voice activity detection methods are based on feature extraction, which represents the distinguishing properties of speech and noise. Feature category of the speech signal will be discriminated with respect to time domain features and frequency domain features [15].

11.5.1.1.1 Time Domain Features

Time domain features are widely used and they are degradable by background noise. The features of speech signal obtained using discrete Fourier transform are as follows:

- Short-term energy
- Zero crossing rate
- Short-time average magnitude

11.5.1.1.2 Frequency Domain Features

It provides the mechanism to obtain the most significant parameters in speech analysis on employing the discrete Fourier transform [DFT] [16]. The features of speech signal are

- Spectral flatness
- Amplitude modulation spectrogram

- Spectral power
- Spectro temporal modulation

11.5.1.2 Normalization

Normalization of speech signal is a process to transform and balance the signal spectrum of the speech signal into a normalized form on the basis of signal threshold. Normalization is considered as a final step in the speech signal processing to reduce the noise issues, such as signal noise distortion and signal channel distortion. Output of normalization of the signal provides an increased signal-to-noise ratio (SNR) value and eliminates or normalizes the signal spectral variation.

11.5.1.2.1 Cepstral Mean Normalization (CMN)

Cepstral mean normalization is a properly employed approach for eliminating or reducing the difference of the speech signal in the cepstral domain. A particular approach is referred to as the simplest normalization technique, which is used for large vocabulary automatic speech recognition applications. Further, it eliminates the convolutional distortions and effects of speech styles caused by speech characteristics in various environments. However, the cepstral mean normalization approach fails to minimize convolutional distortions during time constant and reaches the length of the speech analysis window.

11.5.1.2.2 Cepstral Variance Normalization (CVN)

Cepstral variance normalization is similar to cepstral mean normalization as it computes the variance of every cepstral dimension and normalizes those dimensions into unity values. Cepstral variance normalization is not correlated with any specified type of signal distortion. However, it fails to consider the variability of the speaker and leads to additive noise in the acoustic channel of signal propagation.

11.5.2 Preprocessing of the Video Signals

Preprocessing of video signals is a significant process in the emotion recognition process as it includes multiple operations such as image frame extraction and key frame extraction.

11.5.2.1 Image Frame extraction

Image frame is extracted using video processing libraries such as opencv. Frame extraction parameters such as the number of frames, intervals and frame rate for frame extraction [17] have to be input to capture each frame. Finally, the captured frame is formatted as a separate image file.

11.6 FEATURE EXTRACTION

Feature extraction is to obtain the features of the signals. In this work, multimodal signal has been processed for emotion recognition. In order to enhance recognition efficiency, features from each model have to be extracted. Initially, pre-processed data of visual data in the form of image frame is processed using texture and geometric-based approaches. Next, pre-processed data of speech data is processed using spectral analysis techniques. This feature is considered as the overall shape of the spectral envelope.

11.6.1 Texture Feature Extraction using Grey-level Cooccurrence matrix

Grey-level cooccurrence matrix [18] is a 2D image analysis approach that provides quantitative data on the image textures through spatial–pixel intensity relationship. It considers the relationship between the neighbouring pixels and provides rich texture information on analysing the fine-grained face patterns in the image frames. It is possible to extract 14 features by haralick to characterize the face. Initially, the image frame is transformed into *n*-dimensional Euclidean space to construct the matrix. Further, the co-occurrence matrix is generalized to obtain the features.
Let *I* be a grey-scale image frame and *N* be a total number of grey levels in the image.

In each image, pixel intensity is considered as grey.
Construct grey-level co-occurrence matrix *G* of order *N*.

Occurrence of $(i,j)^{th}$ on the matrix mentions the number of occurrences of a pixel with intensity *i* with the neighbouring pixel with intensity *j*. In other words, it mentions the number of occurrences a pixel *X* with intensity *i* is differentiated from a pixel *Y* with intensity *j* at a distance *k* in a direction *d*. In this distance, *k* is a non-negative integer and distance *d* is mentioned as *d*= {horizontal, Vertical, left diagonal and Right Diagonal}.

Normalized Co-Occurrence Matrix =

$$\frac{\text{Partitioning each element of } G}{\text{Total Number of Co}-\text{Occurence pairs in } G} \tag{11.1}$$

The mathematical model to represent adjacency of the pixel intensity to *N*- dimensional space is represented as

$$Z^n = z^*z^* \ldots\ldots ^*z$$

where Z^n is the Cartesian product of *Z* and *Z* is considered as a set of all integers.

An illustration of the cooccurrence matrix to the image in four dimensions as 4*4 with intensity values 0, 1, 2, 3 Pixel *X* in Z^4 is an 4-tuple of the form $X = (x_{1,} x_{2,} x_3, x_4)$ is represented in Figure 11.4.

Independent direction from *X* in z^3 is

$$\frac{3^3 - 1}{2} \tag{11.2}$$

where x_i ε *Z* and *k* is the distance between *X* and *Y*, and *C* is the count to represent the number of possibilities to form the direction *d*.

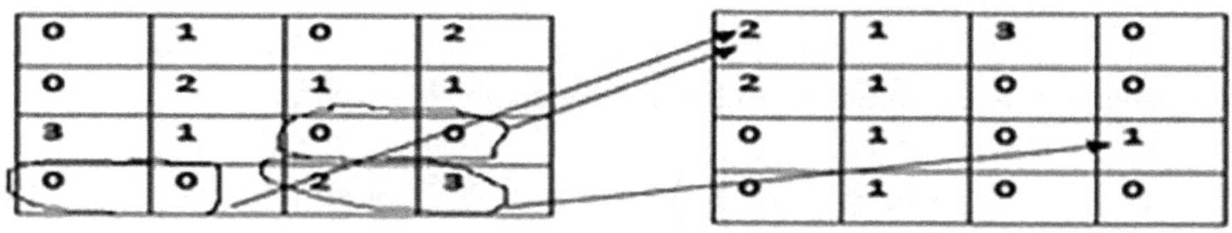

FIGURE 11.4 GLCM matrix computation,

Image *I* is a function from a subset of Z^n to *Z*:

$$\text{If } (X \varepsilon 1)$$

X is assigned to *Y* such that *Y*=*f*(*X*)

where *Y* is the pixel intensity of the image *X*.

Image is a grey scale in form and it is represented in dimensional space $Z^{n.}$

Volumetric data is represented as three-dimensional images. Gray-Level Co-Occurrence Matrix (GLCM) captures the spatial arrangement of texture using three parameters such as distance, angle and number of grey levels for co-occurrence of a specific pixel. On computation of the GLSM, statistical measures that represent the features can be obtained. Those measures characterize the texture and structure of the image. Those texture features are as follows.

11.6.1.1 Contrast

It is considered as grey-level difference in the GLCM matrix on computation of pixel intensity among the neighbours. High contrast value represents large difference among neighbouring pixels. Contrast computation is as follows:

$$\text{Contrast}(d,\theta) = \sum_{i=0}^{N_g-1} X \sum_{j=0}^{N_G-1} Y(i-j)^2 \tag{11.3}$$

11.6.1.2 Energy

It is considered as homogeneity of the image. It value ranges from 0 to 1. A high energy value represents more similar textures. Energy computation is as follows:

$$\text{Energy}(d,\theta) = \sum_{i=0}^{N_g-1} X \sum_{j=0}^{N_G-1} Y * P(i,j)^2 \tag{11.4}$$

11.6.1.3 Homogeneity

It is considered as non-zero elements in the computation and it is inverse of the contrast. Its value ranges from 0 to 1. Homogeneity computation is as follows:

$$\text{Homogeneity} = \sum_{i=0}^{n-1}\sum_{j=0}^{n-1} \frac{1}{1+(i-j)^2} \text{P(i,j)} \tag{11.5}$$

11.6.1.4 Correlation

It is considered as the quantity of energy on measuring the linear dependency between the pixel pairs of the image. Correlation computation is as follows:

$$\text{Correlation} = \sum_{i=0}^{n-1}\sum_{j=0}^{n-1} \frac{1}{1+(i-j)^2} \text{ P(i,j)log(p(i,j)} \tag{11.6}$$

These features help to detect and track the facial landmarks. On feature processing, it is possible to establish landmark facial points.

11.6.1.5 Geometric features

Geometric features are features of the face components with set of the geometric elements such as point, lines, curves and surfaces. Geometric features are extracted using relative positions and shape or sizes of the components with variation such as eyes, nose, mouth and other components of the face. It computes area, angle and Euclidean distance on processing the texture features in each frame. Especially, multiple geometric features are determined to describe facial changes in eyebrow, eyes, nose, mouth and in the other face components [19]. Figure 11.5 represents the architecture of the proposed emotion classification model.

11.6.2 Speech Feature Extraction

The feature extraction technique is employed to extract the multiple types of the spectral speech features in each frame as follows.

11.6.2.1 Pitch estimation using autocorrelation

Pitch is a crucial component in speech signal. Pitch is obtained due to vocal tract and the frequency at which the vocal tracts vibrate is the fundamental frequency. It is a vital feature of the speech signal. Voiced speech segments are recognized through a "pitch step" in the time domain and a "fundamental frequency" in the frequency domain [20]. Autocorrelation function for discrete time signal is mentioned as

$$R_x(m) = \lim n \to \infty (\frac{1}{2N+1}) \sum_{n=-N}^{N} x(n)x(n+m) \tag{11.7}$$

In addition, segmentation of voiced speech against the non-voiced speech is presented through individual pitch epochs. It is significant for multiple high-quality speech synthesis and coding approaches. A frequently employed approach to compute pitch is grounded on identifying the largest value of the autocorrelation function on its ROI. The autocorrelation function of a speech signal is a conversion of the speech signal as it is significant for representing its structure in the waveform [21]. Thus, it is rational to mention a small-time autocorrelation function, which operates on small segments of the speech signal is given as

$$R_x(m) = \frac{1}{N}\sum_{n=0}^{N} \left[x(v+1)w(v)\right] \left[x(v+I+u)w(u+v)\right] \tag{11.8}$$

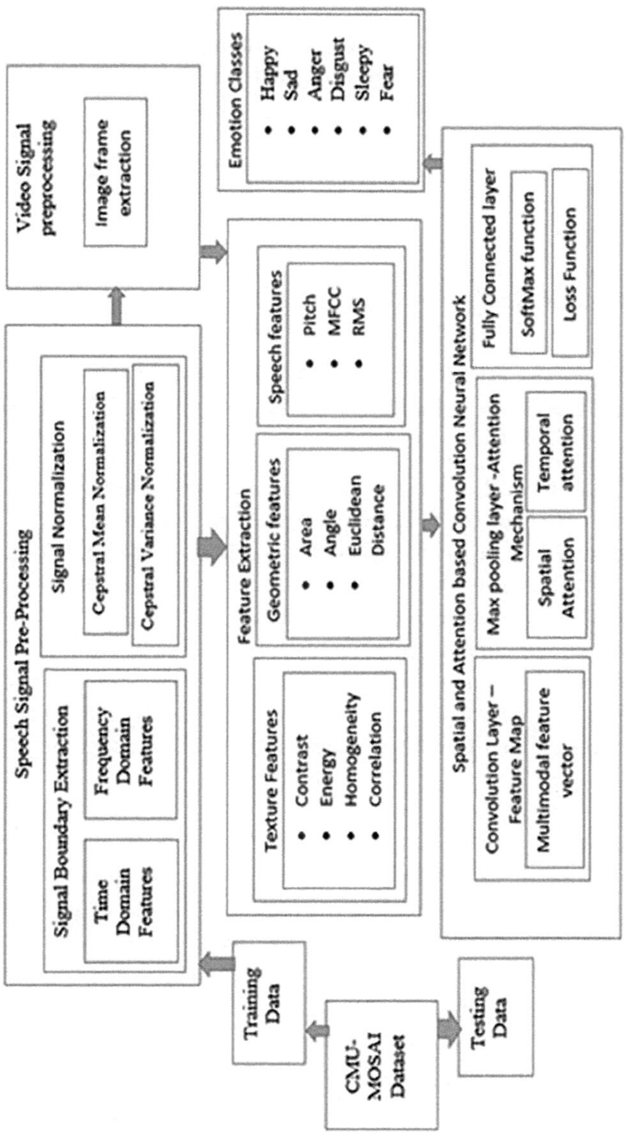

FIGURE 11.5 Architecture diagram.

where $w(v)$ is the window for analysis and N is the selection length which is to be analysed. N^1 is the is the number of speech signal samples gathered in the calculation of $R(u)$ and m_0 is the is the number of autocorrelation points, which require to be computed.

11.6.2.2 Mel Frequency Cepstral Coefficients (MFCC)

MFCC is another feature extraction technique employed to speech signal based on frequency domain to identify the Mel scale. It is employed to extract the spectral features. It is estimated as the frequency domain features. MFCC [22] is computed using the following expression:

$$m = 2595 \log_{10}(\frac{f}{100} + 1) \tag{11.9}$$

where f is the frequency and log is logarithm factor.

11.7 SPATIAL AND TEMPORAL ATTENTION-BASED CONVOLUTION NEURAL NETWORK: MULTIMODAL FEATURE FUSION

Convolution neural network is employed to obtain the multimodal features fusion from the features extracted from the audio signals and video signals. The extracted features are projected to a layer of the convolution neural network for feature fusion and emotion recognition. Attention network, which contains the pretrained ResNet121 convolution neural network architecture [23], processes the feature with respect to spatial and temporal aspects. The extracted feature of the speech and image feature space are processed in the form of a vector in the attention network. Attention network processes the feature vector with spatial attention coefficient and temporal attention coefficient to identify optimal feature for emotion recognition.

- **Convolution layer**

Convolution layers have been incorporated with multiple filters of size 7*7 along dilation factor of size 2*2 to learn the feature vector from the multimodal feature space under multiple spatial and temporal constraints of the architecture with bias as false. A feature map is generated for the multimodal feature vectors. This feature map represents low-level features. Convolution operation with kernel function is represented as

$$F_m \text{ (Multimodal features)} = \text{Conv1D } (F_{S,F}, K_{S,F}) \ \varepsilon \ R^d \tag{11.10}$$

where $K_{S,F}$ is the kernel size of the convolution operation for speech and image modalities and d is the dimension of the low-level multimodal features of speech and image.

Convolution maps of the low-level multimodal features of the speech and image modalities to the dimension d is carried out using dot product in order to enable the positional embedding with temporal information and spatial information to the feature map.

$$H_m \text{ (low level multimodal features)} = F_m \text{ (low-level multimodal features)} + \mathrm{PE}(L_{S,F}, d) \quad (11.11)$$

where $\mathrm{PE}(L_{S,F}, d)$ is the position embedding to each position of the video frame.

- **Attention Mechanism**

Attention mechanism is used to select the weighted feature on refining the feature map to generate the attention map. Weighted feature is computed on the basis of the spatial attention coefficient and temporal attention coefficient to the low-level multimodal features with position embeddings. Attention to the multimodal feature is computed as follows.

Spatial Attention for Facial Features

Attention coefficient is used to calculate the weight according to positional changes of the facial component with reference to the neutral facial components. Weight computation of the spatial attention mechanism to the facial features is calculated with reference to a large variation between the facial component from neutral image. It is represented as follows:

$$\mathrm{WF_F} = \frac{1}{n_k} \sum_{i:x} (w)\ \varepsilon\ \mathrm{Fm} \text{ (low-level multimodal features)} \ldots \quad (11.12)$$

where w represents the variation of the facial component of emotion face and facial component of the neutral face in the feature map. In this computation, high weight represents the facial movement changes from neutral face and low weight represents the no changes or neutral face. Table 11.1 mentions the hyperparameter of ResNet121 architecture.

Temporal Attention for Speech Feature

Attention coefficient is used to calculate the weight according to positional changes of the speech component with reference to the non-voice components in specified

TABLE 11.1
Hyper Parameter of ResNet121Architecture

Hyperparameter	Value
No of epochs	125
Model loss function	Cross Entropy
Model activation function	Sigmoid
Batch size	35
Model learning rate	10^{-6}

time intervals. Weight computation of the speech features is done with reference to pitch changes is as follows:

$$WF_S = \frac{1}{n_k} \sum_{i:x} (i) \varepsilon \, F_m \text{ (low-level multimodal features).} \tag{11.13}$$

where w mentions the variation of the pitch of the emotion speech with pitch of the non-voiced speech in the feature map. In this computation, high weight represents the pitch changes from non-voiced speech beyond the specified threshold and low weight represents the lower changes in pitch in the non-voiced speech.

- **Max pooling layer**

Features with attention weight are processed in the pooling layer to compute the high-level feature of the multimodal feature vector for emotion classification. In this computation, weighted features from the facial and speech feature spaces on the spatial and temporal attention computation are selected. These feature vectors are considered as high-level feature vector. Further, it avoids the vanishing-gradient challenges and increase speed of the multimodal feature propagation into softmax classifier with minimized number of hyperparameters along batch normalization and loss function to handle the over-fitting and under-fitting issues of high-level features. Module is defined in the equation is as follows:

$$HF_m(F_{s,f}) = \frac{1}{F * s} (WF_s + WF_f) \tag{11.14}$$

where F is the feature of the image and S is the feature of the speech.

- **Fully connected layer**

It processes the features from the max pooling in the dense blocks to carry out the fusion task of the features of the speech and facial features in form of a feature vector. It represents the multimodal aggregated feature of the architecture. The architecture processes the multimodal feature from max pooling layer in the dense blocks. Aggregated features are processed to detect the emotion through softmax classifier. The dense block containing reduced features of the multimodal signal is used to obtain the feature map as follows:

$$\text{Feature Map of the Dense Block} = \sigma \left(F_m \sum_{i=0}^{n} I\right) (WF_s + WF_f) \tag{11.15}$$

The feature map extracted is applied with the sigmoid activation function along batch normalization and loss function to obtain the desired emotion classes in the softmax function.

- **Activation layer**

The architecture employs the sigmoid activation function as it improves performance of the training stage and minimizes the errors on removing the non-linearity among the multimodal feature vector. It is activation function represented in an equation:

$$\text{Activation function to the Feature map} = \text{Sigmoid}\left(\left(\sum_{i=0}^{n} \sigma\, w\right)\right) \tag{11.16}$$

where σ represents an activation function and w denotes the weight matrix.

- **Batch Normalization Layer**

Batch normalization is done to provide high learning rate to the training data to provide better classification output. Batch normalization is represented as the transformation of the non-linear multimodal feature into linear multimodal feature on computation of the mean and standard deviation among it. Batch normalization is learned with respect to the epoch and other learning parameters such as weight. Further, it is capable of minimizing the loss of the architecture and it is denoted in the equation as follows:

$$Z_{n=\left(\frac{z-m_z}{s_z}\right)} \tag{11.17}$$

where m_z is the mean of the multimodal features and s_z is the standard deviation of the multimodal features.

- **Loss function**

Loss function employs the cross-entropy function to avoid the over-fitting and underfitting issues of the multimodal feature on the emotion classes. The cross-entropy function [24] to emotion classes is represented as

$$C\,(S, F) = \sum_{X \in \text{Classes}} S(x) \log f(x) \tag{11.18}$$

where $S(x)$ is model true probability distribution of the multimodal features and $f(x)$ is the predicted probability distribution of the multimodal features.

The output of the fully connected layer is mentioned as the classifier output. Further, loss function is incorporated in a fully connected layer to reduce the multimodal feature variance of the emotion. The closest approximation of the testing data from multiple emotion classes is that the minimal residual may be obtained from emotion classes. Softmax function analyses the feature map to compute the emotion class using XGboost classifier [25]. Algorithm 1 represents the multimodal emotion recognition architecture of the emotion classification from the multimodal signals.

- **Flatten layer**

It is a separate layer which is fixed for feature flattening as its aligns the high-level multimodal features on the concatenated vector. It is to avoid the non-linearity and over-fitting challenges of the multimodal features of the emotion classes. Fusion features are projected to the classification function to produce the emotion class.

- **Softmax function**

Softmax function of a fully connected network simultaneously fuses the multimodal features and classifies it with respect to the classifier. Classification reduces the space and time complexity. The classifier processes the concatenated feature into various classes. XGBoost Classifier [25] is incorporated in this chapter. Softmax function is used to generate the different emotion classes. The classification output is generated by integrating the outcomes of the emotion classes based on the voting rule:

$$\text{Softmax(x)j} = \frac{e^{xi}}{\sum_{n-1}^{N} e^{x_n}} \tag{11.19}$$

Derivates of the objective function with respect to the emotion coefficients have been extracted on the computation of error sum of square (SSE). Softmax function uses bagging and aggregation process to compute high-level multimodal fusion features and classifies it into emotion classes. Further, it computes the multiple linear weights of spatial and temporal multimodal features of the fusion vector. In addition, feature weight of the speech and image features is calculated through multiple iterations.

$$\Delta W_i = C(t - \text{net})\, x_i \tag{11.20}$$

where c is the learning rate and x is the feature weight.

On the objective of reducing the SSE and managing loss of the classification outcomes of the multimodal features, gradient boosting will be employed. Algorithm 1 presents the emotion recognition model in the precise recognition of emotion from the video data.

Algorithm 1: Spatial and Temporal Attention Based Convolution Neural Network

Input: Multimodal Video and Speech data
Output: Emotion Class Label C= {Sad, Happy, Disgust, Anger, Sleepiness, Calmness}
Process
  Multimodal Signal Pre-process ()
    Video Pre-Processing ()
      Image Frame Extraction ()
    Speech Preprocessing ()
      Normalization ()
    Feature Extraction _ Image frame ()
      Grey level Cooccurence Matrix (image frames of video)
        Texture Feature = {Contrast, Energy, Homogeneity, Correlation)
      Geometric features = {Facial Components}

```
        Feature Extraction _ Speech frame ()
            Spectral features = {pitch, MFCC}
ResNet121 based spatial and temporal Attention Network ()
    Convolution layer ()
        Set No of  layer =25 and Filter Size= 7*7
            Feature Map Fm  = Low-level Feature (Speech, Image)
                Position Embedding of Fm = H (Low-level Feature (Speech
                ,Image)
Attention Mechanism()
    Set Spatial Attention Coefficient and Temporal Attention Coefficient to
    Feature Map
    Spatial Attention ()
        Compute Weight_Image Features()
                If ( Variation between facial component of the Neutral face and
            Emotion Face is high)
                Set Weight of the face feature = High
            Else
                Set weight of the Face Feature = Low
    Temporal Attention ()
        Compute Weight_Speech Features()
            If ( Variation between Pitch of voice segments and non voice
            segments is high)
                Set Weight of the  pitch feature = High
            Else
                Set weight of the pitch Feature = Low
Max pooling layer()
    Eliminate the features with low weight on the multimodal signals
        Fm ( High level feature) = Weighted Feature of the Speech + Weighted
    Features of the Image
Fully connected layer ()
    Dense Block()
    Feature Fusion Map Fm = Fusion of the Weighted multimodal features
        Activation Function_Sigmoid()
        Batch Normalization ()
            Normalizes with specified emotion features ( Happy, Sad, Anger,
            Disgust , Sleepiness, Fear}
        Flatten Layer ()
                It Transform the 2D feature into 1 dimensional feature vector
            Loss Layer _ Cross Entropy ()
        It eliminates the over fitting and under fitting issues
Softmax()_XgBoost Classifier
    Voting Rule ()
        Classes= Emotions (Happy, Sad, Anger, Disgust , Sleepiness, Fear)
```

11.8 EXPERIMENTAL ANALYSIS

The experimental analysis of the multimodal emotion recognition architecture is evaluated using the CMU-MOSEI dataset [26] in the online Python environment referred Google Colab. Google Colab contains multiple Tensorflow libraries and package to execute the speech signal using deep learning and machine learning architecture on various time and frequency domains

11.8.1 Dataset Description: CMU-Multimodal Opinion Sentiment and Emotion Intensity (CMU-MOSEI) dataset

The dataset consisting of the video clips provides the multilingual vocal and facial expression of the subjects from different cultural backgrounds, that is, Africa, America, Germany, Australia and China. The dataset contains more than 23,500 sentences uttered video from YouTube. Dataset is gender balanced. All the sentence utterances are randomly selected from various topics and it is transcribed and properly punctuated. In this dataset, sentiment annotation and emotion annotation are embedded. Dataset description with dimension and values is represented in Table 11.2.

11.8.2 Performance Metric

The performance of the present multimodal emotion recognition architecture framework is carried out using validation data of the CMU_MOSEI dataset. Validation data is handled using confusion matrix on cross-fold validation. Model validation is done to increase the performance of the emotion recognition with good recognition

TABLE 11.2
Dataset Description

Dimension Name	Dimension Value
Total speech sentences	23,453
Number of the opinion-based speech sentences	18,148
Number of the objective-based speech sentences	5305
Total number of speech video	3228
Total number of distinct speakers	1000
Total number of distinct topics	250
Average number of sentences in a video	7.3
Average length of the sentences	7.28 Seconds
Total number of words in sentences	447143
Number of unique words in sentences	23026
Modalities	Languages, Visuals, Acoustics

accuracy. The performance of the architecture is computed through measures such as accuracy, specificity, sensitivity, peak signal to noise ratio, mean square error and structural similarity index.

11.8.3.1 Sensitivity

It is measure of the multimodal feature of computed classes without exceeding the feature of the high-level features. It calculated using true positive, true negative, false-positive and false-negative parameters of the confusion matrix [27]. Sensitivity is represented as follows:

$$\text{Sensitivity} = \frac{\text{TP}}{TP + FN} \tag{11.21}$$

11.8.3.2 Specificity

It is measure of the relevant features correctly classified into emotion classes among the extracted features. In other words, it is the ratio of the correctly classified features to the number of features extracted. Sensitivity is represented as follows:

$$\text{Specificity} = \frac{\text{TP}}{TP + FN}.. \tag{11.22}$$

11.8.3.3 Peak Signal to Noise ratio

It is measured as a ratio among maximum possible value of multimodal features to the power of arbitrary noise of the multimodal features, which affects the quality class representation of emotions of the multimodal features of different signals [28]:

$$\text{PSNR}= 10.\log_{10}(\text{MAX}^2\ /\text{MSE}) \tag{11.23}$$

MAX is the maximum possible intensity of the frequency at specific intervals.

11.8.3.4 Structural Similarity Index

It is measure of perceptual quantity of the features of the signal. In other words, it is a measure of two windows x and y of the correlation matrix of similar size [29]:

$$\text{SSIM (x,y)} = (2\ \mu_x \mu_y + \text{C1}) \tag{11.24}$$

where μ_x is the spectral feature mean of x and μ_y is the spectral feature mean of y and C is constant.

11.8.3.5 Mean Square Error

It measures the square difference in the spectral features values between the corresponding spectral features of the original speech with noise and final reconstructed spectral features of the speech signal [30]:

$$\mathrm{MSE} = \frac{\sum N * M \left[O(m,n) - r(m,n)\right] 2}{(M * N)} \tag{11.25}$$

where M mentions the row and N mentions the column of the multimodal feature matrix and $O(m,n)$ represents the multimodal feature matrix. In particular, $r(m,n)$ denotes the reconstructed signal.

On utilization of the above measures, classified multimodal features against universal emotion class using multimodal emotion recognition architecture through deep learning models are estimated on partitioning the benchmark dataset from CMU-MOSEI dataset into training and testing data with specified window frames.

11.8.4 Performance Analysis

The performance analysis of the multimodal emotion recognition architecture is done using training data and testing data of the dataset. The performance analysis of the emotion recognition architecture on universal emotion classes is calculated using the above-mentioned metrics and illustrated in Figure 11.7. Spatial and temporal attention-based convolution neural network classifier used for emotion recognition is computed using the sensitivity, specificity and accuracy measures along PSNR, MSE and SSIM. Table 11.3 represents the performance computations of the emotion recognition approaches through multiple performance measures.

On evaluation, it is confirmed that attention-based convolution neural network produces 94.1 percent on sensitivity measure, 93.8 percent on specificity measure, 94.8 percent on accuracy measures, 21.89 db on PSNR 0.710 on SSIM and 2.2 on MSE respectively. Figure 11.6 represents the training loss and training accuracy of the emotion recognition architecture against increasing value of the epoch.

The performance of the convolution neural network produces 90.1 percent on sensitivity measure, 89.8 percent on specificity measure, 90.8 percent on accuracy measures, 26.89 db on PSNR, 0.768 on SSIM and 2.79 on measure square error, respectively. Multimodal emotion recognition architecture outcomes in terms of universal emotion classes through cross-fold validation of the CMU-MOSEI dataset were computed using accuracy measures as represented in Figure 11.7.

Finally, recurrent neural network produces 91.9 percent on sensitivity measure, 90.9 percent on specificity measure, 89.6 percent on accuracy measures, 27.29db on PSNR, 0.781 on SSIM and 2.69 on MSE, respectively. Multimodal emotion recognition architecture outcomes in terms of universal emotion classes through cross-fold validation of the CMU-MOSEI dataset were computed using specificity measure as illustrated in Figure 11.8.

TABLE 11.3
Performance Evaluation of Emotion Recognition Approaches

Emotion Classes	Technique	Accuracy	Sensitivity	Specificity	PSNR	SSIM	MSE
Happy	Spatial and temporal attention-based convolution neural network	93.92	93.48	91.25	21.89	0.708	2.18
	Convolutional neural network	89.08	89.82	88.78	24.69	0.791	2.67
	Recurrent neural network	88.12	88.65	87.42	25.71	0.801	2.72
Sad	Spatial and temporal attention-based convolution neural network	93.19	93.87	91.49	21.92	0.701	2.22
	Convolutional neural network	86.51	86.89	85.12	24.71	0.781	2.77
	Recurrent neural network	87.49	87.83	86.12	24.89	0.7910	2.81
Anger	Spatial and temporal attention-based convolution neural network	94.56	94.12	93.25	21.87	0.718	2.23
	Convolutional neural network	90.10	90.19	89.75	25.89	0.792	2.78
	Recurrent neural network	89.14	89.78	88.21	26.02	0.802	2.71
Fear	Spatial and temporal attention-based convolution neural network	93.9	94.21	93.33	21.96	0.722	2.12
	Convolutional neural network	90.92	91.11	90.63	26.11	0.782	2.69
	Recurrent neural network	89.47	89.98	88.41	27.14	0.774	2.71
Disgust	Spatial and temporal attention-based convolution neural network	93.25	93.45	91.89	22.14	0.732	2.23
	Convolutional neural network	89.61	89.88	88.32	27.15	0.792	2.68
	Recurrent neural network	89.15	89.71	88.52	28.78	0.784	2.72
Sleepiness	Spatial and temporal attention- based convolution neural network	93.69	94.06	93.26	21.89	0.741	2.21
	Convolutional neural network	90.45	89.95	88.96	26.78	0.786	2.78
	Recurrent neural network	90.71	90.89	89.74	27.10	0.798	2.79

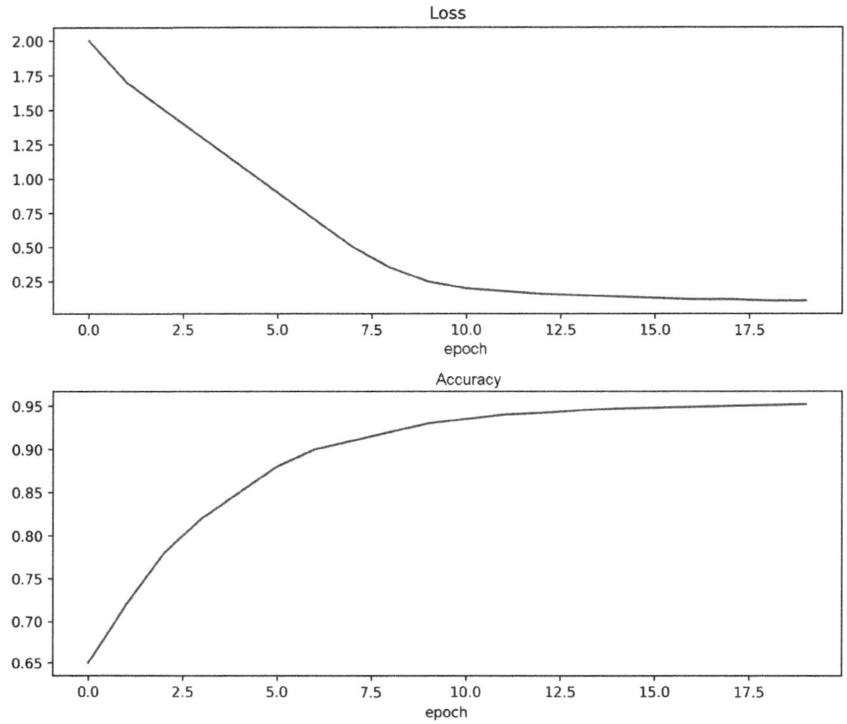

FIGURE 11.6 Training accuracy and training loss on emotion recognition approach.

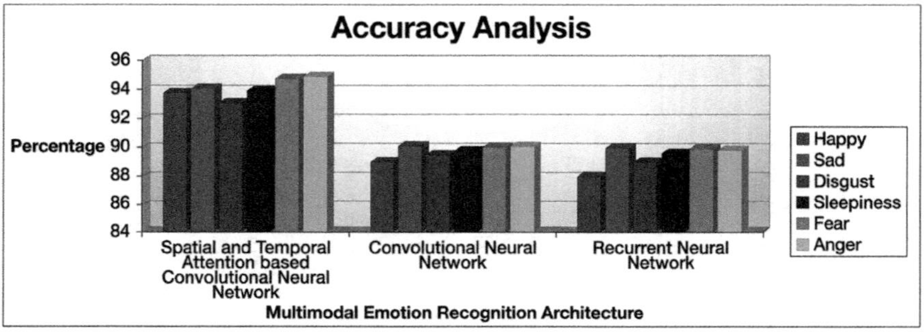

FIGURE 11.7 Accuracy analysis of multimodal emotion recognition architecture,

Multimodal emotion recognition architecture outcomes in terms of universal emotion classes through cross-fold validation of the CMU-MOSEI dataset were computed using specificity measure as illustrated in the Figure 11.9.

Multimodal emotion recognition architecture outcomes in terms of universal emotion classes through cross-fold validation of the CMU-MOSEI dataset were computed using PSNR measure as illustrated in Figure 11.10.

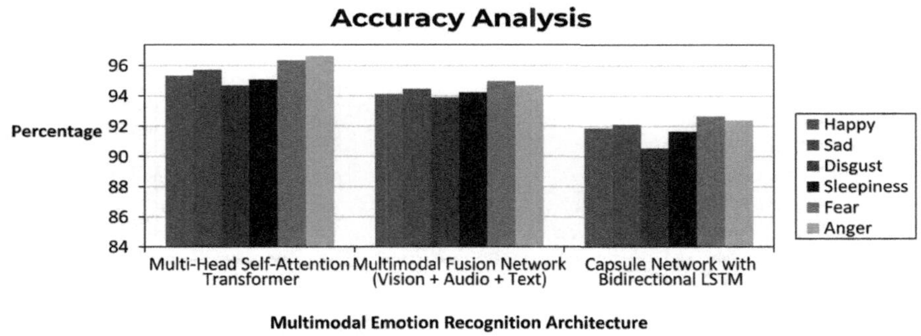

FIGURE 11.8 Sensitivity analysis of multimodal emotion recognition architecture.

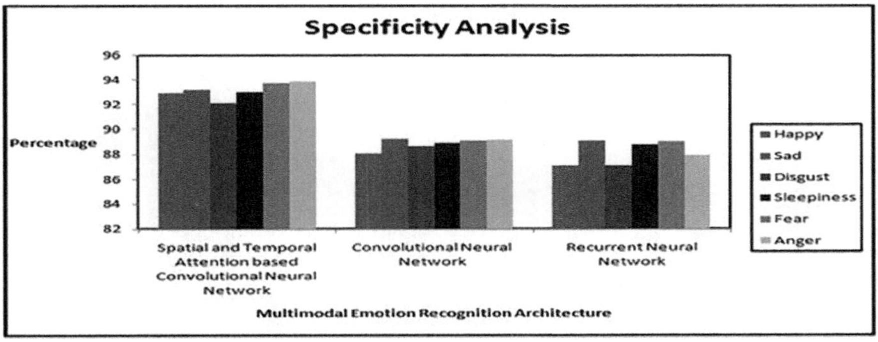

FIGURE 11.9 Specificity analysis of multimodal emotion recognition architecture.

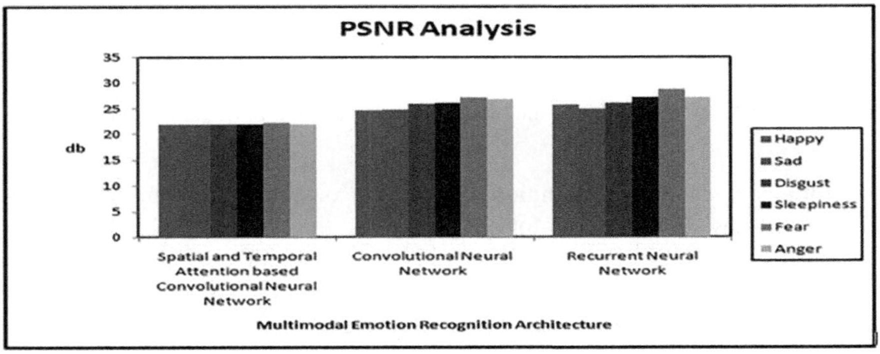

FIGURE 11.10 PSNR analysis of multimodal emotion recognition architecture.

Multimodal emotion recognition architecture outcomes in terms of universal emotion classes through cross-fold validation of the CMU-MOSEI dataset were computed using SSIM measure as illustrated in Figure 11.11.

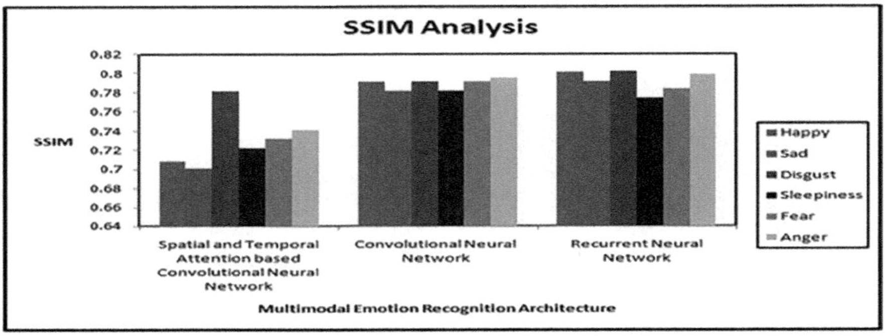

FIGURE 11.11 SSIM analysis of multimodal emotion recognition architecture.

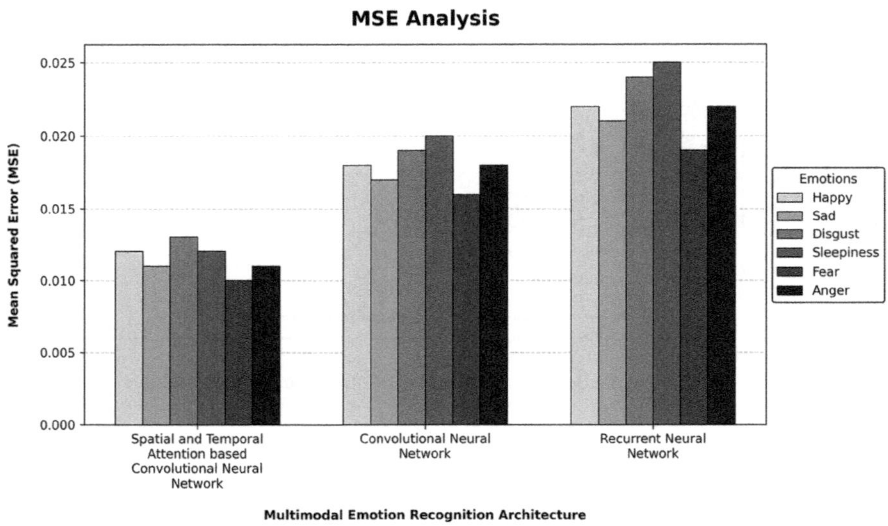

FIGURE 11.12 MSE analysis of multimodal emotion recognition architecture.

Multimodal emotion recognition architecture outcomes in terms of universal emotion classes through cross-fold validation of the CMU-MOSEI dataset were computed using MSE measure as illustrated in Figure 11.12.

Finally, confusion matrix of the cross-fold validation data composed of multimodal signals to different emotion classes is represented in Figure 11.13.

11.9 SUMMARY

The spatial and temporal attention-based convolution neural network has been modelled and implemented to determine the emotion classes according to different weighted features of the multimodal signals effectively. The features extracted from texture analysis and spectral analysis to the image and speech data are processed as a tensor in the layer of the attention-based convolutional neural network. Attention

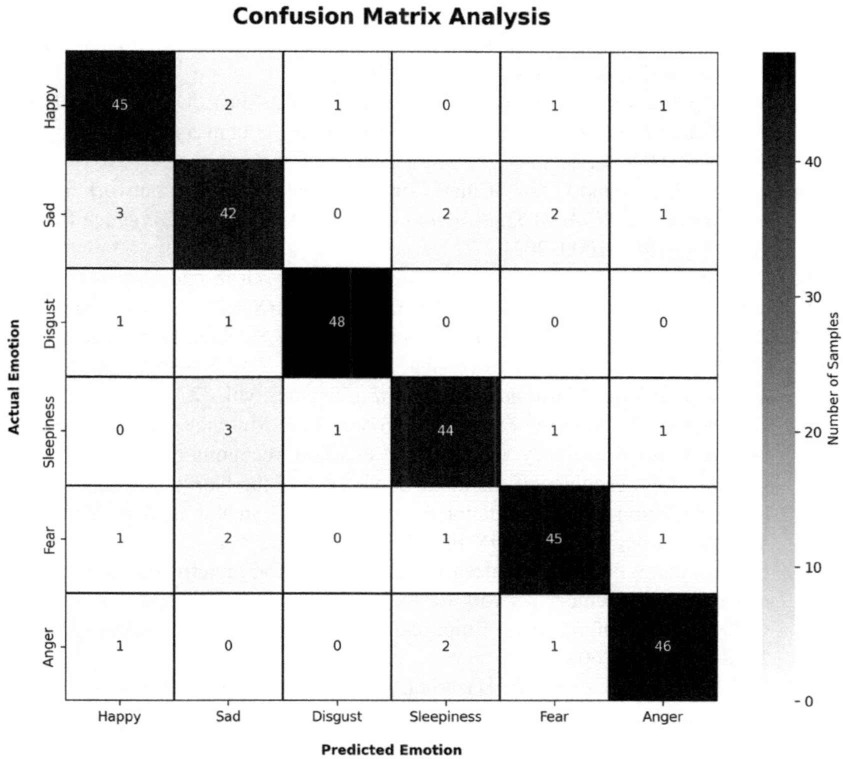

FIGURE 11.13 Confusion matrix.

mechanism uses attention coefficient to process the video and audio features in the form of tensor to obtain the weighted feature, which represents the emotions. Finally, the weighted features from attention mechanism in the max pooling layer is projected to fully connected layer to simultaneously carry out multimodal feature fusion and recognition using softmax function. Softmax function use XGboost classifier to classify the fusion feature into the emotional classes. Experimental and performance analysis of the emotion recognition architectures has been performed using CMU-Multimodal Opinion Sentiment and Emotion Intensity (CMU-MOSEI) dataset and its outcomes are measured using multiple metrics to evaluate its performance.

REFERENCES

[1] J. Deng, Z. Zhang, F. Eyben, and B. Schuller, "Autoencoder-based unsupervised domain adaptation for speech emotion recognition," *IEEE Signal Processing Letters*, vol. 21, no. 9, pp. 1068–1072, 2014.

[2] W. Zheng, M. Xin, X. Wang, and B. Wang, "A novel speech emotion recognition method via incomplete sparse least square regression," *IEEE Signal Processing Letters*, vol. 21, no. 5, pp. 569–572, 2014.

[3] M. Chen, X. He, J. Yang, and H. Zhang, "3-d convolutional recurrent neural networks with attention model for speech emotion recognition," *IEEE Signal Processing Letters*, vol. 25, no. 10, pp. 1440–1444, 2018.
[4] Y. Mao, Q. Sun, G. Liu, X. Wang, W. Gao, X. Li, and J. Shen, "Dialoguetrm: Exploring the intra-and inter-modal emotional behaviors in the conversation," arXiv preprint arXiv:2010.07637, 2020.
[5] Z. Lian, B. Liu, and J. Tao, "Ctnet: Conversational transformer network for emotion recognition," *IEEE/ACM Transactions on Audio, Speech, and Language Processing*, vol. 29, pp. 985–1000, 2021.
[6] W. Shen, J. Chen, X. Quan, and Z. Xie, "Dialogxl: All-in-one xlnet for multi-party conversation emotion recognition," arXiv preprint arXiv:2012.08695, 2020.
[7] Z. Yang, Z. Dai, Y. Yang, J. Carbonell, R. R. Salakhutdinov, and Q. V. Le, "Xlnet: Generalized autoregressive pretraining for language understanding," *Advances in Neural Information Processing Systems*, vol. 32, pp. 5753–5763, 2019.
[8] D. Hazarika, S. Poria, A. Zadeh, E. Cambria, L.-P. Morency, and R. Zimmermann, "Conversational memory network for emotion recognition in dyadic dialogue videos," in Proceedings of the 2018 Conference of the North American Chapter of the Association for Computational Linguistics: Human Language Technologies, Volume 1 (Long Papers), 2018, pp. 2122–2132.
[9] D. Hazarika, S. Poria, R. Mihalcea, E. Cambria, and R. Zimmermann, "Icon: interactive conversational memory network for multimodal emotion detection," in Proceedings of the 2018 Conference on Empirical Methods in Natural Language Processing, 2018, pp. 2594–2604.
[10] N. Majumder, S. Poria, D. Hazarika, R. Mihalcea, A. Gelbukh, and E. Cambria, "Dialoguernn: An attentive rnn for emotion detection in conversations," in Proceedings of the AAAI Conference on Artificial Intelligence, vol. 33, 2019, pp. 6818–6825.
[11] D. Zhang, L. Wu, C. Sun, S. Li, Q. Zhu, and G. Zhou, "Modeling both context-and speaker-sensitive dependence for emotion detection in multi-speaker conversations." *IJCAI*, 2019, pp. 5415–5421.
[12] X. Lu, L. Zhu, Z. Cheng, L. Nie, and H. Zhang, "Online multi-modal hashing with dynamic query-adaption," in SIGIR. ACM, 2019, pp. 715–724.
[13] Y. Xu, Y. Xu, T. Lv, L. Cui, F. Wei, G. Wang, Y. Lu, D. Florencio, C. Zhang, W. Che et al., "Layoutlmv2: Multi-modal pre-training for visually-rich document understanding," arXiv e-prints, pp. arXiv–2012, 2020.
[14] S. Pramanik, P. Agrawal, and A. Hussain, "Omninet: A unified architecture for multi-modal multi-task learning," arXiv e-prints, pp. arXiv–1907, 2019.
[15] K. Santhakumar, B. R. Kiran, T. Gauthier, S. Yogamani et al., "Exploring 2d data augmentation for 3d monocular object detection," arXiv preprint arXiv:2104.10786, 2021.
[16] G. Chen and X. Zeng, "Multi-modal emotion recognition by fusing correlation features of speech-visual," *IEEE Signal Processing Letters*, pp. 1–1, 2021.
[17] C. Zheng, L. Zhu, S. Zhang, and H. Zhang, "Efficient parameter-free adaptive multi-modal hashing," *IEEE Signal Processing Letters*, vol. 27, pp. 1270–1274, 2020.
[18] S. Mai, H. Hu, and S. Xing, "Modality to modality translation: An adversarial representation learning and graph fusion network for multimodal fusion," in Proceedings of the AAAI Conference on Artificial Intelligence, vol. 34, no. 01, 2020, pp. 164–172.
[19] S. Mohla, S. Pande, B. Banerjee, and S. Chaudhuri, "Fusatnet: Dual attention based spectrospatial multimodal fusion network for hyperspectral and lidar classification," in Proceedings of the IEEE/CVF Conference on Computer Vision and Pattern Recognition Workshops, 2020, pp. 92–93.

[20] Y. Wang, W. Huang, F. Sun, T. Xu, Y. Rong, and J. Huang, "Deep multimodal fusion by channel exchanging," *Advances in Neural Information Processing Systems*, vol. 33, 2020.

[21] Z. Huang, X. Wang, L. Huang, C. Huang, Y. Wei, and W. Liu, "Ccnet: Criss-cross attention for semantic segmentation," in Proceedings of the IEEE/CVF International Conference on Computer Vision, 2019, pp. 603–612.

[22] X. Shu, L. Zhang, G.-J. Qi, W. Liu, and J. Tang, "Spatiotemporal coattention recurrent neural networks for human-skeleton motion prediction," *IEEE Transactions on Pattern Analysis and Machine Intelligence,* 2021, vol. 22, pp. 90–102.

[23] K.-H. Lee, X. Chen, G. Hua, H. Hu, and X. He, "Stacked cross attention for image-text matching," in Proceedings of the European Conference on Computer Vision (ECCV), 2018, pp. 201–216.

[24] S. Xing, S. Mai, and H. Hu, "Adapted dynamic memory network for emotion recognition in conversation," IEEE Transactions on Affective Computing, 2020.

[25] D. Ghosal, N. Majumder, S. Poria, N. Chhaya, and A. Gelbukh, "Dialoguegcn: A graph convolutional neural network for emotion recognition in conversation," in Proceedings of the 2019 Conference on Empirical Methods in Natural Language Processing and the 9th International Joint Conference on Natural Language Processing (EMNLP-IJCNLP), 2019, pp. 154–164.

[26] C. Busso, M. Bulut, C.-C. Lee, A. Kazemzadeh, E. Mower, S. Kim, J. N. Chang, S. Lee, and S. S. Narayanan, "Iemocap: Interactive emotional dyadic motion capture database," *Language Resources and Evaluation*, vol. 42, no. 4, p. 335, 2008.

[27] Y.-H. H. Tsai, S. Bai, P. P. Liang, J. Z. Kolter, L.-P. Morency, and R. Salakhutdinov, "Multimodal transformer for unaligned multimodal language sequences," in Proceedings of the conference. Association for Computational Linguistics. Meeting, vol. 2019. NIH Public Access, 2019, p. 6558.

[28] X. Ju, D. Zhang, J. Li, and G. Zhou, "Transformer-based label set generation for multi-modal multi-label emotion detection," in Proceedings of the 28th ACM International Conference on Multimedia, 2020, pp. 512–520.

[29] S. Poria, E. Cambria, D. Hazarika, N. Majumder, A. Zadeh, and L.- P. Morency, "Context-dependent sentiment analysis in user-generated videos," in Proceedings of the 55th annual meeting of the association for computational linguistics (volume 1: Long papers), 2017, pp. 873–883.

[30] A. Zadeh, P. P. Liang, N. Mazumder, S. Poria, E. Cambria, and L.-P. Morency, "Memory fusion network for multi-view sequential learning," in Thirty-Second AAAI Conference on Artificial Intelligence, 2018.

12 Real-Time Sign Language Recognition and Grammatically Correct, Coherent Sentence Formation Using Deep Learning

Krithik S., Geraldine Bessie Amali D., Mathew Mithra Noel, and Muhammad Rukunuddin Ghalib

12.1 INTRODUCTION

Sign language, as a natural and essential means of communication for individuals who are deaf and mute, has been extensively studied in linguistics and cognitive science. It primarily uses hand gestures and body expressions to express meaning. A sign can either be a blue-collar element or a non-blue-collar element. Body language and facial expressions are examples of non-blue-collar components, whereas palm configurations, movements, and orientations make up blue-collar components. The act of signing can be either done by the movement of the entire body or just the hands. Hand action can be executed with one hand or both; however, depending on the user's hand preference, the dominant hand will complete most of the movement [1].

While sign language is an effective mode of communication within the deaf community, it presents challenges in communicating with non-signers, particularly in contexts such as online meetings and conferences where real-time interpretation is necessary [2], particularly when it comes to translating the gestures into grammatically complete and coherent sentences. Many prepositions, conjunctions, and other linguistic elements that are vital for constructing full sentences in spoken language are absent in sign language. For example, if a person wants to say, "Nice to meet you" in sign language, signs for "Nice," "meet," and "you," are available. But the preposition "to" is missing. This leads to translation to merely "Nice meet you", which is not a coherent sentence. The existing solutions concentrate only on accurately

 DOI: 10.1201/9781003678359-12

recognizing the signs and not on generating full coherent sentences [3,4]. Thus, identifying approaches that can be used to solve the problem is of essential importance.

Advancements in deep learning (DL) techniques and natural language processing (NLP) have revolutionized the field of computer vision and language processing [5,6]. These technologies enable machines to recognize patterns in visual data, such as hand gestures in sign language, and to understand and generate human language with increasing accuracy and fluency. This chapter will investigate and propose a framework that interprets hand signs captured through a live web camera using a Mediapipe holistic model followed by a stacked long short-term memory (LSTM) model to identify keywords and use a T5 base model to generate meaningful sentences [7].

12.2 LITERATURE SURVEY

12.2.1 Survey of the Existing Work

Recent advancements in DL techniques have been used to recognize sign language [8]. Kothadiya et al. [9] employed a deep learning model that used LSTM and gated recurrent unit (GRU) architectures, a feedback-based learning model, to identify signs. It was used for the recognition of isolated Indian Sign Language (ISL) gestures. The model reached approximately 97 percent accuracy across 11 distinct signs. However, the model's performance depended on the brightness levels, which required frame enhancement for angular and non-stable input data. The paper "Sign Language Recognition System for Communicating to People with Disabilities" by Obi et al. [10] focuses on using CNNs to classify American Sign Language (ASL) gestures. It achieved an accuracy of 96.3 percent for ASL alphabet letters. Nonetheless, the study's exclusive focus on alphabets neglects other signs, leading to delays when forming complete words. Pigou et al. [11] present a dual-CNN architecture that uses Microsoft Kinect to extract spatial and motion-based features from both the hands and the upper body. The model attained an accuracy of 95.68 percent. The model relies on Kinect for depth maps. This might limit its applicability, especially in the outdoor setting. Camgoz et al. [12] present a transformer-based architecture that integrates both sign language recognition and translation in an end-to-end framework using connectionist temporal classification (CTC) loss. This approach eliminates the need for frame-level annotation. However, further developments are needed to enhance the model's understanding of linguistic connections between sign articulators for effectiveness. Bertasius et al. [13] undertake a critical literature review of machine learning methods. First introduced in the Facebook AI Research paper "Is Space-Time Attention All You Need for Video Understanding?" by Adeyanju et al. [14], the model offers a novel, convolution-free approach to video classification by employing self-attention mechanisms across both spatial and temporal dimensions. The model outperforms alternative attention schemes achieving high accuracy on several benchmarks, including Kinetics-400 and Kinetics-600. TimeSformer offers improved training and inference efficiency compared to conventional 3D convolutional neural networks (3D CNNs), while maintaining competitive accuracy and supporting longer video sequences exceeding one minute. Shrawankar and Dixit [15] proposed a methodology for sentence generation that uses pseudo two-dimensional hidden Markov models (P2DHMM) for feature extraction, and look ahead left to right (LALR) parsing and part-of-speech (POS) tagging. However, language design errors may impact the model's accuracy. More recently, Jeyasheeli and Indumathi [16] presented a comparative analysis of two sentence generation methods for ISL: one using traditional NLP-based parsing and another employing a bidirectional LSTM (Bi-LSTM) model. Their findings indicated that the Bi-LSTM model achieved higher accuracy (95.6 percent) than the traditional method (90.6 percent). Tao et al. [17] cover static/dynamic SL translation and generation, and transformer-based video models, while Baihan et al. [18] propose CNNSa-LSTM (CNN + self-attention + LSTM) for dynamic sign language recognition with improved temporal modelling. A large pretrained video-language model tailored to sign language using CLIP-style vision-text pretraining for downstream SLR/SLT tasks by Luqman [19] and a MobileNet-based SLR pipeline for ASL alphabet by Abd Al-Latief et al. [20] suggest extension to continuous/dynamic sign sequences.

12.2.2 Gaps Identified from the Survey

Current research in sign language recognition reveals several limitations. Firstly, there is a lack of real-time translation capabilities [21]. Most of the existing studies focus only on the recognition of signs from recorded video inputs. The models treat the task as a classification problem rather than enabling live interaction. Thus, it cannot be deployed in a real-time setting. Secondly, many systems do not consider non-alphabetic signs. They predominantly focus on recognizing individual letters rather than the full range of vocabulary. This also limits the system's applicability for real-world communication, which often relies on gestures that convey whole words. Thirdly, dependency on specific hardware such as Microsoft Kinect is also a constraint [22,23]. These specialized devices are ineffective in outdoor settings as they rely on environmental factors such as lighting variations, complex backgrounds, and background noise.

12.3 DATASET CREATION

12.3.1 Dataset for Sign Language Recognition

The sign language recognition dataset has seven glosses, each containing 54 videos. These glosses represent common sign language expressions, like "Hello", "Bye", "House", "Nice", "No", "Thank you", "Yes". A "no-action-made" class is also added to detect no movements. Each video contains 30 frames, capturing the hand movements associated with the corresponding sign language expressions. Later, to test the performance on sentence formation, 20 videos of glosses including "Father", "Food", "Go", "Hello", "House", "How", "I", "Want", "You" are formed.

12.3.2 Data for Complete Sentence Formation

The sentence formation dataset consists of 310 rows of data organized into two columns. The first column consists of sets of glosses and the second column includes

	A	B
1	Glosses	Complete Sentence
2	I, like, ice cream	"I like to eat ice cream."
3	She, take, dog, walk, park	"She takes her dog for a walk in the park."
4	We, watch, movie, theater	"We watch a movie at the theater."
5	They, visit, museum, yesterday	"They visited the museum yesterday."
6	He, play, guitar, friend, house	"He plays the guitar at his friend's house."
7	She, study, library, afternoon	"She studies at the library in the afternoon."
8	We, cook, dinner, kitchen	"We cook dinner in the kitchen."
9	They, swim, beach, summer	"They swim at the beach in the summer."
10	I, read, book, bed	"I read a book in bed."
11	He, drive, car, work, every, morning	"He drives his car to work every morning."
12	She, dance, party, last, night	"She danced at the party last night."
13	They, play, soccer, field, weekend	"They play soccer on the field every weekend."
14	We, eat, dinner, restaurant, celebrate, anniver	"We eat dinner at the restaurant to celebrate our anniversary."
15	He, ride, bike, school, every, day	"He rides his bike to school every day."
16	She, cook, delicious, meal, family	"She cooks a delicious meal for her family."
17	They, go, hike, mountain, last, summer	"They went on a hike in the mountains last summer."
18	I, attend, conference, city, next, week	"I will attend the conference in the city next week."

FIGURE 12.1 Sample dataset for complete sentence formation.

complete sentences formed from these glosses. The dataset covers a large range of sentence types, including declarative, interrogative, exclamatory, imperative, simple, and compound sentences. This enables the NLP model to generate grammatically correct sentences from the input glosses. A sample of the dataset is presented in Figure 12.1.

12.4 EXPERIMENTAL SETUP AND COMPARISON OF VARIOUS DEEP LEARNING ARCHITECTURES FOR SIGN LANGUAGE RECOGNITION

Various DL architectures were implemented and tested for the recognition of the signs. The models are compared based on their accuracy of classification and inference time. This was essential to determine if the model was suitable for real-time applications. A two-stage strategy was utilized to ensure that the sign language recognition model is applicable in real time. Firstly, for faster classification, a video classification model optimized for low-inference time was trained. This is done by processing a specific number of frames representing a single sign and quickly retrieving the corresponding gloss. To ensure uniform dimensionality in the number of frames, videos exceeding a predefined maximum frame count are truncated, while those with fewer frames are zero-padded. This normalization is critical for feeding data into the fixed-architecture neural network. Secondly, a stream classification method was explored.

12.4.1 A Hybrid Convolutional Neural Network (CNN)-Gated Recurrent Unit (GRU) network Deep Learning Architecture

This methodology uses a combination of CNN and GRU in a sequential manner to effectively model the spatial and temporal features in sign language recognition [24]. In this model, CNN layers extract the spatial features from each video frame, while

Layer (type)	Output Shape	Param #
input_layer_1 (InputLayer)	(None, 224, 224, 3)	0
true_divide (TrueDivide)	(None, 224, 224, 3)	0
subtract (Subtract)	(Nane, 224, 224, 3)	0
inception_v3 (Functional)	(Norte, 2048)	21,802,784
Total params: 21,802,784 (83.17 MB)		
Trainable params: 21,768,352 (83.04 MB)		
Non-trainable params: 34,432 (134.50 KB)		

FIGURE 12.2 Model summary of the CNN.

the GRU captures temporal dependencies across frame sequences. This approach helps in the model learning both the physical gestures and their evolution over time.

For spatial feature extraction, a pretrained InceptionV3 architecture is used. Inception V3 performed better than AlexNet, VGGNet, ResNet-152 pretrained models. InceptionV3 matched AlexNet in training time while using less memory and achieving 9 percent higher accuracy. The CNN model summary is presented in Figure 12.2.

The extracted spatial features are sent to the sequential recurrent GRU layers. These layers understand the sequential patterns and accurately capture the temporal features. GRU layers are presented in Figure 12.3.

With test data, the model achieved an accuracy of 88.5 percent. The average inference time on the test data was 2.57 seconds. The sample output achieved from the model for the various gestures is presented in Figure 12.4.

12.4.2 A Hybrid CNN–LSTM–GRU Deep Learning Architecture

This hybrid model includes a CNN layer to extract spatial features and a two-stage sequential LSTM layer followed by a GRU layer for capturing complex temporal dependencies [25]. The LSTM layer captures long-range temporal patterns, while the GRU layer refines these temporal features, improving both model convergence and generalization. A pretrained InceptionV3 architecture is followed by the LSTM–GRU layers that learn temporal dependencies across the video sequence. This enhances the model's ability to interpret both short- and long-duration gestures, thereby improving its performance in the real world. The model architecture is presented in Figure 12.5. With test data, the model's accuracy was 85.7 percent. The average inference time on the test data was 2.28 seconds. The sample output achieved from the model is shown in Figure 12.6.

12.4.3 TimeSformer Transformer Architecture

TimeSformer is a transformer-based architecture that processes spatiotemporal data in videos. It applies spatial and temporal attention mechanisms independently within each transformer block at the frame-level. TimeSformer model applies a frame sampling strategy for short-length videos. It extracts eight frames randomly ensuring

Layer (type)	Output Shape	Param #	Connected to
input layer 8 (Inputlayer)	(None, 30, 2048)	0	
input_layer_9 (InputLayer)	(None, 30)	0	
gru_5 (GRU)	(None, 30, 16)	99,168	input_layer_8[0] input_layer_9[8]
gru_6 (GRU)	(None, 8)	624	gru_5[0][0]
dropout 3 (Dropout)	(None, 8)	0	gru_6[0][0]
dense 6 (Dense)	(None, 8)	72	dropout_3[0][0]
dense 7 (Dense)	(None, 7)	63	dense_6[0][0]

Total params: 299,783 (1.14 MB)
Trainable params: 99,927 (390.34 KB)
Non-trainable params: (0.008)
Optimizer params: 199,856 (780.69 KB)

FIGURE 12.3 Model summary of the sequential model (GRU layers).

FIGURE 12.4 Sample output of hybrid CNN – GRU DL architecture.

temporal diversity while preserving computational efficiency. These selected frames are processed using the VideoMAE image processor, which formats the input data for the TimeSformer architecture. The model is fine-tuned using custom sign language datasets through the Hugging Face Transformers library. This fine-tuning process helps the model adapt to various unique visual and temporal characteristics of sign language. By using the model's robust spatiotemporal feature extraction capabilities, the proposed system demonstrates an improved accuracy of 89.7 percent and an

Layer (type)	Output Shape	Param#	Connected to
input_layer_2 (InputLayer)	(None, 30, 2048)	0	
input_layer_3 (InputLayer)	(None, 30)	0	
1stm (LSTM)	(None, 30, 16)	132,160	input_layer_2[0]... input_layer_3[0]..
gru (GRU)	(None, 8)	624	1stm[0][0]
dense (Dense)	(None, 8)	72	gru[0][0]
dropout (Dropout)	(Nane, 8)	0	dense[0][0]
dense_1 (Dense)	(None, 7)	63	dropout [0] [0]

Total params: 398,759 (1.52 MB)
Trainable params: 132,919 (519.21 KB)
Non-trainable params: 0 (0.00 B)
Optimizer params: 265,840 (1.01 MB)

FIGURE 12.5 Model summary of the sequential model (LSTM-GRU layers).

FIGURE 12.6 Sample output of hybrid CNN – LSTM-GRU DL architecture.

average inference time on the test data is 2.08 seconds. The sample output achieved from the model is in Figure 12.7.

12.4.4 A Hybrid Mediapipe and Stacked LSTM Deep Learning Architecture

This methodology uses a combination of Mediapipe and stacked LSTM architectures sequentially. In this approach, Mediapipe's ability to extract landmarks representing

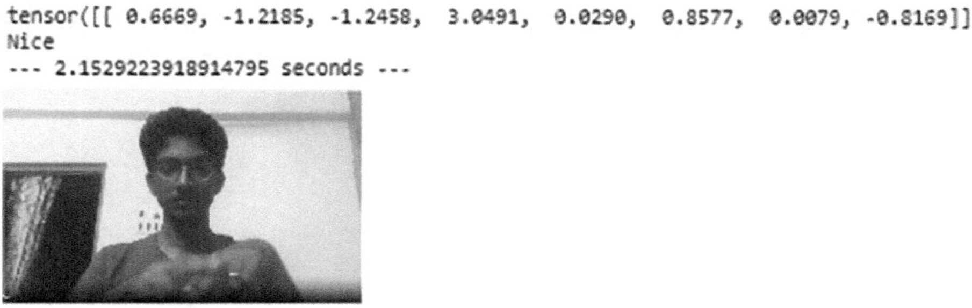

FIGURE 12.7 Sample output of the TimeSformer model.

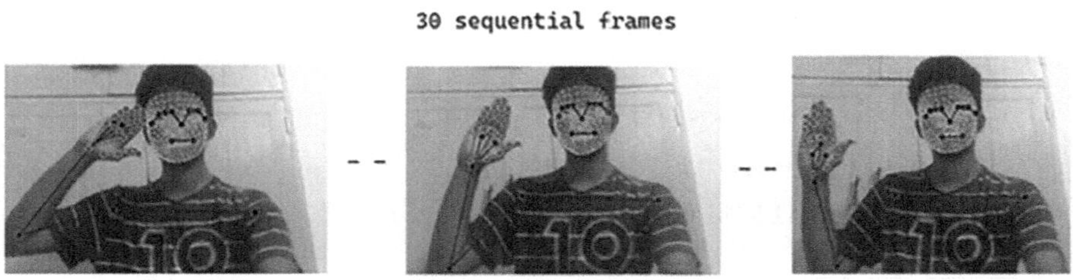

FIGURE 12.8 Sequence of frames with Mediapipe landmarks.

key points on the hands and face from sign language videos is used. To model the temporal dynamics of sign gestures, a stacked LSTM network was trained using the extracted landmark sequences [26,27]. By stacking multiple LSTM layers, the model was able to learn more complex temporal structures, thereby improving its ability to discriminate between similar gestures. The sample landmark sequence extracted using Mediapipe is presented in Figure 12.8.

Mediapipe selects the key landmarks such as facial, left-hand, and right-hand from each video frame to maximize the model's generalizability across signers with different physical characteristics. As a result, each input sample was structured as a tensor of shape *(30, 372)*, corresponding to 30 consecutive frames and 372 holistic key points per frame (Figure 12.9).

To capture temporal context and inter-frame dependencies, a sliding window of 30 sequential frames was used as input to the LSTM model. Each input sequence was mapped to a corresponding gloss label, representing the sign language gesture being performed. Mediapipe's compatibility with streaming data ensures efficient real-time landmark extraction, while the LSTM model's low-inference latency allows for near-instantaneous classification of gestures. The combined pipeline demonstrates high efficiency and responsiveness. Experimental evaluation on test datasets yielded an accuracy of 96.7 percent, indicating the model's strong performance in recognizing isolated sign language gestures. Sample output achieved from the model can be seen in Figure 12.10.

```
Layer (type)                 Output Shape              Param #
=================================================================
lstm_6 (LSTM)                (None, 30, 64)            111872

lstm_7 (LSTM)                (None, 30, 128)           98816

lstm_8 (LSTM)                (None, 64)                49408

dense_4 (Dense)              (None, 64)                4160

dense_5 (Dense)              (None, 32)                2080

dense_6 (Dense)              (None, 10)                330

=================================================================
Total params: 266,666
Trainable params: 266,666
Non-trainable params: 0
```

FIGURE 12.9 Stacked LSTM model for sequence detection.

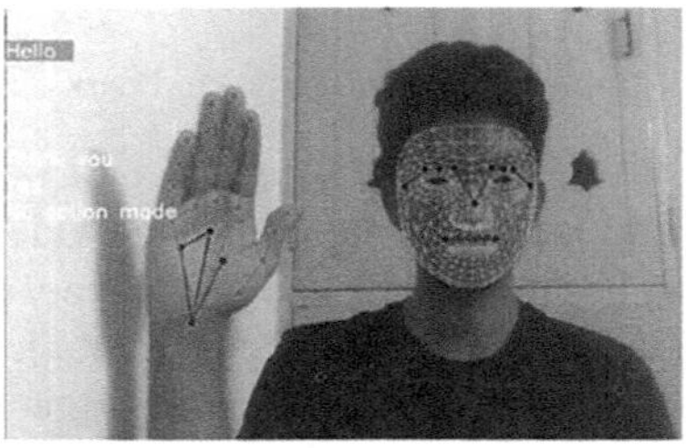

FIGURE 12.10 Sample output of hybrid Mediapipe–Stacked LSTM deep learning architecture.

12.4.5 A Hybrid Mediapipe and Bi-Directional LSTM Deep Learning Architecture

This model uses the Mediapipe library to extract and filter landmarks corresponding to pose, face, left hand, and right hand from sign language videos. For gesture classification, a bidirectional long short-term memory (BiLSTM) network was employed instead of the previously used stacked LSTM architecture. The BiLSTM processes input sequences in both forward and backward directions, allowing it to capture temporal dependencies from both past and future contexts. This bidirectional structure is particularly beneficial in tasks where understanding the full context of a sequence is essential for accurate sentence formation.

Despite its architectural advantages, the BiLSTM model achieved a classification accuracy of only 12.1 percent on the test data. This result might be due to several contributing factors. One possible explanation lies in the nature of the dataset itself. If the sequences predominantly exhibit patterns that progress in a forward manner without requiring backward temporal information, a unidirectional LSTM may be better suited to the task. Another important consideration is model complexity. The BiLSTM architecture, by design, introduces a larger number of parameters and increased computational demands. In cases where the dataset is limited in size, this added complexity can lead to overfitting, thereby hindering the model's ability to generalize effectively to new data. Furthermore, BiLSTM networks often require longer input sequences to fully utilize their bidirectional capabilities. When the input consists of relatively short sequences as in our dataset, the model may not gain sufficient contextual benefit from bidirectional processing. These findings suggest that the effectiveness of bidirectional models in sign language recognition is highly contingent upon the characteristics of the dataset and sequence length.

12.4.6 MoViNets A0 Stream Deep Learning Architecture

Mobile video networks (MoViNets) are designed for online inference and streaming video applications. 3D CNNs have been traditionally used for video-based recognition tasks. But their high computational demands and significant memory usage hinder their practicality for real-time deployment. A simpler alternative involves the use of 2D frame-based classifiers, which offer both simplicity and computational efficiency by analyzing individual frames or entire videos in real time. However, these classifiers do not capture temporal context, resulting in limited accuracy and incorrect predictions across consecutive frames. To overcome these limitations, recurrent neural networks (RNNs), which propagate state information forward through time, can be used. Ensembling techniques can be explored to enhance accuracy without significantly compromising computational efficiency. For this study, A0-stream variant MoViNets was utilized. This takes as input both an RNN state and a video frame, outputting an updated state alongside frame-level predictions. Preprocessing involved wrapping the ASL dataset with a frame-generator class that yields individual video frames on demand. From each video, 15 frames were sampled at regular intervals. Padding was applied to videos shorter than the required length.

The MoViNet-A0-stream model achieved an accuracy of 40.8 percent on the test dataset while supporting real-time recognition capabilities. The key factor that may have contributed to this moderate performance is the model's relatively small size and reduced parameter count. This may have limited its capacity to model the complex spatial-temporal patterns inherent in sign language gestures. Insufficient diverse data might also have constrained the model's generalization ability.

Extensive experimentation on the various deep learning models for sign language recognition indicate that the hybrid Mediapipe with stacked LSTM deep learning architecture provided the best categorical accuracy with good generalizability.

12.5 EXAMINING SUITABLE MODELS FOR SENTENCE FORMATION

12.5.1 Few-Shot Prompting with Lang Chain and GPT4

Large language models (LLMs) possess zero-shot capabilities, enabling them to perform various tasks without prior task-specific training. However, their performance

```
examples = [
    {
        "query": "Glosses: [I, want, food]",
        "answer": "Complete sentence: \"I want to eat food\""
    },
    {
        "query": "Glosses: [I, want, go, home, eat]",
        "answer": "Complete sentence: \"I want to go home and eat food\""
    }
]
```

FIGURE 12.11 Few-shot learning prompt.

declines when faced with more complex tasks in a purely zero-shot setting. To overcome this limitation, few-shot prompting has emerged as an effective technique. It involves providing explicit examples within the prompt, thereby providing the model with additional context information. These examples provided within the prompt help the model understand the task and guide the model's subsequent outputs, thereby improving accuracy and relevance.

In this study, the examples used to condition the model consist of glosses paired with their corresponding complete sentences, as shown in Figure 12.11.

Using the PromptTemplate method from the LangChain package, a prompt template was created. Additionally, a prefix statement is prepended to every prompt to further guide the model's behavior. The prefix reads as follows:

> Imagine you're a proficient sign language translation expert. You possess the ability to construct grammatically complete and meaningful sentences when presented with glosses from a sign language user. For example:

An LLM chain was constructed using the FewShotPromptTemplate() function and an open-source GPT-4 model. When provided with a query such as glosses [I, want, meet, You, food], the entire prompt, including the prefix and few-shot examples, is tokenized and passed to the model. The response generated follows the demonstrated pattern, for example:

> Imagine you're a proficient sign language translation expert.
> You possess the ability to construct grammatically complete and meaningful sentences when presented with glosses from a sign language user. For example:
>
> User: Glosses: [I, want, food]
> AI: Complete sentence: "I want to eat food"
>
> User: Glosses: [I, want, go, home, eat]
> AI: Complete sentence: "I want to go home and eat food"
>
> User: Glosses: [I, want, meet, You, food]
> AI: Complete sentence: "I want to meet you for a meal at your house."

Despite the effectiveness of this approach, the average inference time for the LLM chain to generate a response is approximately 40 minutes, rendering it unsuitable for real-time applications.

12.5.2 Prompt Chat Method Using Gemini-API

The Gemini API was utilized to explore the prompt chat method to enhance task context for the model. This method involves engaging the model in a conversational manner by providing examples progressively and discussing the task before presenting the actual query. This helps in familiarizing the model with the task and providing a rich context through dialogue, which facilitates more informed and accurate responses to subsequent queries. The Google generative AI package was employed to access the model's API, and the chat history was structured as follows:

```
convo = model.start_chat(history=[
  {
  "role": "user",
  "parts": ["you are a sign language translation expert. You can make grammatically complete, meaningful sentences when you are given the glosses from a sign language user. For example: glosses - {I, want, food}, complete sentence - \"I want to eat food\"; glosses - {I, want, go, home, eat}, complete sentence - \"I want to go to home and eat food\". Given the glosses, make perfect English sentences by adding prepositions at the necessary places."]
  },
  {
  "role": "model",
  "parts": ["Certainly! Please provide the glosses, and I'll help you create the corresponding English sentences with appropriate prepositions."]
  },
  {
  "role": "user",
  "parts": ["glosses - {Mother, I, go, home, you}"]
  },
  {
  "role": "model",
  "parts": ["I go home with you, Mother"]
  },
])
```

Following this initialization of the chat history, the model can be queried by sending messages formatted as "Glosses – {glosses…}" and is expected to generate grammatically complete sentences via the API. For example, the following query:

```
convo.send_message("glosses - {hi, how, you, can, we, play, football, tomorrow}")
print(convo.last.text)
```

produced the output:

"Hi, how are you? Can we play football tomorrow?"

The major disadvantage of the system arises from the latency of the API calls, which have an average inference time of approximately three seconds per request. This delay significantly affects the responsiveness and overall efficiency of the pipeline. Inference time also changes with network speed and server load, resulting in inconsistent and unpredictable processing durations. Furthermore, the model occasionally generates responses unrelated to the gloss-to-sentence conversion task, undermining precision and potentially compromising the system's reliability and effectiveness.

12.5.3 Fine-tuning a T5-base Model

Google developed an advanced language model known as the text-to-text transfer transformer (T5). The distinguishing feature of T5 lies in its "text-to-text" framework, which unifies a wide range of NLP tasks encompassing translation, summarization, and classification. In this study, the T5 model was fine-tuned for the specialized task of generating grammatically complete sentences from glosses in sign language data. The fine-tuning process involved several key steps. First, the model and its associated tokenizer were initialized using the pretrained "t5-base" checkpoint. A preprocessing function was then defined to tokenize dataset examples, which consist of paired glosses and corresponding complete sentences. This function converts the text inputs into token ids and attention masks compatible with the model's input requirements. Following this, both the training and validation datasets were tokenized in batches using the preprocessing function. Training arguments were specified, including the output directory for storing trained models, evaluation strategy, logging configurations, learning rate, batch size, weight decay, number of epochs, and other relevant hyperparameters. The model was then configured for training with these parameters, enabling efficient monitoring of performance metrics and checkpointing of the best-performing model states. Finally, the training was executed using the tokenized datasets and specified configurations. Throughout this process, the model's progress was continuously evaluated and logged, facilitating iterative improvements. This fine-tuning approach allowed the T5 model to better adapt to the specific task of sentence formation from glosses, thereby improving its performance on this domain-specific challenge.

The training loss is decreased to 0.05 (Figure 12.12), and the validation loss is decreased to 0.18 (Figure 12.13).

The sample inputs and outputs generated by the model can be seen in Figure 12.14. The average inference time of the fine-tuned model is 0.43 seconds, making it suitable for real-time conversion of glosses to complete sentence formation.

12.6 OVERVIEW OF THE PROPOSED DEEP LEARNING ARCHITECTURE

The proposed architecture is a combination of Mediapipe, LSTM, and T5 Base models, as shown in Figure 12.15. It operates in two main phases: the sign language recognition phase and the sentence formation phase. In the recognition phase, the system accesses the user's camera to capture real-time hand gestures. Mediapipe is used for extracting spatial and temporal data from the body, hands, and face in the live video frames. The data is stored in a stream buffer until 30 frames are collected. Then, the

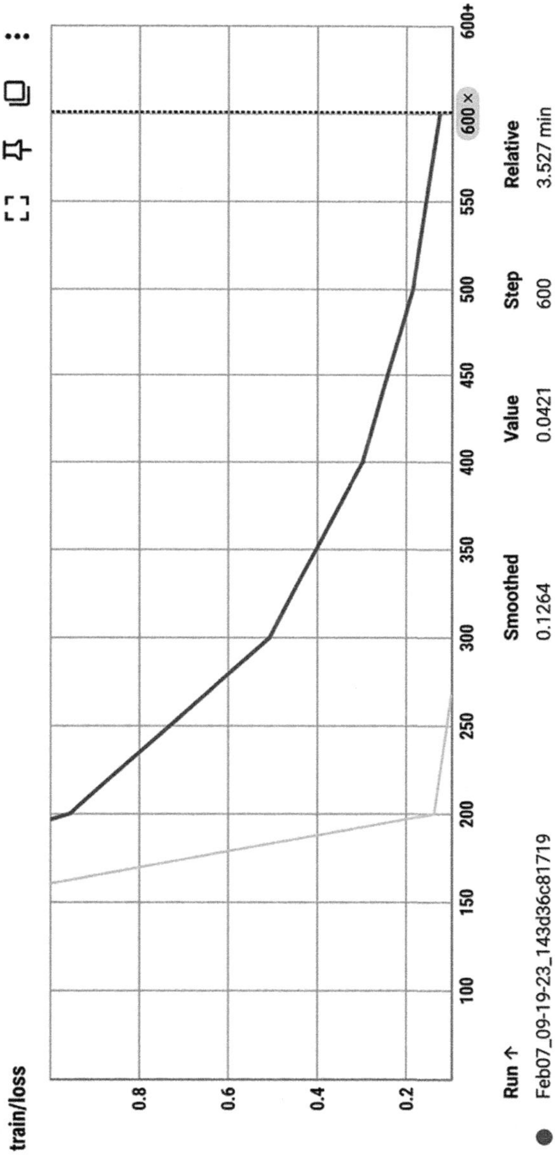

FIGURE 12.12 Training loss of T5 base.

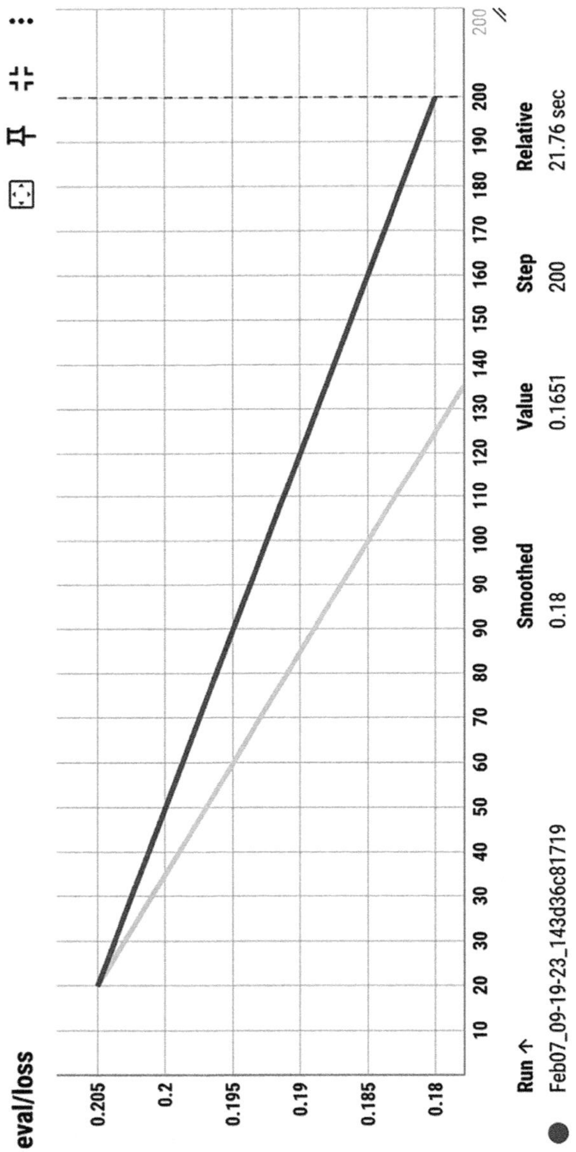

FIGURE 12.13 Validation loss of T5 base.

```
glosses - He, skate, rink, ice, hockey, game
target - "He skates at the ice rink for the hockey game."
generated - "He skates at the rink for the ice hockey game."
****************************************
glosses - Who, provide, assistance, necessary, event, unexpected, occurs
target - "Who will provide the necessary assistance if an unexpected event occurs?"
generated - "Who provided the assistance necessary for the event unexpected occurs?"
****************************************
glosses - He, cook, breakfast, kitchen, every, morning
target - "He cooks breakfast in the kitchen every morning."
generated - "He cooks breakfast in the kitchen every morning."
****************************************
glosses - When, it, appropriate, use, humor, social, gathering
target - "When is it appropriate to use humor in a social gathering?"
generated - "When is it appropriate to use humor at a social gathering?"
****************************************
glosses - We, travel, Europe, vacation, last, year
target - "We traveled to Europe for vacation last year."
generated - "We traveled to Europe for vacation last year."
****************************************
glosses - How, they, manage, stay, productive, despite, challenges
target - "How do they manage to stay productive despite facing challenges?"
generated - "How did they manage to stay productive despite challenges?"
****************************************
Average Rough Score: 0.6504814776205694
```

FIGURE 12.14 Sample output of fine-tuned T5-base.

LSTM model processes these sequences, capturing dependencies in the movement patterns. If the model's confidence exceeds the threshold, the output gloss is stored in a gloss buffer and classified. After the glosses are collected, they are forwarded to the sentence formation phase. When the user signals the end of a phrase, the T5 Base model interprets the classified glosses and generates coherent sentences. Finally, the sentence is displayed to the user, forming a robust pipeline for real-time sign language processing. The proposed approach integrates computer vision and NLP for effective communication between signers and non-signers, making interactions more inclusive.

12.7 RESULTS AND DISCUSSIONS

Based on the comprehensive experimentation and analysis, the integration of Mediapipe for landmark extraction, an LSTM-based model for temporal gesture recognition, and the T5-base model for sentence generation emerges as an effective and practical pipeline for sign language recognition and sentence formation. Mediapipe provides a lightweight and efficient framework for extracting keypoint data from video frames, enabling robust real-time tracking of hand and body movements. LSTM networks, with their inherent capability to model temporal dependencies, are well-suited to interpret the sequential patterns in sign gestures. The T5-base model, fine-tuned for gloss-to-sentence transformation, converts recognized glosses into grammatically coherent and contextually meaningful English sentences. This modular architecture addresses several core challenges in sign language recognition,

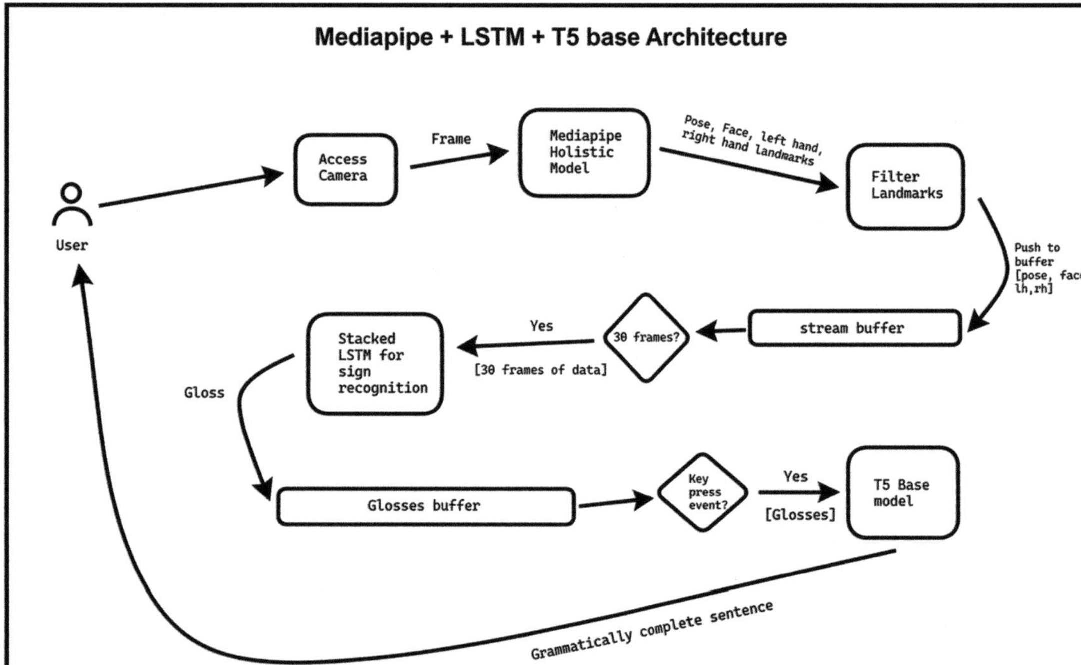

FIGURE 12.15 Proposed deep learning architecture for sign recognition and sentence formation.

TABLE 12.1
Summarization of Models' Performance in Sign Language Recognition

Methodology	Categorical Accuracy	Supports real time
CNN + GRU	88.5	2.87s delay
CNN + LSTM + GRU	85.7	2.28s delay
TimeSformer (trained-600-kinetics)	89.7	2.08s delay
Mediapipe + Stacked LSTM	96.7	Yes
Mediapipe + Stacked Bi-directional LSTM	12.1	Yes
MoViNet (A0-stream)	40.8	Yes

TABLE 12.2
Summarization of Models' Performance in Complete Sentence Generation

Methodology	Time taken
LangChain with GPT	40 mins
Gemini API – Prompt chat	3 secs
T5 - Base	0.43 secs

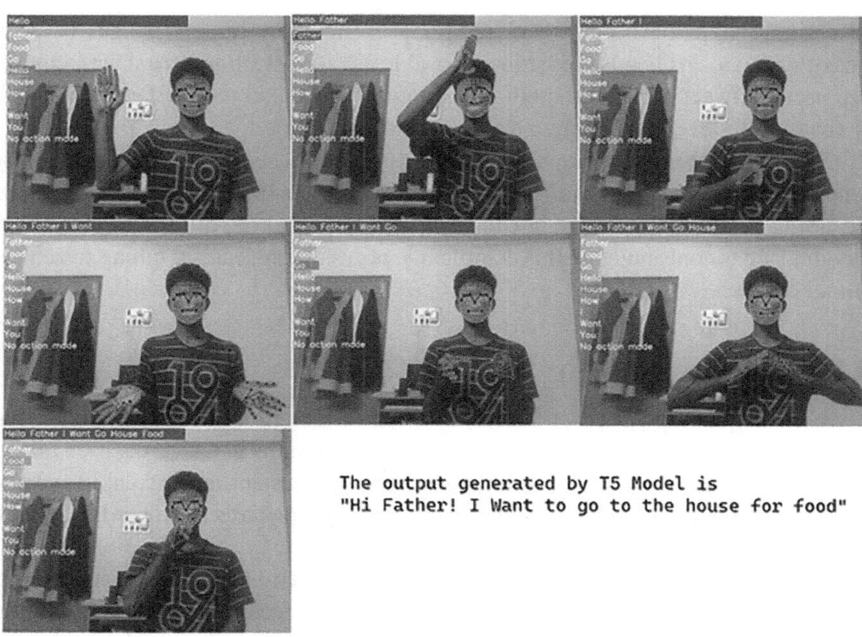

FIGURE 12.16 Sample output 1 by Mediapipe + LSTM + T5-base model architecture.

including real-time processing, sign transition handling, left-hand dominance, and sentence fluency with a minimal inference time.

The approaches for recognizing sign language, their classification accuracy, and their support for real-time processing are compiled in Table 12.1. The table displays the amount of time the model takes to generate glosses in cases when it is not appropriate for real-time processing.

Table 12.2 summarizes the methodologies used for complete sentence formation along with the time taken by them to produce results.

The conclusive architecture consists of Mediapipe + LSTM + T5 base for sign language recognition and complete sentence formation. Figures 12.16 and 12.17 are the results obtained from the proposed architecture. While the model generally performs adequately, it occasionally produces erroneous results during transitions between different signs.

Several challenges were encountered and addressed during the development of the sign language recognition system. One significant issue was the lack of sufficient training data. To mitigate this, various fundamental operations and data augmentation techniques—such as random cropping, rotation, scaling, and color jittering were applied. These methods effectively increased the diversity and volume of training samples, thereby enhancing the model's generalization capabilities.

Another challenge involved handling left-hand dominance among signers. The available training data predominantly represented right-hand dominant individuals. To accommodate left-handed users, a selection mechanism was introduced, allowing users to indicate hand dominance. Subsequently, video frames were horizontally inverted when necessary, ensuring consistent model input without impacting performance.

Real-time sign transition posed additional difficulties. Rapid changes between signs often led to erroneous predictions. To address this, a waiting period was required from users between signs, although this solution could lead to user fatigue and was not always effective in preventing transition-related errors.

Finally, hyperparameter optimization—particularly regarding frame quantity, frame selection strategies, and model architecture—remained a complex and time-consuming process. Variability in camera frame rates and the differing durations of individual signs contributed to this complexity, necessitating careful tuning to achieve optimal results.

12.7.1 Security and Safety Standards

Strong security and privacy are becoming essential design requirements rather than an afterthought due to the quick spread of AI-driven systems in real-time, user-facing, and safety-critical applications. Building reliable, consistent, and scalable systems requires adherence to established security and privacy standards as AI models process more sensitive visual and behavioral data.

From a security and privacy perspective, the suggested architecture is in line with well-known standards including the General Data Protection Regulation (GDPR), ISO/IEC 27001, and the NIST Cybersecurity Framework [28], with conditional application to Health Insurance Portability and Accountability Act (HIPAA) in

FIGURE 12.17 Sample output 2 by Mediapipe + LSTM + T5-base model architecture.

healthcare-related installations. The technology complies with data minimization and privacy-by-design principles by drastically reducing the exposure of raw video data by landmark extraction on the user's device using Mediapipe. Additionally, fine-grained access control, safe deployment, and compliance with enterprise-grade security and privacy standards are made possible by the modular separation of the perception, temporal modelling, and language generation components (Figure 12.18).

By incorporating platforms like Enkrypt AI, which offer ongoing risk assessment, privacy monitoring, and compliance reporting for machine learning systems, AI governance and security may be further reinforced. Throughout the model lifespan, Enkrypt AI may function as an external governance layer in the suggested architecture, keeping an eye on model behavior, identifying security flaws, and guaranteeing compliance with legal requirements like GDPR, NIST, and ISO/IEC 27001.

12.8 CONCLUSION AND SCOPE OF FUTURE WORK

The study explored various deep learning architectures for sign recognition and sentence formation. The proposed model using a Mediapipe holistic model followed by a stacked LSTM model and a T5 base model achieved an accuracy of 96.7 percent with real-time support and no delays. The hybrid model with Mediapipe and Bidirectional LSTM performed poorly due to the inherent pattern of the dataset. The other hybrid models performed fairly in terms of accuracy but incurred delay, which made it unsuitable for real-time applications. The performance of sign language recognition systems was influenced by hyperparameter choices, such as the number of input frames. Thirty frames were reasonable considering the sign gesture time. However, this approach introduces challenges, particularly during transitions between signs, where frames may capture overlapping gestures from consecutive signs, potentially leading to misclassification. Dynamic frame selection that can adaptively choose frames based on gesture motion or velocity can be considered. Temporal context modelling through recurrent or attention-based mechanisms allows the model to consider adjacent frames, improving prediction continuity across sign boundaries. Additionally, feedback mechanisms, such as confidence scoring or user input, may refine predictions in uncertain cases, enhancing overall accuracy.

Improvements in complete sentence generation from sign language glosses can be achieved by incorporating emotional context and memory mechanisms. Integrating emotional modelling via emotion-specific embeddings or conditioning inputs enables

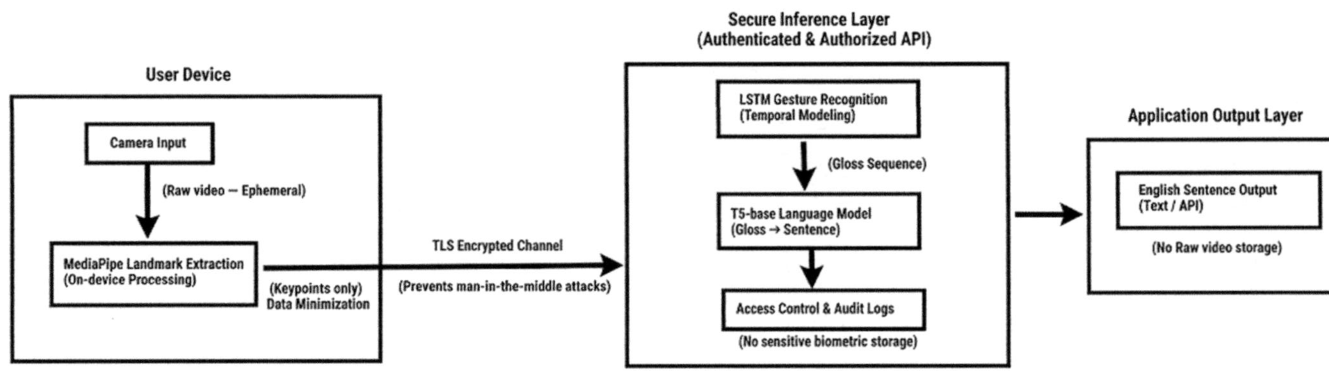

FIGURE 12.18 High-level security architecture diagram.

the generation of sentences with expressive, contextually appropriate emotional tone. Using pretrained emotion classifiers can further guide this process. Introducing memory components, such as recurrent or transformer-based architectures with attention, supports temporal coherence by retaining contextual information across sentence sequences.

REFERENCES

[1] R. Shrivastava, A hidden Markov model-based dynamic hand gesture recognition system using OpenCV, in: Proc. 3rd IEEE Int. Adv. Comput. Conf. (IACC), Ghaziabad, India, 2013, pp. 947–950.

[2] X. Lu, B. Qi, H. Qian, Y. Gao, J. Sun, J. Liu, Kinect-based human finger tracking method for natural haptic rendering, *IEEE Sensors Council, Entertainment Computing*, 33 (2020) Art. no. 100335.

[3] F.-S. Chen, C.-M. Fu, C.-L. Huang, Hand gesture recognition using a real-time tracking method and hidden Markov models, *Image Vis. Comput.,* 21 (8) (2003) 745–758.

[4] Y. Yang, D. Ramanan, Articulated human detection with flexible mixtures of parts, *IEEE Trans. Pattern Anal. Mach. Intell.*, 35 (12) (2013) 2878–2890

[5] M. A. Ahmed, B. B. Zaidan, A. A. Zaidan, M. M. Salih, M. M. B. Lakulu, A review on systems-based sensory gloves for sign language recognition state of the art between 2007 and 2017, *Sensors*, 18 (2018) 2208.

[6] L. Zheng, B. Liang, A. Jiang, Recent advances of deep learning for sign language recognition, in: Proc. Int. Conf. Digit. Image Comput., Techn. Appl. (DICTA), 2017, pp. 1–7.

[7] S. Stoll, N. C. Camgoz, S. Hadfield, R. Bowden, Text2Sign: Towards sign language production using neural machine translation and generative adversarial networks, *Int. J. Comput. Vis.*, 128 (2020) 891–908.

[8] L. Zheng, B. Liang, A. Jiang, Recent advances of deep learning for sign language recognition, in: Proc. Int. Conf. Digit. Image Comput., Techn. Appl. (DICTA), 2017, pp. 1–7.

[9] D. Kothadiya, C. Bhatt, K. Sapariya, K. Patel, A. B. Gil-González, J. M. Corchado, Deepsign: Sign language detection and recognition using deep learning, *Electronics*, 11 (11) (2022) 1780.

[10] Y. Obi, K. Claudio, V. Budiman, S. Achmad, A. Kurniawan, Sign language recognition system for communicating to people with disabilities, *Procedia Computer Science,* 216 (2023) 13–20.

[11] L. Pigou, S. Dieleman, P. J. Kindermans, B. Schrauwen, Sign language recognition using convolutional neural networks, in: L. Agapito, M. M. Bronstein, C. Rother (Eds.), Computer Vision—ECCV 2014 Workshops, vol. 8925, Springer, 2015, pp. 572–578.

[12] N. C. Camgoz, O. Koller, S. Hadfield, R. Bowden, Sign language transformers: Joint end-to-end sign language recognition and translation, arXiv preprint, arXiv:2003.13830, 2020.

[13] G. Bertasius, H. Wang, L. Torresani, Is space-time attention all you need for video understanding?, arXiv preprint, arXiv:2102.05095, 2021.

[14] A. Adeyanju, O. O. Bello, M. A. Adegboye, Machine learning methods for sign language recognition: A critical review and analysis, *Intell. Syst. Appl.*, 12 (2021) 200056.

[15] U. Shrawankar, S. Dixit, Framing sentences from sign language symbols using NLP, *Int. J. Adv. Res. Comput. Commun. Eng.*, 5 (1) (2016) 15–19.
[16] P. G. Jeyasheeli, N. Indumathi, Sentence generation for Indian sign language using NLP, *Webology*, 18 (S1) (2021) 196–210.
[17] Tao, T., Zhao, Y., Liu, T., & Zhu, J. Sign language recognition: A comprehensive review of traditional and deep learning approaches, datasets, and challenges. *IEEE Access*, 12 (2024) 75034–75060.
[18] Baihan, A., Alutaibi, A. I., Alshehri, M., & Sharma, S. K. Sign language recognition using modified deep learning network and hybrid optimization: a hybrid optimizer (HO) based optimized CNNSa-LSTM approach. *Sci. Rep.* 14(1) (2024) 26111.
[19] Luqman, H. SignVLM: A pre-trained large video model for sign language recognition. *PeerJ Comp. Sci.* 11 (2025) e3112.
[20] Abd Al-Latief, S. T., Yussof, S., Ahmad, A., & Khadim, S. Deep learning for sign language recognition: A comparative review. *J. Smart Internet Things*, 2024(01) (2024) 77–116.
[21] T. Starner, J. Weaver, A. Pentland, Real-time American Sign Language recognition using desk and wearable computer-based video, *IEEE Trans. Pattern Anal. Mach. Intell.*, 20 (12) (1998) 1371–1375.
[22] G. Marin, F. Dominio, P. Zanuttigh, Hand gesture recognition with Leap Motion and Kinect devices, in: IEEE Int. Conf. Image Process. (ICIP), Paris, France, 2014, pp. 1565–1569.
[23] X. Lu, B. Qi, H. Qian, Y. Gao, J. Sun, J. Liu, Kinect-based human finger tracking method for natural haptic rendering, IEEE Sensors Council, 2025.
[24] P. Witoonchart, P. Chongstitvatana, Application of structured support vector machine backpropagation to a convolutional neural network for human pose estimation, *Neural Netw.*, 93 (2017) 1–10.
[25] S. He, Research of a sign language translation system based on deep learning, in: 2019 Int. Conf. Artif. Intell. Adv. Manuf. (AIAM), Dublin, Ireland, 2019.
[26] E. J. Escobedo Cárdenas, L. Ramirez Cerna, G. Chávez, Dynamic sign language recognition based on convolutional neural networks and texture maps, in: Proc. Int. Conf. Comput. Vision Appl., 2019, pp. 265–272.
[27] Y. Liao, P. Xiong, W. Min, J. Lu, Dynamic sign language recognition based on video sequence with BLSTM-3D residual networks, *IEEE Access*, 7 (2019) 38044–38054.
[28] Lanz, J. The Updated NIST Cybersecurity Framework. *CPA J.*, *94*(5/6) (2024) 70–72.

13 Emotion Detection across Modalities

A Deep Dive into Multimodal Systems

Vijaya Kumar T., Krishnamoorthi M., and Prabha S.U.

13.1 INTRODUCTION

The faculty of multimodal emotion detection is based on the fact that human affect, however, does not express itself in a single medium but rather cooccurs in terms of voice tone, facial muscles, gaze, body posture, language usage, and even such a change of physiological states as heart rate or skin conductance. Each of these capture a complementary aspect of the identical internal configuration: speech provides evidence of prosody and rhythm, sight provides evidence of micro-expressions and gesticulation, text provides evidence of semantic and syntactic information, and biosignals contain evidence of subconscious reactions that cannot be voluntarily controlled. Systems based on each of these streams alone are weak (e.g. face recognition suffers by being occluded or having low light, and speech models in high-noise acoustics do the same). The blend of modalities thus enables the emotion recognition systems to estimate the richness, redundancy, and contextual sensitivity of human perception, which is essential in the reliable affect-conscious applications that have been implemented in the wild [1].

The state-of-the-art multimodal emotion systems usually take the sequence signal acquisition, preprocessing, feature extraction, and fusion, and then classifying or regressing to discrete emotions or continuous affect dimensions including valence and arousal. Recent progress in deep learning has changed the emphasis on handcrafted features towards learned ones, in which convolutional, recurrent, and transformer-based networks collectively capture the temporal dynamics and cross-modal relationships [1]. Architectures could handle each of the modalities using dedicated subnetworks and then fuse the modalities in a combined latent space, or they can use attention mechanisms that dynamically bias modalities depending on their instantaneous reliability and relevance. This flexibility of design has allowed significant performance improvements in benchmark datasets, as well as is subject to inquiry regarding interpretability, generalization, and fairness across the various groups of people regarding demographics and cultural values.

DOI: 10.1201/9781003678359-13

In spite of their potential, the multimodal emotion detection system has huge issues with data, modeling, and deployment. Techniques used to gather and label multimodal emotion datasets in a time-synchronized and high-quality manner are costly and time-consuming in nature and classification is typically noisy, caused by subjectivity, cross-cultural differences, and context effects in perceiving and expressing emotions [2]. The data streams that are in the real world may be incomplete, corrupted, or asynchronous, which means that models must deal with missing modalities and misaligned time gracefully [3]. Moreover, cross-dataset and cross-domain generalization is still a challenge: models that have been trained on controlled, lab environments tend to perform poorly when it comes to in-the-wild interactions, such as interactions between different languages, accents, devices, and environments. To ensure these challenges can be met, tough combination plans, domain customization methods, and stringent assessment plans are necessary exceeding closed-set benchmarking [4].

Reliable multimodal emotion detection has numerous implications in areas including adaptive tutoring systems, mental health monitoring, driver state assessment, the social robot and the conversational agent sensitive to the effects of its user. Meanwhile, the capabilities of detecting emotional conditions raise serious ethical and social issues regarding privacy, consent, bias, and manipulation, as well as the fact that affective profiling may be severely abused. The development of these systems requires visible design of models, thorough management of data gathering and data storage and correspondence with societal laws and cultural standards of emotions surveillance. The chapter hence does both provide the current status of the art and the algorithms grounded in the multimodal emotion recognition, and previews to the human-centered, ethical, and practical issues that should inform future study and implementation.

13.2 THEORETICAL FOUNDATIONS OF EMOTION REPRESENTATION

Human emotions can be represented through three main models: categorical, dimensional, and appraisal-based. Categorical models (e.g., Ekman's six basic emotions) are easy to interpret but lack nuance. Dimensional models VAD (Valence–Arousal–Dominance) capture intensity and blends but are complex to annotate [5]. Appraisal models (e.g., component process model) provide context-sensitive insights but require high-level semantic reasoning. Each model has distinct strengths and limitations depending on the application domain. In subsequent sections, these models are mapped to deep learning formulations, where categorical theories define discrete output classes, dimensional theories inform regression targets in VAD space, and appraisal theories inspire context-aware architectures and auxiliary tasks.

13.3 CATEGORICAL MODELS

Emotions are perceived in categories and can be identified universally and categorically in categorical models, which in turn include the basic emotion theory by Paul Ekman as one of the most effective approaches in the field. Ekman introduced six core emotions of happiness, sadness, anger, fear, disgust, and surprise that he considers to be biologically hardwired and culturally invariant [6,7]. All these emotions are commonly identified with specific facial expressions and physical reactions, which makes the model easy and convenient in the work of the emotion recognition. Categorical models are simple and easy to interpret and this has made them very popular. Nonetheless, they tend to be weak in the representation of the nuances, combinations, and the variations of intensity of more complex emotional experiences. These simple emotions commonly constitute the output classes of softmax layers of facial, speech, or multimodal classifiers in deep learning systems, an explicit instance of categorical theory.

13.4 DIMENSIONAL MODELS

Dimensional models describe emotions as the objects of a continuous multi-dimensional space, not as distinct categories. The VAD model is the most popular. Deep models apply this perspective by training deep regression heads and appropriately defined loss functions (e.g. CCC or MSE) to predict continuous valence, arousal, and dominance, in many cases with categorical outputs in a multitask configuration as given in Table 13.1.

- Valence refers to the positivity or negativity of an emotion (pleasant → unpleasant).
- Arousal indicates the intensity of emotion (calm → excited).
- Dominance measures the degree of control or influence (submissive → empowered).

This model allows for the representation of complex and blended emotions, making it suitable for applications requiring fine-grained emotional analysis.

13.5 APPRAISAL MODELS

The appraisal theories are concerned with the process through which people appraise the occurrences in the surrounding and the result of the appraisal being associated

TABLE 13.1
Comparison of Emotions in VAD Space

Emotion	Valence	Arousal	Dominance
Joy	High	High	High
Anger	Low	High	High
Sadness	Low	Low	Low
Fear	Low	High	Low
Surprise	Neutral	High	Moderate
Disgust	Low	Medium	High

TABLE 13.2
Comparison of Emotion Representation Models

Model Type	Key Feature	Advantages	Limitations	Common Use Cases
Categorical	Discrete emotion classes	Simple, interpretable, easy annotation	Inflexible, lacks nuance	Facial expression recognition, chatbots
Dimensional	Continuous emotion space (VAD)	Captures intensity and blends	Complex labeling, may lack clarity	Affective computing, music emotion analysis
Appraisal	Cognitive evaluation of stimuli	Context-sensitive, theory-driven	Hard to model, requires semantic reasoning	Cognitive models, advanced HCI systems

with emotion [8]. The component process model (CPM) by Klaus Scherer is one well-known model.

The key appraisal dimensions in CPM include the following:

- Novelty: How new or unexpected the event is
- Pleasantness: How enjoyable the event is
- Goal relevance: How important the event is to personal goals
- Coping potential: Ability to deal with the event
- Norm compatibility: Consistency with social norms or personal values

Appraisal dimensions can be encoded as auxiliary variables or latent factors, encouraging attention mechanisms, graph structures, or high-level semantic encoders to focus on novelty, goal relevance, and coping potential. It provides a way to incorporate the appraisal theory into contemporary multimodal transformers and graph neural network (GNN)-based fusion as given in Table 13.2. The descriptive models of appraisal provide an understanding of more of the contextual and cognitive relation of emotions, yet are difficult to execute in practice in real-time systems because of the requirement of the high-level semantic perception [9][10].

13.6 OVERVIEW OF EMOTION MODALITIES

Emotions are expressed through facial expressions, speech, text, physiological signals, and body gestures [11]. Each modality contributes unique cues:

- *Facial expressions* are intuitive but sensitive to lighting/occlusion.
- *Speech* captures arousal effectively but is affected by noise and accent.
- *Text* provides semantic depth but lacks non-verbal context.
- *Physiological signals* reveal unconscious states but require intrusive sensors.
- *Body gestures* complement other cues but are ambiguous in isolation.

TABLE 13.3
Comparative Summary of Emotion Modalities

Modality	Emotional Clues	Sensing Method	Strengths	Limitations
Facial expressions	Visual (face muscles)	Camera/video	Intuitive, strong for basic emotions	Affected by lighting, occlusion
Speech	Vocal tone, prosody	Microphone/ audio	High arousal sensitivity, natural cues	Noisy environments, accent variation
Text	Semantic and sentiment	Text input/ Transcripts	Good for long-term mood/context analysis	Ambiguity, lacks non-verbal context
Physiological	Biometric signals	Wearable sensors	Reliable for unconscious emotions	Intrusive, hardware-dependent
Body Gestures	Movement/ posture	Camera/Motion sensors	Complements other cues well	Hard to interpret in isolation

Multimodal systems leverage the complementarity of these modalities to achieve robustness and higher accuracy as given in Table 13.3.

13.7 CRITICAL ANALYSIS OF FUSION TECHNIQUES

Multimodal emotion detection systems use the data collected across multiple modalities (facial expressions, speech, text, physiological signals, and body gestures) [11]. All these modalities give complementary cues on emotions. Nevertheless, the real strength of multimodal systems does not only lie in the ability to gather this data but to successfully combine it with the ability to derive a more accurate and stronger inference of emotional states. Table 13.4 discusses the three significant types of fusion paradigm—early fusion, late fusion, and hybrid fusion—some emerging strategies such as attention-based fusion and graph-based methods, their architectures, merits and shortcomings:

- The early fusion allows a good interaction between inter-modes; it is noise-sensitive and needs to be synchronized. It is moderately accurate and moderately low in reliability in the real-world scenario.
- Late fusion: Late fusion is resilient to missing/noisy modalities and can be used in real time but it lacks cross-modal synergy, resulting in intermediate accuracy.
- Hybrid fusion combines feature-level with decision-level integration and is perhaps more highly accurate, although at a significant computational price, making it less realistic at high frequencies.
- Attention-based fusion dynamically assigns modality weights, which can provide the state-of-the-art accuracy and interpretability at the expense of large datasets and training resources.

TABLE 13.4
Comparative Insights

Fusion Strategy	Strengths	Weaknesses	Accuracy	Computational Cost	Real-Time Suitability	Use Case Examples
Early Fusion	Strong inter-modality interaction; rich low-level correlations	Sensitive to noise; requires synchronization	Moderate–High	Moderate	Limited to synchronized environments	VR/AR systems
Late Fusion	Robust to missing/noisy modalities; modular	Weak cross-modal synergy	Moderate	Low	High	Surveillance, chatbots
Hybrid Fusion	Balances interaction & robustness; higher accuracy	High complexity; latency issues	High	High	Limited	Healthcare, education
Attention-based Fusion	Context-aware; interpretable; dynamic weighting	Requires large datasets; training-intensive	Very High	High	Moderate	Tutoring systems, dialogue agents
Graph-based Fusion	Models structured relationships	Extremely resource-intensive	Very High	Very High	Low	Research-focused systems

- Graph-based fusion specifically represents inter- and intra-modality relations and provides extremely high research accuracy at prohibitive computational cost.

Summary: Early fusion suits controlled environments, late fusion fits modular systems, hybrid fusion improves accuracy but increases latency, attention-based fusion enhances adaptability, and graph-based fusion remains experimental but promising.

13.8 CRITICAL ANALYSIS OF DEEP LEARNING ARCHITECTURES

13.8.1 Modality-Specific Networks

Each modality such as vision, speech, or text has unique data characteristics that are best handled by specialized architectures [12]:

- The convolutional neural network (CNNs) are good in cases when it is about vision (facial recognition, gestures) and poor at being able to pick out their temporal patterns.
- Recurrent neural networks and long short-term memories (RNNs/LSTMs) work well on sequential data such as speech, but have problems with scalability and vanishing gradient problems.
- Transformers are better than RNNs in both short- and long-range dependencies, which are both captured effectively with self-attention. They allow parallel sequence processing, enhancing accuracy and speed of training. To illustrate that, transformers are more successful in matching textual sentiment and facial expression compared to LSTMs.
- GNNs explicitly capture inter- and intra-modality dependencies, which enable them to be promising to structured fusion. Their computational complexity, however, limits them to research uses only in most cases.

Summary: CNNs and RNNs remain strong for modality-specific tasks, but transformers and GNNs represent the next stage of multimodal fusion with higher accuracy and interpretability, albeit at the cost of higher computation.

13.8.2 Multistream and Parallel Architectures

In multistream networks, each modality is processed in a separate stream using modality-appropriate layers (e.g., CNN for vision, LSTM for audio). The outputs are then concatenated or fused before final prediction as in Figure 13.1.

13.8.3 Attention-Based Fusion Models

Attention mechanisms allow models to focus on the most informative features or modalities dynamically. There are multiple forms:

- Intra-modality attention: Identifies the most relevant features within a single modality (e.g., important facial regions or speech segments).

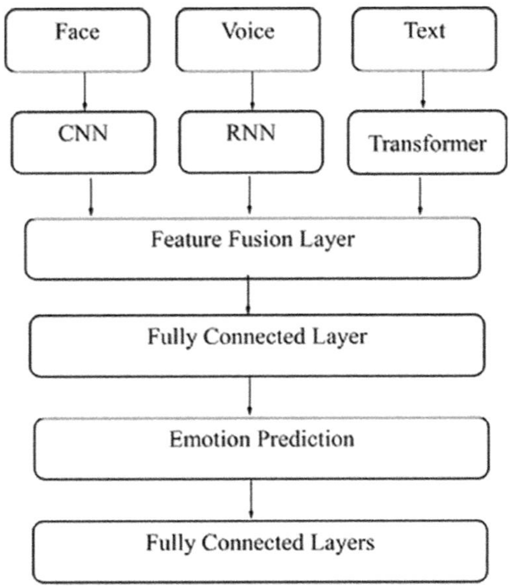

FIGURE 13.1 Multistream and parallel architecture diagram.

- Cross-modal attention: Allows one modality to attend to another (e.g., text attending to facial features).
- Modality attention: Assigns importance weights to different modalities based on their relevance in a given context.

13.8.4 Multimodal Transformers

Transformers are revolutionizing multimodal learning due to their powerful self-attention mechanisms. These models can jointly model and align different modalities in a unified architecture.

- Types:
  - Early fusion transformers: Combine input embeddings from all modalities at the start (e.g., VisualBERT).
  - Late fusion transformers: Use separate encoders for each modality, followed by fusion (e.g., FLAVA).
  - Cross-modal transformers: Employ cross-attention layers where each modality attends to others (e.g., CM-Attention, MERLOT).

13.8.5 Graph Neural Networks (GNNs) for Structured Fusion

In GNN-based architectures, each modality or even subcomponent (e.g., facial region, speech segment) can be modelled as a node, and relationships among them are represented as edges [13].

- Graph construction: Nodes = modalities or features; Edges = statistical or learned relationships.
- Graph convolution: Propagates information through connected nodes to capture interactions.

13.8.6 Multitask and Multilabel Architectures

Some models treat emotion detection as a multitask problem, where each modality contributes to separate but related tasks (e.g., valence prediction, arousal detection) as illustrated in Figure 13.2.

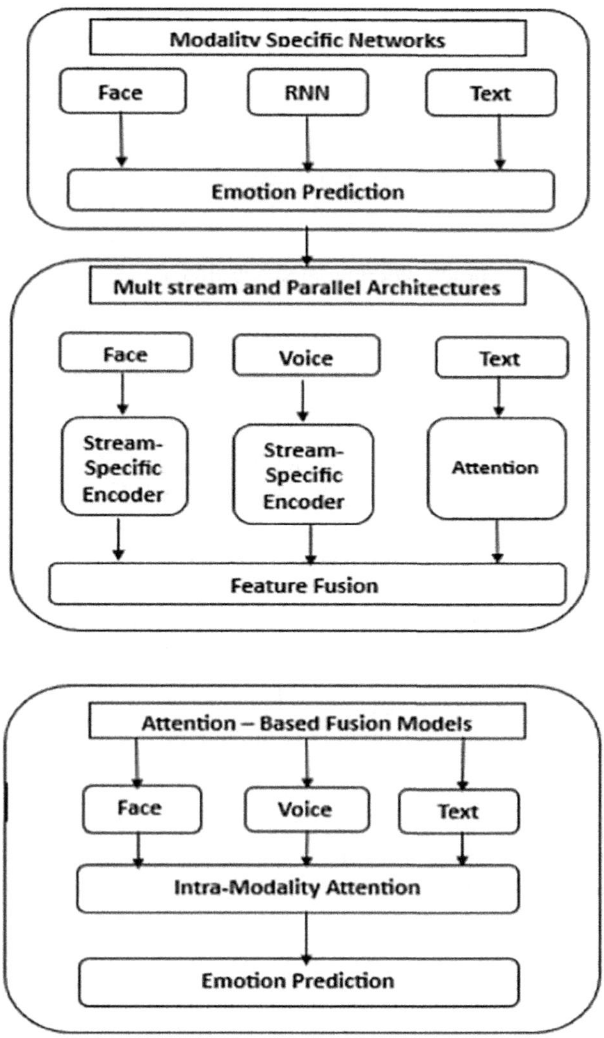

FIGURE 13.2 Deep learning fusion architectures.

- Architecture: Shared feature extraction → Task-specific heads.
- Benefits: Enhances generalization and captures different emotional dimensions simultaneously.

13.9 BENCHMARK DATASETS AND EVALUATION METRICS

Datasets such as IEMOCAP, MELD, SEMAINE, DEAP, CMU-MOSEI, and RECOLA provide multimodal benchmarks with categorical and dimensional labels as given in Table 13.5 [14]. Evaluation uses classification metrics (Accuracy, F1-score, Precision/Recall), regression metrics (MSE, RMSE, PCC, CCC), and temporal metrics (DTW, time-windowed F1). These enable fair model comparison but also highlight issues with dataset imbalance, synchronization, and cultural bias [15,16].

13.10 CHALLENGES AND RESEARCH OPPORTUNITIES

Among the major challenges, such aspects are the data quality, the absence of modalities, cultural variability, real-time requirements and the insufficient general applicability across domains [17]. Privacy, consent, and algorithmic bias represent ethical issues that are also a hindrance. The use of self-supervised learning, domain adaptation, lightweight edge deployment, federated learning, and explainable AI techniques are opportunities [16].

13.10.1 Gap in Quality Data

No evidence of inconsistency in data quality was found. Noisy, incomplete or missing modalities are one of the main issues of multimodal systems. An example is that part of the face could be covered, speech distorted by noise in the background, or the physiology sensors could not effectively record correct signals because of movement. Such discrepancies may worsen the output of models.

Research Opportunity: The creation of modality-agnostic frameworks, imputation approaches and powerful data fusion schemes capable of managing missing or corrupted data without the system itself being affected.

13.10.2 Diversity in Display of Emotions

The expression of emotions is highly diverse in terms of cultures, sex, age, and personality of a person. A smile could imply happiness in one situation and in another, it could be a social disguise of an uncomfortable situation. Emotion modelling is complex by virtue of this diversity.

Research Opportunity: Working with cross-cultural datasets, adaptive learning models and personalized emotion recognition systems able to work across the user populations with some ability to customize.

TABLE 13.5
Available Datasets

Dataset	Modalities Covered	Emotion Labels Type	Total Samples	Description
IEMOCAP	Speech, video (facial expressions), text	Categorical (e.g., happy, sad, angry)	~12 hours of recordings	A widely used dataset featuring dyadic conversations by actors with detailed annotations for both categorical and dimensional emotions.
MELD	Audio, video, text	Categorical (7 emotions)	13,000+ utterances	Derived from the TV series *Friends*, it supports multimodal emotion and sentiment analysis in a multi-speaker context.
SEMAINE	Audio, video, facial action units	Dimensional (Valence, Arousal)	~6 hours of video	Contains spontaneous emotional interactions between humans and virtual agents, labeled with continuous emotion dimensions.
DEAP	EEG, physiological signals, facial video	Dimensional (Valence, Arousal, Dominance)	32 participants, 40 videos each	Designed for affective computing using physiological and visual cues in response to emotional stimuli (music videos).
CMU-MOSEI	Text, audio, video	Categorical and Dimensional	23,453 annotated sentences	Includes sentiment, emotion, and personality labels across six emotions and large-scale speaker diversity.
RECOLA	Audio, video, ECG, EDA	Dimensional (Valence, Arousal)	23 subjects	Focused on collaborative tasks in French, annotated in real-time for continuous emotions.

13.10.3 Real-Time Processing and System Efficiency

Real-life uses, like in-vehicle monitoring or telemedicine, need emotion detection systems to run in real time and with low latency whilst being able to utilize very few computational resources. Deep learning architectures and multimodal data fusion tend to add to the complexity of the processing.

Research Opportunity: The investigation of lightweight models, edge computing services, and lean fusion approaches that can commit to real-time inference without compromising performance.

13.10.4 Limitations on the Dataset: Standardization Problems

The datasets that are publicly available usually are too small, do not contain equal representation of emotion, or are multimodally synchronized. Also, the annotation schemes and labelling procedures are diverse and it is hard to make fair comparisons between models.

Research Opportunity: Developing large, balanced and well-annotated multimodal images with standardized emotion taxonomies and evaluation measures to achieve reproducibility and benchmarking.

13.10.5 Generalizability and Domain Adaptation

These trained models that have been trained using controlled datasets usually do not work well in unknown conditions or real-world situations because of domain shift. This restricts usage in solutions such as classrooms, hospitals or in public environments.

Research Opportunity: Investigating domain adaptation and transfer learning algorithms and zero-shot learning and enhancing model robustness in different working conditions.

13.10.6 Ethical, Legal, and Privacy Concerns

The process of gathering and processing sensitive emotional data brings up various issues of consent of the users, data privacy, algorithmic biasness, and emotional manipulation. These systems are likely to be abused with irresponsible governance and corruption of trust.

Research Opportunity: Promoting ethical principles, privacy-saving computation (e.g., federated learning), and explainable AI (XAI) to bring the use of emotion recognition systems into a transparent, fair, trustworthy, and accountable state.

13.10.7 Interpretability and Transparency

Multimodal deep learning models can be black-box models. This restricts their use in areas that demand accountability and interpretability in areas like healthcare or education.

Research Opportunity: The development of interpretable fusion architectures and attention-driven models that can provide an insight into the decisions and allow human-in-the-loop systems.

13.11 APPLICATIONS OF MULTIMODAL EMOTION DETECTION

Applications span across domain [18–20]:

- *Healthcare*: Monitoring and assistive patient care, mental health.
- *Education*: Adaptive emotive intelligent tutoring systems.
- *Customer Service*: Sentiment-reasonable assistants.
- *Automotive*: Fatigue/stress driver monitoring systems.
- *Entertainment and gaming*: Adaptive storytelling by emotion of player.

Comparative Impact: Healthcare and education have an increased societal impact because they have a direct effect on well-being and learning outcomes, whereas gaming and entertainment are more focused on engagement and personalization.

13.12 CONCLUSION

Multimodal emotion recognition has been an important move in the direction of affective computing in which classical categorical, dimensional, and appraisal-based theories are being incorporated into deep learning models and fusion configurations. Transformers and GNNs offer better performance in cross-modal alignment than CNNs and RNNs, although they have higher initiatives. Fusion strategies provide trade-offs (accuracy, robustness and real-time feasibility) and this necessitates a good choice depending on the application environment. Multimodal systems have a huge potential of constructing emotionally intelligent and socially responsible applications of AIs by resolving the existing data quality, ethical, and efficiency problems in real-time. Future studies are encouraged to further incorporate theoretical model of emotion in designing models, loss functions and assessment metrics with which deep learning systems will be psychologically reasonable when they become used in more complex environments.

REFERENCES

1. Wu, You, Qingwei Mi, and Tianhan Gao. "A comprehensive review of multimodal emotion recognition: Techniques, challenges, and future directions." *Biomimetics* 10, no. 7 (2025): 418.
2. Lian, Hailun, Cheng Lu, Sunan Li, Yan Zhao, Chuangao Tang, and Yuan Zong. "A survey of deep learning-based multimodal emotion recognition: Speech, text, and face." *Entropy* 25, no. 10 (2023): 1440.
3. Mamieva, Dilnoza, Akmalbek Bobomirzaevich Abdusalomov, Alpamis Kutlimuratov, Bahodir Muminov, and Taeg Keun Whangbo. "Multimodal emotion detection via attention-based fusion of extracted facial and speech features." *Sensors* 23, no. 12 (2023): 5475.
4. Dai, Wei, Dequan Zheng, Feng Yu, Yanrong Zhang, and Yaohui Hou. "A novel approach to for multimodal emotion recognition: Multimodal semantic information fusion." *arXiv preprint arXiv:2502.08573* (2025).

5. Lai, Songning, Xifeng Hu, Haoxuan Xu, Zhaoxia Ren, and Zhi Liu. "Multimodal sentiment analysis: A survey." *Displays* 80 (2023): 102563.
6. Krishna, D. N., and Ankita Patil. "Multimodal emotion recognition using cross-modal attention and 1D convolutional neural networks." In *Interspeech*, pp. 4243–4247. 2020.
7. Tsai, Yao-Hung Hubert, Shaojie Bai, Paul Pu Liang, J. Zico Kolter, Louis-Philippe Morency, and Ruslan Salakhutdinov. "Multimodal transformer for unaligned multimodal language sequences." In *Proceedings of the Conference. Association for Computational Linguistics. Meeting*, vol. 2019, p. 6558. 2019.
8. Gu, Xin, Yinghua Shen, and Jie Xu. "Multimodal emotion recognition in deep learning: a survey." In *2021 International Conference on Culture-oriented Science & Technology (ICCST)*, pp. 77–82. IEEE, 2021.
9. Abdullah, Sharmeen M. Saleem Abdullah, Siddeeq Y. Ameen Ameen, Mohammed AM Sadeeq, and Subhi Zeebaree. "Multimodal emotion recognition using deep learning." *Journal of Applied Science and Technology Trends* 2, no. 01 (2021): 73–79.
10. Yang, Xiaocui, Shi Feng, Yifei Zhang, and Daling Wang. "Multimodal sentiment detection based on multi-channel graph neural networks." In *Proceedings of the 59th Annual Meeting of the Association for Computational Linguistics and the 11th International Joint Conference on Natural Language Processing (Volume 1: Long Papers)*, pp. 328–339. 2021.
11. Patlar Akbulut, Fatma. "Hybrid deep convolutional model-based emotion recognition using multiple physiological signals." *Computer Methods in Biomechanics and Biomedical Engineering* 25, no. 15 (2022): 1678–1690.
12. Dixit, Chhavi, and Shashank Mouli Satapathy. "Deep CNN with late fusion for real time multimodal emotion recognition." *Expert Systems with Applications* 240 (2024): 122579.
13. Xie, Jinbao, Jiyu Wang, Qingyan Wang, Dali Yang, Jinming Gu, Yongqiang Tang, and Yury I. Varatnitski. "A multimodal fusion emotion recognition method based on multitask learning and attention mechanism." *Neurocomputing* 556 (2023): 126649.
14. Hu, Guimin, Yi Xin, Weimin Lyu, Haojian Huang, Chang Sun, Zhihong Zhu, Lin Gui, Ruichu Cai, Erik Cambria, and Hasti Seifi. "Recent trends of multimodal affective computing: A survey from NLP perspective." *arXiv preprint arXiv:2409.07388* (2024).
15. Simić, Nikola, Siniša Suzić, Nemanja Milošević, Vuk Stanojev, Tijana Nosek, Branislav Popović, and Dragana Bajović. "Enhancing emotion recognition through sfederated learning: A multimodal approach with convolutional neural networks." *Applied Sciences* 14, no. 4 (2024): 1325.
16. Liu, Chenyu, Xinliang Zhou, Yihao Wu, Yi Ding, Liming Zhai, Kun Wang, Ziyu Jia, and Yang Liu. "A comprehensive survey on EEG-based emotion recognition: A graph-based perspective." *arXiv preprint arXiv:2408.06027* (2024).
17. Ramaswamy , Manju Priya Arthanarisamy, and Suja Palaniswamy. "Multimodal emotion recognition: A comprehensive review, trends, and challenges." *Wiley Interdisciplinary Reviews: Data Mining and Knowledge Discovery* 14, no. 6 (2024): e1563.
18. Li, Jiang, Xiaoping Wang, Guoqing Lv, and Zhigang Zeng. "GraphMFT: A graph network based multimodal fusion technique for emotion recognition in conversation." *Neurocomputing* 550 (2023): 126427.

19. Liu, Chenyu, Xinliang Zhou, Yihao Wu, Ruizhi Yang, Zhongruo Wang, Liming Zhai, Ziyu Jia, and Yang Liu. "Graph neural networks in EEG-based emotion recognition: a survey." *arXiv preprint arXiv:2402.01138* (2024).
20. Lu, Nannan, Zhiyuan Han, Min Han, and Jiansheng Qian. "Bi-stream graph learning based multimodal fusion for emotion recognition in conversation." *Information Fusion* 106 (2024): 102272.

14 Multimodal Disentangled Representation Learning for Enhanced User Behavior Analysis in Recommendation Systems

M.K. Kavitha Devi, K. Indira, Raja Lavanya, S. Jeniba, and V. Dharsana

14.1 INTRODUCTION

The flooding of user-generated content following the increase of online platforms, and it presents both opportunities and challenges to the recommendation algorithms. Effective user behavior analysis is a prerequisite of individualized experiences that could result in an enhanced user pleasure and engagement. Such methods, however, are largely based on single model data sources, and this has always led to inaccurate representations of user patterns and preferences.

The proposed chapter presents the multimodal disentangled representation learning as a combination of multiple modes of data to develop a unified knowledge of the behavior of the user, which eliminates the gap created by traditional methods. The challenges in representing the user behavior that occurs currently, the literature review, the implementation of the proposed system, and the results of the conducted experiments, as well as the implications of the findings, will be discussed in the sections presented in the following sections. A number of approaches to analyzing user behavior with regards to recommendation systems have been suggested, as can be observed through the available literature. The traditional methods, such as the popular matrix factorization and collaborative filtering, may become generally accepted because they are easy to employ and highly efficient at measuring interactions of the user-item data. Nonetheless, the impact of such methods is usually

DOI: 10.1201/9781003678359-14

not optimal and does little to incorporate the additional contextual data provided in the multimodal data sets.

Recently, the combination of multiple types of data, including textual, visual, and behavioral data, has been investigated. Studies on deep learning techniques have demonstrated potential in finding complex relationships within multimodal data and particularly applying neural networks. Disentangled representation learning has the potential to be more efficient and interpretable with the advantageous property of enabling segregation and representation. In spite of these changes, a challenge exists as far as the effectiveness of the integration of multimodal data and the surety of generalizability across numerous users is concerned. In proposing a distinctive method that gives the best of multimodal data and learning disentangled representations, this chapter aims to address this gap.

The online market has grown at such a high rate that the user-generated content has become so much so that we address the crippling user-generated content volume. Although this information is useful in understanding user taste, it is a great challenge to the conventional recommendation that is mainly based on uni-modality data like explicit ratings, or the history of purchases. Such traditional models as matrix factorization (MF) and collaborative filtering (CF) tend to be weak on the aspects of personalization, context-awareness, and interpretation.

Contemporary customers form their preferences in a diverse medium—in reviews, images, clicks, purchasing patterns, and even social activities. The multimodal information is important to capture and process this data to come up with viable, appealing, and meaningful suggestions.

In order to deal with this deficiency, this chapter introduces multimodal disentangled representation learning (MDRL) as a new framework that mixes textual, visual, and behavioral data and decomposes user behavior into segments that can be interpreted. MDRL is trained to learn disentangled representations of the various dimensions of user preference without combining with others, which is not only better at predicting the actual preferences, but it is also more explainable in terms of what the recommendations are basing themselves on.

14.2 MULTIMODAL DISENTANGLED REPRESENTATION LEARNING (MDRL)

14.2.1 Overview of the MDRL Framework

With the digital conversion, the increasing amount of user data, be it clicks or reviews, visual interactions, demands highly potent recommendation systems capable of processing such user information intelligently and incorporating the same in the system. Conventional recommendation systems consider solely one type of data such as ratings of users or browsing records. Nevertheless, this can lead to poor personalization, in particular when the users communicate via an array of contents. In order to address this drawback, the multimodal disentangled representation learning (MDRL) framework is proposed.

MDRL aims not only to integrate these modalities but also to factorize the factors of user behavior. An example of such would be that the user might prefer a smartphone because of its camera (feature appreciation), its sleekness (aesthetic preference), or it is the current trend (temporal interest). The disentanglement mechanism facilitates the system to process these factors in an independent and identical manner.

Modality-specific encoders, like long short-term memory (LSTM) to encode text, convolutional neural network (CNN) to encode images, and recurrent neural networks (RNNs) to encode sequential data about the behavior, are used by MDRL technically. Every encoder produces a latent representation, which encodes the information peculiar to that modality. These representations are then fed through a layer of disentanglement to learn independent latent variables, which construe discrete behavioral factors.

Such latent variables are then transferred to a fusion mechanism that optimizes the disentangled outputs into a singular vector. This fusion may be either a simple concatenation, or a more advanced attention-based technique. The recommendation layer then inputs this fused representation and reliably gives ranked results based on user interest.

Such a layered design has several advantages as it can support superior personalization, robustness to missing data and improved interpretability of the MDRL framework. As an example, the system will be able to recommend products despite the fact that no image data is provided, although the system will rely on behavioral and textual information.

The MDRL framework that has been proposed is a set of various components that should be able to capture and represent the behavior of a user across diverse modalities:

Data modalities: In the framework, three kinds of data are integrated: behavioral data (the history of the interaction of a user), visual data (images of products), and textual data (accepted by a user). In the case of a smartphone, the textual information will be the user reviews of the features, the visual information will be a set of device images, and the behavior information will be all the user clicks and purchases as well as rating. MDRL is a significant neural framework that combines three major types of data: behavioral, visual and textual data.

14.2.1.1 Behavioral Data

Behavioral data consists of user activities like clicks or purchases, ratings, and amount times on specific content. Behavioral data indicates the explicit and implicit information conversion of the interaction of a user with a digital platform. This involves action like clicks on object, buying history, rating of products, browsing duration, adding to cart, watching history (in cases of media), and even scrolling sequences. Such interactions can be seen as the direct manifestation of user intentions, interest, and their preference, so behavioral data becomes one of the most important input in recommender systems.

Behavioral data, in relation to MDRL, gives temporal and contextual cues. An example is a user who is always touched by the high-resolution camera phones having a preference in the hardware of a camera phone. Likewise, regular contact at times-of-day-based demands or in times of sale can get one to know the buying trends or availability preferences. Often, behavioral data is sequential, and thus models, such as RNNs or transformers with attention, are normally applied to analyze it.

Notably, behavioral information can also ensure filling in the blanks that are created because of the lack of visual or textual data. In the event that a product image or review is not in place, user interaction data continues to bear enough sign to give the recommendation models intelligent predictions.

14.2.1.2 Visual Data

Visual data usually comprises product image or even thumb images, which are used to provide a visual expression to a product. Visual data is the images or thumbnails of the goods used to be displayed with the goods in online stores or the design schemes. In online shopping, the images are used by users in order to determine the physical aspect of a product, including its color, shape, texture, or beauty. As an example, a person who has his/her interest in furniture may pay attention to the grain of wood or the pattern of the upholstery, whereas in fashion, the contrast of the color or the way the material drapes of the item may be more important.

In order to derive meaningful details of a visual information, CNNs are used to identify high-level characteristics. Other than being instrumental in classification of items, these features are also instrumental in the modeling of preferences. An example is that a user could engage in recurrent buying of products that are dark in color implying a visual preference which a system can learn and take action on it.

The data in the form of visual representations is even more prominent in cases where information presented in written forms is scarce. Reviews will often be eschewed by the user in multilingual or global platform environments because of language barriers, yet the analytic to be learned (derived using visual input) will contain useful learning information. Hence, visual modality can be a strong supplement in multimodal learning.

14.2.1.3 Textual Data

Text data consist of what is written by the user such as a review or a comment. Textual data are made up of user reviews, comments, ratings with commentary as well as questions or a discussion about a product. Those writings are semantically rich texts that describe subjective views, emotions, and ratings on detailed product reviews.

Textual data, in case of the absence of behavioral data, often describes why something was done as opposed to what was done by showing the behavioral data. The review that says, "Battery lasts all day and charges fast," also mentions a particular feature (battery life) that was valued by the customer. This data is run through text encoders like LSTM, BERT, or transformers in order to infer a context, sentiment polarity, and keyword focus.

A special opportunity of textual data is that the recommendation can be explained. A system would recommend a product and give the justification, "On your interest to buy the fastest charging phones we have found this product in our records and we can refer it as, fast charging phones as that is what you like," quoted literally some of the review alerts and this will be accompanied with the reference that the user is interested in.

Disentangled representation learning: The system core is a neural network that can learn disentangled representations, that is, it can split multiple aspects of user behavior. Encoder–decoder networks are applied to this, with each modality being input, independently, to the encoder, and then concatenated with learned information. Once the model reads through the reviews, it can decide as to whether the sentiment is positive or negative. These image processing aspects such as color and design would be of interest to image-processing networks and the usage trend would be recorded by the behavioral data network.

The most important concept in the MDRL framework is the disentangled representation learning concept that is concerned with recovering independent factors of variation of multimodal inputs. The conventional representation learning algorithms tend to create tangled features that identify, where each of the latent variables may depict a combination of factors. This makes it non-interpretable and rigid. Disentangled representation learning addresses this issue such that each latent variable has different semantic meanings.

Architecture is usually encoder–decoder of each modality. These networks also learn latent factors that identify particular elements of the data, independently of each other. As an example, a text review encoder may detect the sentiment, product features being mentioned, or comparisons. At the same time, a visual encoder will recognize the color interests, patterns or kind of product preferred, whereas the behavioral encoder will note the sequences of usage and frequency.

The encoder outputs are given to a disentanglement layer, which compels the latent variables to encode independent aspects of behavior of the user. This is usually by means of methods like $beta$-VAE (variational autoencoders), InfoGAN, or mutual information constraints. Separating these factors out, the model may single-out and decode the likes of the aesthetic, brand loyalty, or the engagement pattern.

A significant benefit to a disentangled learning method is its interpretability. Considering that each latent factor is connected with a particular behavior trait, system designers and end-users can gain a stronger idea as to why particular recommendations are made. Having an example, when the recommendation is motivated by a couple of factors, such as a camera quality, then it is much easier to justify and believe in the results of the system.

In addition, disentangled representations are resistant to missing data. In case one of the modalities is absent (e.g., an image or text is not available), the system can use other modalities to come up with informed predictions. In real-world problems, it is essential where missing data is the rule.

Integration mechanism: The fusion layer tries to increase the ability of the model to utilize complementary information via combining the representations that were learned with several modalities. The multi-level integration lends itself to more specific recommendations on the basis of personal preferences. Under this method, the sentiments of the review and the combination of these sentiments with the visual input or the usage pattern such as high engagement in photo applications would be allowed.

After being encoded and disentangled, the second transformation moment that is vital to the MDRL pipeline is related to the integration mechanism. In this component, there is the integration of the latent factors learned in multiple modalities in order to create a single user representation by the recommendation engine.

The MDRL framework employs a two-level fusion approach, in which feature-level fusion of the modality-specific representations followed by refinement on them via modeling of the interactions. These features are either concatenated by a fusion layer in a simple way, in an element-wise manner or in an attention-based way. The attention mechanisms enable the model to ensure that various modalities have an important weight depending on the task or the context of the user.

As a concrete example, a user seeking fashion items would have a high visual bias, but one seeking electronics would be willing to place a higher emphasis on text-based opinions. The fusion layer dynamically adapts in order to accentuate the most informative modality.

The resulting fused vector is then fed to a recommendation layer, which is usually a feed-forward neural network, or potentially a ranking model, predicting a list of personalized recommendations. The predicted relevance scores, which are based on the fused representation, rank this output.

One of them is multi-level integration that enables the interaction between information at early and late fusion stations. This is such that low-level information as well as the high-level abstractions are able to be captured in the final user profile.

Such integration renders the system being very flexible and sensitive to user intent. As an example, a user who regularly utilizes smartphones with a high review score discussing the topic of camera quality and the one that spends more time surfing through pictures of smooth camera modules can appreciate the tool of MDRL, which will not occur to a user that has not actively searched this type of merchandise.

A combination of various sources of information reduces dependence on one modality with MDRL and thus allows them to create more accurate and context-sensitive recommendations, as well as more fine-grained personalization.

14.2.2 Recommendation Systems

Recommendation systems aim to find out items, which could be items, videos, or articles, of user interest or insight. These systems have since entered into the fabric of digital platforms (e-commerce sites, streaming services, news aggregators, etc.). They have traditionally used two main strategies, namely, content-based filtering and collaborative filtering.

Content-based filtering is concerned with suggesting the similar items that the user liked before. It applies item attributes—such as genre, brand or specifications—in developing a profile of consumers' likes and dislikes. By way of example, if the user looks at a lot of sci-fi films, the system will indicate further sci-fi films, no matter what some other users might be watching. This approach is effective in managing new users, which however creates a situation of so-called filter bubble in which the recommendations are too narrow and redundant. The resultant filter bubble produces a biased account, the reduction of which would result in a narrowing of the users interests due to further exposure to new or heterogeneous material. Besides, content-based systems need rich metadata, and they can be ineffective when the attributes of the items are scarce or ambiguous.

Collaborative filtering (CF), in turn, uses the matrices of user-item interaction. It presupposes that those users that had similarity in the preferred items will continue to do so. An example can be given where should User A and User B have rated a series of goods, or items in the similar amount, items loved by User A, but not viewed by User B will be suggested to User B. It is a scalable and successful model but is subject to a significant drawback referred to as a cold start problem in which a system lacks sufficient information about new users or items to make any suggestions.

Matrix factorization (MF) is one of the common methods in collaborative filtering. It breaks up the interaction matrix into latent user and item dimensions. It has an inner product of these vectors that predicts the rating of an item by the user. In spite of being elegant as a theory, in real-life situations, user behavior is much more subtle and MF fails to address that. It also finds it difficult to support contextual or multimodal content such as text, images or time.

14.2.3 Limitations of Traditional Systems

Cold start: None of its instances are able to make recommendations to new users/items.

Sparsity: Data turns out to be sparse because most users utilize a fairly small part of the items.

The absence of context: They do not look at the contextual clues such as time, visual stimulation, or sentiments.

In order to curb these difficulties, recent systems have neural architectures, graph-based systems, and reinforcement learning to learn more complicated sets. Multimodal input integration has been of great necessity especially since the interaction with the user is no longer based on clicks and rating but the reviews, likes, shares, and media consumption.

MDRL can remedy shortcomings in the conventional recommendation systems. It brings contextual understanding, makes cold start issues less severe with cross-modal guesses, and enhances accuracy of predictions using multimodality richness. Besides, considering the delinking of latent variables, MDRL enables system designers to know the reason to make a recommendation, which enhances clarity.

14.3 THE MDRL FRAMEWORK

14.3.1 Core Concept

The MDRL model combines modality-wise data and seeks disentangled variables that represent different behavioral dimensions of users quite independently. The MDRL framework proposes a complex and structured data strategy of integration and analysis of data associated with individual users based on multiple modalities.

At its essence, it embraces the fusion of behavioral, visual, and textual inputs in a series of deep learning-based steps that pull out, disentangle, compose, and

TABLE 14.1
MDRL Framework

Modality	Source	Preprocessing Techniques
Textual data	User reviews, comments	Tokenization, stop-word removal, embeddings (Word2Vec, BERT)
Visual data	Product images, thumbnails	Normalization, resizing, augmentation, feature extraction (ResNet)
Behavioral data	Clickstreams, purchases, browsing logs	Time encoding, sequence structuring, vectorization

interpret user behavior. That allows producing exceptionally accurate and explicable recommendations. Each of the steps in the pipeline is explained in detail as seen below:

It runs through a well-organized pipeline:

1. Textual, visual, and behavioral preprocesses
2. Deep learning encoders in feature extraction
3. The disentanglement layer is aimed to learn interpretable latent factors
4. Modality combination combining the fusion layer
5. The output of ranked results by the recommendation layer

Preprocessing of textual, visual, and behavioral data

The preprocessing module is the basis of the MDRL framework as it performs operations on each modality that prepares it for further processing. Raw data on users is usually untidy, unformatted, and in a wide variety of forms, thus needing careful transformation. Table 14.1 shows the preprocessing techniques used for textual, visual, and behavioral data in the MDRL framework. These techniques convert raw data into structured formats that can be effectively used for learning and recommendation tasks.

14.3.2 Feature Extraction Using Deep Learning Encoders

At this step of the pipeline, each preprocessed modality is fed through deep learning-based encoders to obtain high-level encoded representations that could be subsequently fused to bring out the recommendations. In the textual data case, through a text encoder (LSTM, GRU, or transformer-based models), the feature extraction is done weighing the textual data as sentiment, keywords, and contextual embeddings, expressing the opinion of the user. In terms of visual presentation, any image encoder (e.g., CNN or vision transformer, e.g., ResNet) is able to capture the visual aesthetic, shapes, and colors that could sway preference. Lastly, in behavioral data, behavioral encoder, which is an RNN or attention network-based explores patterns in the engagement and temporal patterns of use. Table 14.2 summarizes three deep learning encoders for text, image, and behavioral data. Using transformers, CNNs, and RNNs, they extract modality-specific features to create rich multimodal representations for intelligent systems.

TABLE 14.2
Feature Extraction Using Deep Learning Encoders

Encoder Type	Architecture Used	Output Features
Text encoder	LSTM / GRU / Transformers	Sentiment, keywords, contextual embeddings
Image encoder	CNN / ViT (e.g., ResNet)	Visual aesthetics, shapes, colors
Behavioral encoder	RNN / Attention Networks	Engagement patterns, temporal usage trends

TABLE 14.3
Disentanglement Layer to Learn Interpretable Latent Factors

Latent Factor	Interpretation
z1	Preference for visual aesthetics
z2	Sentiment toward specific features
z3	Frequency of engagement

14.3.3 Disentanglement Layer to Learn Interpretable Latent Factors

Disentangled representation learning is an effective strategy that makes the factor of user interest represented as a latent vector in a model one-to-one perfection an intelligible and beneficial aspect in the features of representation. Constraints in the learning process can be enforced to encourage the independence of distinct factors by using special models to do so, like β-VAE (beta variational autoencoder) or InfoGAN. This implies that every dimension (or latent factor) is promoted to represent distinct, mutually exclusive trait, for example, one dimension may say how sensitive a user is to the visual beauty of products, another encodes how a user feels about specific features of the products, and another captures such behavioral patterns as frequency of use. This kind of disentanglement does not only increase the transparency of the model but also allows more accurate and predictable recommendations since it recommends against specific factors that are of interest to individual users.

Disentangled representation learning enhances interpretability by attempting to guarantee that the latent variables represent an independent and semantically interpretive feature of user behavior (e.g., visual preference, sentiment, engagement pattern, and so forth). This separation allows reasoning precisely as to the reason behind the generation of recommendations unlike entangled embeddings. Moreover, disentangling also makes the system more robust, as it isolates the effect of missing or noisy modalities and makes dependable recommendations even in the case of partial multimodal information. Table 14.3 presents a disentanglement layer that decomposes user data into three interpretable latent factors, each capturing a distinct aspect of user behavior. These factors represent visual aesthetic preferences, sentiment toward specific features, and engagement frequency, enabling more transparent and explainable recommendations.

14.3.4 Fusion Layer for Modality Combination

Once disentangled latent variables that constitute particular areas of user interests are extracted, these variables should be combined in an effective way to comprise a comprehensive user profile. This is achieved in the fusion layer where multiple modalities, that is, visual, textual, and behavioral data, are being fused. The fusion strategy is also important since it dictates the fusion of information of the various sources. An example is early fusion, which combines features prior to the proceeding training of the models so that all modalities learn together.

The late fusion methodologies deal with every modality separately and after combining their results, which is convenient in case the sources of data are heterogeneous.

More sophisticated techniques such as attention-based fusion naturally operationalize context-specific priorities on individual models and thus offer versatile and user-contextually aware suggestions. The attention-based fusion dynamically allocates the weights to modalities depending on the context of the user, which allows to fine-tune personalization, and gated fusion explicitly regulates the modality contribution to avoid the contribution of noisy or unreliable modalities. Whereas attention is more context aware in terms of recommendation, gated fusion is more stable in cases of shallow or incomplete information. MDRL builds on such adaptive fusion strategies by an optimum integration of multimodal information.

In the meantime, gated fusion is trained to activate the most productive modalities selectively relying on the downstream task, thus enabling the network to adapt itself to paying attention to the most distinctive characteristics. Using the proper fusion strategy makes the multidimensional interests of a user captured and used in a precise manner. Table 14.4 presents four fusion strategies for combining multimodal features: Early Fusion, Late Fusion, Attention-Based Fusion, and Gated Fusion. These methods enable effective, flexible, and context-aware integration of information from different modalities.

TABLE 14.4
Fusion Layer for Modality Combination

Fusion Strategy	Description
Early fusion	Combines modality features before model training
Late fusion	Fuses results after each modality is processed independently
Attention-based fusion	Assigns dynamic weights to modalities based on user context
Gated fusion	Learns to activate relevant modalities based on task-specific learning

14.4 RECOMMENDATION LAYER TO OUTPUT RANKED RESULTS

The last step of the recommendation pipeline takes in the overall profile developed in the preceding steps, which consists of a combination of information in any modality as well as disentangled latent factors, and interprets it to recommend action. This gets accomplished by a cascade of well-planned neural network modules. The fully connected layers are used to project the user representation that has been fused to item ranking scores, learning higher order relationships between user interests, and item characteristics. A ranking loss function is then used, which optimizes the network to produce a high-quality recommendation by directly targeting ranking goals using a framework such as Bayesian personalized rank (BPR), hinge loss, or cross-entropy loss. This will make the most pertinent elements pertinent to every user to be ranked first in the output of the model.

The final step is output—a model trained to select and show only the top-N ranked items that provide to each user a personalized, prioritized list of recommendations that closely matches their personal tastes. This multiple stage turnaround ensures that the interest area of users is diversified and subtle and is richly captured and reflected back in final recommendations being as accurate as personalized as possible. Table 14.5 describes the recommendation layer that converts fused multimodal representations into personalized ranked results. Fully connected layers generate ranking scores, ranking loss functions optimize recommendation quality, and the output layer produces the top-N items tailored to each user.

14.4.1 System Architecture

The design of the multimodal disentangled representation learning (MDRL) structure is meant to be a complex design that stages learning between multiple data modalities. The initial theory is preprocessing. This is where normalization and extraction of features is performed over all forms of data including textual user review and visual picture of the product as well as user history activity. The text input is cleaned, tokenized, and embedded using pretrained models such as Word2Vec, GloVe, or BERT. Such models provide vectors that denote the semantic meaning of the words, for example, 0.1, 0.2, 0.3, denotes love. Visual data may include resizing of images, normalization of images, and extraction of features of images, which are already trained CNNs to generate feature vectors such as [0.5, 0.6, 0.7] and the like.

TABLE 14.5
Recommendation Layer to Output Ranked Results

Component	Description
Fully connected layers	Maps fused representation to item ranking scores
Ranking loss function	Optimizes for ranking quality (e.g., BPR, Hinge, Cross-Entropy)
Output layer	Produces top-N ranked item list personalized for each user

Any history of contact makes a feature vector that gives an account of the behavioral patterns of the user like ratings and stamps like [2, 4.5] that shows that there have been two contacts with an average rating of 4.5. Specific encoding phase is then applied to each of the modalities where the interaction data is fed to an RNN, which learns to understand sequential user activity, text data to an LSTM or Transformer that is able to capture contextual relationships, and images to CNNs that learn visual features.

Isolating variables that determine the actions of a user, the framework isolates specific aspects of user preferences, for example, sentiment, aesthetic appeal, and behavioral trends in the process of disentangling. It decomposes, which means more interpretability and customization to ascertain why people would wish to see high-quality photos or specific features of a product. The last features then construct a fusion layer, which chiefly employs attention mechanisms or concatenation to combine latent representations of the two modality inputs to a sole, integrative representation (vector). Lastly, with the ranks of product interest, the personalized recommendation is, in fact, created by the recommendation jolt through deep learning. The softmax layer yields an output as the probability of the preferences of the user which lists out a set of well identified recommended products as (Product A, Product B, and Product C). This multimodal approach simplifies providing very specific and personal recommendations depending on the detailed profile of the user and product characteristics.

The MDRL system has a properly designed architectural process that works in a pipeline combining multimodal disentangled representation learning that analyzes user behavior. The phases are as follows in details:

1. Preprocessing Stage

There are three types of data prepared to train this model at this first tier namely textual, visual and behavioral data.

- *Textual data*: Cleaned and tokenized and then embedded with models, such as Word2Vec, GloVe, or BERT in order to derive the semantics.
- *Visual data*: The image is resized into an acceptable shape and utilized as input in pretrained CNNs such as ResNet in retrieving image representations.
- *Behavioral data*: A set of sequences e.g. [2, 4.5] to denote the number of actions and average rating.

2. Encoding Stage

The modality of each of these engages a specific deep encoding:

- *Text*: Embedded with LSTM or Transformers attention-based embeddings.
- *Images*: Processed through CNNs to get the visual semantics.
- *User activity*: Learned as a sequence of behavior with RNNs.

3. Disentanglement Phase

The stage divides latent variables into dimensions that represent various behavior dimensions (e.g., sentiment, aesthetic preference). Such interpretable models such as VAE can decipher such characteristics.

4. Fusion Layer

A mechanism of combining (fusion) such as attention or concatenation compacts modality-specific latent features into a single user profile.

5. Recommendation Layer

The softmax or sigmoid layer runs deep neural models to produce ranked lists of items in personalized form, e.g., ["Product A", "Product B"].

6. Evaluation Metrics

Performance is assessed using:

- *Precision*: How many recommended items are relevant.
- *Recall*: How many relevant items are recommended.
- *F1-Score*: Harmonic mean of precision and recall.
- *MAP* (mean average precision): Measures average precision at each correct item.
- *NDCG* (normalized discounted cumulative gain): Evaluates ranking quality based on position relevance.

The performance is measured using several metrics:

- Precision
- Recall
- F1-score
- Mean average precision (MAP)
- Normalized discounted cumulative gain (NDCG)

14.4.2 Practical Uses of MDRL Online Retail and E-Commerce

Through the integration of textual data (user reviews, product descriptions), visual data (product images), and behavioral data (clicks, cart additions, purchases), MDRL improves tailored product suggestions in e-commerce platforms such as Amazon, Flipkart, or Alibaba. The algorithm can determine whether a user favors particular features, functionality, or brand dependability by deciphering these characteristics, and then accurately tailoring results to their interests.

Platforms for online streaming: Using text metadata, visual clues (such as thumbnails and scene styles), and behavioral interactions, MDRL is used by platforms such as Netflix, Spotify, and YouTube to suggest material that aligns with a user's tastes.

14.4.2.1 Personalization on Social Media

By examining post captions and comments (text), images or videos (visual), and user involvement patterns (behavioral), social media sites use MDRL to suggest posts, advertisements, or new relationships.

14.4.2.2 MOOCs and Online Education

By assessing learning resources (text and videos), peer interactions (forums), and user behavior (completion rates, quiz scores), MDRL facilitates individualized learning on educational platforms.

14.4.2.3 Platforms for Health and Wellbeing

In order to provide individualized care and health recommendations, MDRL is used in telemedicine and wellness platforms using textual notes, medical image analysis, and behavioral patterns.

14.4.3 Future Directions and Open Challenges

14.4.3.1 Scalability

MDRL models face difficulty handling increasing users, modalities, and real-time interaction. Techniques such as model compression, distributed training, and edge AI deployment are potential solutions.

14.4.3.2 Handling Missing Modalities

In real-world data, some users/items may not have all modalities. MDRL must address this via modality dropout, zero-shot learning, or synthetic data generation.

14.4.3.3 Interpretability and Transparency

Though MDRL offers disentangled representations, understanding and explaining recommendations to end-users remains challenging. Explainable AI techniques are required.

14.4.3.4 Concerns about Ethics and Privacy

When MDRL processes personal behavioral and visual data, privacy considerations must be addressed. Strategies for mitigating bias, federated learning, and differentiated privacy are essential.

14.4.3.5 Time-Based Modeling and Changing Preferences

Over time, user preferences change. To adjust to these changes, future MDRL models should use temporal learning strategies like time-LSTM or real-time reinforcement learning.

14.5 DATASET AND EXPERIMENTAL SETUP

14.5.1 Dataset Used

The Amazon Multimodal Product Review Dataset employed in this research includes about 50,000 user-item interactions, which consists of 12,000 users and 9,000 products with available textual reviews, product images and ratings. The data has a slightly lopsided rating pattern with more positive ratings, which is a real-world user behavior. Experiments are conducted on the Amazon Multimodal Product Review Dataset, containing:

- Textual reviews
- Product images
- Ratings and user interaction history

14.5.2 Preprocessing

The main problems of preprocessing were the lack of modalities, variable length reviews, and uneven image quality. Text data were tokenized and padded to constant lengths, images were resampled and normalized and behavioral sequences temporally encoded. There was a pre-processing standardization of all the steps to be reproducible.

- *Text*: Tokenized, embedded using Word2Vec/BERT
- *Images*: Resized, normalized, features extracted via ResNet/CNN
- *Behavioral data*: Encoded via sequences of ratings and timestamps

14.5.3 Evaluation Metrics

Metrics of evaluation measure the recommendation systems in various aspects. Precision indicates accuracy, that is, how many of the items recommended are truly relevant to the user (going by the recommendation). Recall is a percentage expression of all pertinent items that the system has been able to retrieve demonstrating completeness. NDCG (normalized discounted cumulative gain) is used to measure the quality of ranking based on where the relevant items the user wants to see are located on the recommendation list. Averaged with the scores, MAP (mean average precision) gives a general precisional score. Finally, F1-score is a unification of precision and recall value and, in other words, it makes up a harmonic value of precision and recall, providing an accurate yet balanced measurement of the accuracy and coverage of such a system. Collectively, these measures give an in-depth picture of the effectiveness of recommendations. Table 14.6 presents evaluation metrics used to assess recommendation performance. Precision, Recall, NDCG, MAP, and F1-score measure the relevance, ranking quality, retrieval effectiveness, and overall accuracy of the recommended items.

TABLE 14.6
Evaluation Metrics

Metric	Description
Precision	% of relevant items among recommended
Recall	% of relevant items successfully retrieved
NDCG	Ranking quality of recommendations
MAP	Mean precision across all users
F1-score	Harmonic mean of precision and recall

14.5.4 Results

With regard to these five metrics, Precision, Recall, NDCG, F1, and MAP, Table 14.7 helps to compare the three different recommendation models, including the matrix factorization (MF), collaborative filtering (CF), and the proposed modified deep RNN locally (MDRL). The greater it is, the more effective the performance of the model. The moderate-withhold sample results of MF and CF are relatively close, with significant differences being observed between the two models, MF, and CF each on all metrics in their comparison. As an example, MDRL gets a precision of 0.74 and a recall of 0.69, as opposed to 0.61 and 0.55 with MF. Likewise, the NDCG and F1 of MDRL is significantly greater with a higher ranking and balanced recommendations. Overall ranking quality plunges, as shown by the MAP score, and is also significantly better in MDRL. This sharp rise in values proves that the MDRL provides much more accurate, relevant, and effective suggestions than classical ones. Its adaptive fusion strategy and disentangled latent representations hypothesized with healthy interpretability and fault-resistance results in the performance gains of MDRL. The application of MDRL is always superior to matrix factorization and collaborative filtering because it is capable of simultaneously modeling multimodal user data and splitting independent preference dimensions. Unlike the baseline methods which only have user-item interactions, MDRL will leverage both textual reviews, visual features, and behavioral patterns, and result in more precise ranking and better personalization. The statistical significance of the performance gains made by MDRL is confirmed with the help of paired t-tests with the 95 percent confidence level ($p<0.05$) that the gains made are not the result of random variation.

Table 14.7 is a bar chart analysis of Precision, Recall, NDCG, F1-score, and MAP compared with Precision and Recall, NDCG, F1-score, and MAP of MF, CF, and MDRL models, which gives a clear visual explanation of the better performance of MDRL.

The developed results can be easily shown to prove the effectiveness of the new MDRL model compared with conventional matrix factorization and collaborative filtering methods. With the help of sophisticated methods, MDRL does not only refine the quality of predicted items recognized in terms of their relevancy (which can be observed through the increased precision and recall) but enriches the quality of ranking (as can be witnessed by the high NDCG and MAP values).

TABLE 14.7
Evaluation Metrics

Model	Precision	Recall	NDCG	F1	MAP
Matrix factorization (MF)	0.61	0.55	0.58	0.58	0.50
Collaborative filtering (CF)	0.63	0.56	0.60	0.59	0.52
MDRL (Proposed)	0.74	0.69	0.71	0.71	0.64

Moreover, its overall performance in all the measures such as the F1 score is close to the one of the other models thus implying that MDRL is more suitable in covering the vast and multifaceted user preferences contributing to more individualistic and fulfilling recommendations. These enhancements support the fact that MDRL has a major potential to enhance overall user engagement and overall effectiveness in a recommendation system.

14.6 ADVANTAGES AND INTERPRETABILITY

- *Explainability*: Every latent factor is translatable (e.g., prefers red-colored items).
- *Robustness*: Acts better when data are sparse or missing owing to multimodal redundancy.
- *Personalization*: Models more sophisticated and situational interests of a user.
- *Scalability*: Performs well on large-sized dataset with heterogeneous input.

14.7 APPLICATIONS AND USE CASES

- *E-Commerce*: By ensuring the visualization of the products and the reviews, the products can be recommended.
- *Streaming sites*: Recommendations by show/movie thumbnails and description as well as viewing history.
- *Healthcare*: Individual considerations of the treatment in the usage of the medical images, reports, and history of behavior.
- *Education*: (Adaptive learning platforms, based on text logs, interaction behavior, and visuals).

14.8 CONCLUSION AND FUTURE ENHANCEMENTS

The MDRL architecture proposed achieves a substantial improvement in user behavior analysis in recommendation systems, and synthesizes textual, visual, and behavioral data. This approach is multimodal and gives a clearer definition of what the user likes as opposed to traditional early methods of matrix factorization and collaborative filtration. Through its disentanglement of latent factors, it separates specific interests of different users and it allows more personalized, more transparent, and more interpretable recommendations. The experiments focus on one representative multimodal dataset to demonstrate the effectiveness of MDRL compared to traditional MF and CF methods. While ablation studies, cross-validation, and comparisons with additional recent multimodal baselines are not included in this version; they are planned as part of future work to further strengthen the experimental evaluation.

Going forward, to enhance the user representations, other modalities such as audio data or even signals on social media may be added to enable the system to respond to complex and dynamic user persuasions. All in all, MDRL is an excellent and versatile model of next-generation recommendation systems, which can provide highly topical and catchy personalized suggestions.

REFERENCES

[1] Liu, Fan, et al. "Disentangled multimodal representation learning for recommendation." *IEEE Transactions on Multimedia* 25 (2022): 7149–7159.

[2] Hu, Haibing, et al. "Modality-disentangled feature extraction via knowledge distillation in multimodal recommendation systems." *IEEE Transactions on Computational Social Systems* 12(6) (2025): 4517–4531. DOI: 10.1109/TCSS.2025.3555653.

[3] Khalafaoui, Yasser, et al. "CADMR: Cross-attention and disentangled learning for multimodal recommender systems." *arXiv preprint arXiv:2412.02295* (2024).

[4] Yang, Dingkang, et al. "Disentangled representation learning for multimodal emotion recognition." *Proceedings of the 30th ACM International Conference on Multimedia.* 2022: 1642–1651. DOI: 10.1145/3503161.3547754

[5] Cui, Zeyu, et al. "Disentangled item representation for recommender systems." *ACM Transactions on Intelligent Systems and Technology (TIST)* 12(2) (2021): 1–20. DOI: 10.1145/3445811.

[6] Khalafaoui, Yasser. "Advancements in multimodal and unsupervised learning with application to recommender systems." (2024). arXiv:2412.02295. vol. 2412.02295.

[7] Ning, Hailong, et al. "Disentangled representation learning for cross-modal biometric matching." *IEEE Transactions on Multimedia* 24 (2021): 1763–1774.

[8] Zhang, Xiaokun, et al. "Beyond co-occurrence: Multimodal session-based recommendation." *IEEE Transactions on Knowledge and Data Engineering* 36, no. 4 (2023): 1450–1462.

[9] Liu, Huiting, et al. "Unbiased meta reinforcement learning for interactive recommender systems." *IEEE Transactions on Multimedia* 27 (2025): 5185–5197. DOI: 10.1109/TMM.2025.3543045.

[10] Yuan, Xu, et al. "Cross-modal feature alignment and fusion with contrastive learning in multimodal recommendation." *Knowledge-Based Systems* 326 (2025): Article 114020. DOI: 10.1016/j.knosys.2025.114020

15 Generative AI in Multimodal Biological Data

Transformations, Techniques, and Future Directions

Tomojit Ghosh and Caleb Hendren

15.1 INTRODUCTION

The rapid advancement of Omics technologies—including genomics, transcriptomics, proteomics, and metabolomics—has fundamentally transformed modern biomedical research. These platforms enable comprehensive profiling of biological systems across multiple molecular layers, providing unprecedented insight into cellular mechanisms, disease etiology, and therapeutic response. However, the complexity of Omics data—characterized by high dimensionality, sparsity, and small sample sizes—poses significant challenges for conventional computational and statistical methods, which often struggle to capture nonlinear relationships and hidden biological structures.

Generative artificial intelligence (GenAI) has recently emerged as a transformative approach for modeling complex biological data. By learning the underlying distributions that govern biological variability, generative models can synthesize realistic data, impute missing values, and denoise experimental measurements. These capabilities are particularly valuable for high-dimensional, low-sample-size (HDLSS) datasets, where the scarcity of samples can hinder robust inference and model generalization [15]. Through data augmentation and manifold learning, GenAI enables deeper exploration of biological heterogeneity while improving the reliability of downstream analyses.

Among existing architectures, generative adversarial networks (GANs) [9], variational autoencoders (VAEs) [13], and denoising diffusion probabilistic models (DDPMs) [12] have shown strong potential in biological applications. GAN-based frameworks can generate synthetic gene expression or single-cell profiles to expand limited datasets and enhance classifier training. VAEs are effective for learning compact, interpretable latent representations that capture key biological variations and support dimensionality reduction. Diffusion models, a more recent innovation, excel

DOI: 10.1201/9781003678359-15

at denoising and data reconstruction, providing robust solutions for incomplete or noisy Omics measurements. Collectively, these models have enabled advances in areas such as single-cell clustering, trajectory inference, biomarker discovery, and disease modeling.

The integration of generative AI with biological data analysis represents a notable shift toward more flexible and data-efficient computational biology [5,18]. Beyond improving data quality and interpretability, these models pave the way for cross-modal data fusion and simulation of cellular processes under hypothetical perturbations—offering opportunities for hypothesis generation and experimental design [15]. As the field evolves, future research must focus on improving model interpretability, addressing data bias, and ensuring the ethical use of synthetic biological data. This chapter provides a comprehensive overview of the role of generative AI in biological discovery. We begin by outlining the theoretical foundations of modern generative models, followed by applications across Omics modalities, including synthetic data generation, imputation, and cross-modal alignment. The chapter concludes with a discussion of current limitations, open challenges, and emerging directions shaping the next generation of data-driven biology.

15.2 FOUNDATIONS OF GENERATIVE MODELS FOR BIOLOGICAL DATA

Generative artificial intelligence (GenAI) models have advanced rapidly over the past decade, offering new frameworks for data-driven discovery in both natural and life sciences. In biological research, these models address the analytical limitations of high-dimensional, sparse, and heterogeneous *Omics* data, where classical statistical methods often fail to capture nonlinear dependencies or hidden biological structure. Unlike discriminative models that map inputs to labels, generative models learn the underlying probability distribution of the data, allowing them to produce realistic synthetic samples and infer latent representations that encode biologically meaningful variation. This section provides a conceptual overview of three major families of generative models—generative adversarial networks (GANs), variational autoencoders (VAEs), and denoising diffusion probabilistic models (DDPMs)—that form the theoretical backbone of modern generative AI in computational biology.

15.2.1 Generative Adversarial Networks (GANs)

GANs, first introduced by Goodfellow et al. in 2014 [9], revolutionized the field of generative modeling by framing data synthesis as a game-theoretic competition between two neural networks. A GAN consists of a generator (G) and a discriminator (D). The generator attempts to create synthetic samples that resemble true data, while the discriminator evaluates whether a given sample originates from the real data distribution. Through this adversarial interplay, the generator gradually learns to approximate the underlying data manifold, producing highly realistic outputs.

Mathematically, the generator transforms random noise vectors $z \sim p_z(z)$—typically sampled from a standard normal or uniform distribution—into synthetic data points $G(z)$. The discriminator receives both real samples $x \sim p_{\text{data}}(x)$ and generated samples $G(z)$ and outputs the probability $D(x)$ that the input is real. The joint objective of both networks is expressed as a mini-max optimization problem:

$$\min_{G}\max_{D} V(D,G) = \mathbb{E}_{x\sim P_{data}(x)}\left[\log D(x)\right] + \mathbb{E}_{z\sim P_z(z)}\left[\log\left(1-D(G(z))\right)\right] \quad (15.1)$$

Here, the discriminator seeks to maximize the log-likelihood of correctly classifying real and fake samples, while the generator minimizes the same objective by producing samples that the discriminator misclassifies as real. When the game reaches equilibrium, the generated distribution $p_g(x)$ ideally converges to the true data distribution $p_{data}(x)$.

Several architectural variants have been developed to improve GAN training stability and sample quality, including the Wasserstein GAN (WGAN), LeastSquares GAN (LSGAN), and conditional GAN (CGAN). Conditional GANs, for example, introduce class or condition labels into both G and D, enabling conditional sample generation $G(z, y)$ that captures cell-type-specific or disease-specific patterns.

In the biological domain, GANs have been applied to expand limited gene expression datasets, generate synthetic single-cell RNA-seq profiles, and remove experimental batch effects [17]. By synthesizing realistic yet diverse data, GANs can augment training sets for classification and clustering, assist in rare-cell identification, and facilitate the benchmarking of computational pipelines. The flexibility of GANs also allows them to model nonlinear relationships across modalities such as transcriptomics and proteomics, providing a versatile tool for multi-Omics data generation.

15.2.2 Variational Autoencoders

VAEs, introduced by Kingma and Welling [13], extend classical autoencoders through a probabilistic framework that integrates variational inference with deep learning. VAEs learn a latent variable model in which observed data x are assumed to arise from unobserved latent variables z. The model approximates the intractable posterior distribution $p_\theta(z|x)$ using a tractable variational distribution $q_\phi(z|x)$ parameterized by a neural encoder network. This probabilistic formulation enables VAEs to both compress high-dimensional data into a low-dimensional latent space and generate new samples consistent with the learned distribution.

The encoder network $q_\phi(z|x)$ outputs the mean $\mu_\phi(x)$ and variance $\sigma^2_\phi(x)$ of a Gaussian distribution over latent variables z, while the decoder network $p_\theta(x|z)$ reconstructs input data from z. The optimization objective maximizes the marginal log-likelihood of the data, which is approximated by the evidence lower bound (ELBO):

$$\mathcal{L}(\theta,\phi;x) = E_{q_\phi(z|x)}\left[\log p_\theta(x|z)\right] - D_{KL}\left(q_\phi(z|x)\,|\,p_\theta(z)\right) \quad (15.2)$$

The first term represents the expected reconstruction log-likelihood, encouraging accurate data reconstruction, while the second term is the Kullback–Leibler divergence that regularizes the approximate posterior toward the prior $p_\theta(z)$, typically $\mathcal{N}(0,I)$.

To allow gradient propagation through the stochastic latent variables, VAEs employ the reparameterization trick:

$$z = \mu_\phi(x) + \sigma_\phi(x) \odot \epsilon, \qquad \epsilon \sim \mathcal{N}(0,I) \tag{15.3}$$

where $\odot$ denotes element-wise multiplication. This formulation expresses random sampling as a differentiable operation, enabling end-to-end training via stochastic gradient descent.

In biological applications, VAEs are particularly valuable for learning smooth, interpretable latent representations of *Omics* data. In single-cell transcriptomics, VAEs can project tens of thousands of gene-expression dimensions into a compact latent space, facilitating visualization, clustering, and trajectory inference [23]. The probabilistic nature of VAEs also supports uncertainty quantification, making them useful for denoising noisy sequencing data and imputing missing gene counts. Variants such as CVAEs and hierarchical VAEs further extend this framework to capture structured dependencies across multiple biological conditions or cell types.

15.2.3 Denoising Diffusion Probabilistic Models

Denoising diffusion probabilistic models (DDPMs), commonly known as diffusion models [12], represent a more recent and highly expressive class of generative frameworks. Unlike GANs or VAEs, which generate samples directly from a latent code, diffusion models generate data by gradually transforming random noise into structured samples through a sequence of learned denoising steps. The approach is inspired by nonequilibrium thermodynamics [28], where data generation is viewed as the reversal of a diffusion (noise) process.

The forward process $q(x_t|x_t-1)$ defines a Markov chain that progressively corrupts the original data x_0 by adding Gaussian noise at each timestep t∈ [1, T]. It is expressed as:

$$q\left(x_t|x_{t-1}\right) = \mathcal{N}!\left(x_t; \sqrt{1-\beta_t}\,x_{t-1}, \beta_t\mathbb{I}\right) \tag{15.4}$$

where $\beta_t \in (0,1)$ controls the noise variance schedule. After many diffusion steps, the data distribution approaches an isotropic Gaussian. A closed-form solution gives the noisy sample at any step t as:

$$q\left(x_t|x_0\right) = \sqrt{\overline{\alpha_t}}\,x_0 + \sqrt{1-\overline{\alpha_t}}\,\epsilon, \qquad \epsilon \sim \mathcal{N}(0,I), \tag{15.5}$$

where $\alpha_t = 1-\beta_t$ and $\overline{\alpha_t} = \prod_{i=1}^{t} \alpha_i$. This forward process is fixed; learning occurs during the reverse phase.

The reverse process $p_\theta\left(x_{t-1} \middle| x_t\right)$ learns to invert the diffusion by predicting the noise component ε_t that was added during the forward step. A neural network parameterized by θ is trained to approximate this reverse conditional distribution. The model is optimized by minimizing a mean-squared error between the true and predicted noise:

$$\mathscr{L}_\theta = \mathbb{E}_{t,x_0,\epsilon}\left[\|\epsilon - p_\theta\left(x_t, t\right)\|^2\right] \tag{15.6}$$

Once trained, the model generates new data by sampling Gaussian noise and iteratively applying the learned denoising transformations from $t{=}T$ down to $t{=}\ 1$. Each reverse step refines the sample, reconstructing complex structures such as gene-expression patterns or molecular interaction profiles.

Diffusion models offer several advantages over GANs and VAEs. Their likelihood-based formulation avoids adversarial instability and mode collapse, leading to more stable training and consistent sample diversity. They also excel at capturing fine-grained structural details, making them ideal for tasks that require precise quantitative fidelity. In computational biology, DDPMs can be applied to synthesize realistic *Omics* data for rare diseases, enhance data quality via denoising, and integrate heterogeneous modalities by modeling shared latent diffusion trajectories. Emerging research further combines diffusion processes with graph or sequence encoders to extend their applicability to biological networks, protein conformations, and molecular dynamics simulations.

15.3 GENERATIVE MODELING FRAMEWORKS FOR BIOLOGICAL DATA

The recent adoption of generative artificial intelligence (GenAI) in computational biology has accelerated data-driven discovery across transcriptomics, epigenomics, proteomics, and metabolomics. Generative models alleviate the limitations of high dimensionality, sparsity, and sample scarcity that characterize most biological datasets. By learning the underlying distribution of complex molecular measurements, these models can synthesize realistic biological samples, recover missing information, and reveal latent structures representing cellular or disease processes. The following sections outline representative applications of three key model families—GANs, VAE), and DDPMs—with emphasis on architectures that have demonstrated utility in biological data analysis.

15.3.1 GAN-Based Models for Biological Data

GANs have been widely explored for tasks ranging from single-cell data augmentation to cross-omics translation and dimensionality reduction. Their adversarial framework enables the synthesis of highly realistic samples, even from small, noisy datasets.

DR-A: Lin et al. [17] proposed DR-A, an adversarial variational autoencoder with dual matching (AVAE-DM) for dimensionality reduction of scRNA-seq data. The model employs two discriminators to align both latent and reconstructed distributions with the real data distribution, while the generative process models count data using a zero-inflated negative binomial likelihood. When evaluated on five benchmark scRNA-seq datasets, DR-A demonstrated superior clustering performance, achieving normalized mutual information (NMI) scores that were 10–15 percent higher than those obtained with t-SNE, UMAP, and scVI.

cscGAN: The conditional single-cell GAN (cscGAN) [22] learns gene–gene dependencies across multiple cell types and generates synthetic single cells conditioned on cell-type labels. Using the Wasserstein distance with gradient penalty mitigates mode collapse and stabilizes training. On a PBMC dataset with 68 k cells, cscGAN generated realistic expression profiles whose synthetic clusters overlapped the real cell populations, improving downstream classification accuracy.

scDEC: scDEC [18] jointly performs dimensionality reduction and clustering for scATAC-seq data by coupling two symmetric GANs that transform data between latent and observed spaces. Continuous latent variables (z) represent style, while discrete variables (c) capture cluster identity. Adversarial and round-trip losses ensure consistency between domains. On benchmark datasets, scDEC achieved higher clustering performance than scABC, SCALE, and cisTopic, accurately identifying rare subpopulations.

LSH-GAN: Proposed by Lall et al. [14], LSH-GAN addresses the high-dimensional small-sample (HDSS) problem in scRNA-seq data. By integrating locality sensitive hashing into the generator input, it selects representative subsamples from real data, combining them with noise to guide realistic synthetic-cell generation. Benchmarking on multiple scRNA-seq datasets showed that LSH-GAN produced more biologically coherent cells and improved clustering accuracy (adjusted Rand Index $\approx$0.9) compared with cscGAN, SUGAR, and Splatter.

scIGANs: Xuet al. [33] developed scIGANs for dropout imputation in scRNAseq data. Instead of imputing from observed neighbors, it generates synthetic cells using a conditional GAN and imputes missing values via k-nearest neighbors among generated samples. The method recovers biologically meaningful variation without oversmoothing, outperforming 11 competing imputation algorithms and maintaining rare-cell structure.

omicsGAN: Ahmed et al. [1] proposed omicsGAN, a cascaded WassersteinGAN framework to integrate two omics layers (e.g., mRNA and miRNA) using their interaction network. Iterative updates propagate information across modalities, producing synthetic datasets with enhanced predictive power for disease phenotypes. On TCGA cancer cohorts, synthetic data improved AUC scores for subtype prediction and patient survival analysis compared to original measurements.

scPreGAN: Wei et al. [32] introduced scPreGAN to predict single-cell transcriptional responses to external perturbations. By coupling an autoencoder with dual GAN modules, the framework learns shared latent representations and condition transformations. Evaluations on interferon beta (IFN-β), helminth infection, and LPS stimulation datasets showed that scPreGAN most accurately reproduced real perturbed states, outperforming scGen, trVAE, and stVAE.

Overall, GAN-based approaches have shown promise for enhancing limited biological datasets, recovering incomplete information, and capturing transitions between cellular or experimental states. Nevertheless, their training process can be sensitive to hyperparameter choices and prone to instability. Moreover, because the learned representations are not explicitly probabilistic, model interpretability remains limited. These challenges have motivated the development of hybrid architectures that integrate adversarial learning with probabilistic or variational frameworks to achieve greater stability and biological transparency.

15.3.2 VAE-Based Models for Biological Data

VAE-derived models form the backbone of probabilistic generative modeling in biology because of their stable optimization and interpretable latent spaces.

scVAE: Gronbech et al. [10] proposed scVAE to model raw scRNA-seq count data using Poisson, negative binomial, and zero-inflated variants rather than normalized expressions. This direct likelihood modeling preserves rare-gene signals and allows clustering without preprocessing. On PBMC and TCGA datasets, scVAE identified biologically relevant immune subtypes and outperformed Seurat in cluster accuracy.

mtVAE: Gomari et al. [8] developed mtVAE for metabolomics data, employing a conditional VAE to extract nonlinear latent features. Training on 4,500 human blood samples, mtVAE captured metabolite correlations linked to Type 2 diabetes, leukemia, and schizophrenia, outperforming PCA and kernel PCA in reconstruction error and biological interpretability.

CAMPA: Spitzer et al. [29] introduced the conditional autoencoder for multiplexed pixel analysis (CAMPA) for spatial proteomics. A conditional VAE learns condition-invariant latent representations of pixel-level molecular profiles, allowing clustering into consistent subcellular landmarks (CSLs) across perturbations. CAMPA achieved superior cross-condition alignment and imputation accuracy on high-resolution imaging datasets.

PaCCMann^{RL}: Born et al. [3] combined conditional VAEs with reinforcement learning to design anticancer compounds conditioned on transcriptomic profiles. Two VAEs (for gene expression and SMILES strings) are integrated with a drug-sensitivity critic. The framework generated novel molecules predicted to exhibit high efficacy across breast, lung, and prostate cancers, many resembling FDA-approved drugs.

MassGenie: Shrivastava et al. [27] developed MassGenie, a transformer-VAE hybrid translating mass-spectrometry peaks into chemical structures. Trained on synthetic spectra from FragGenie, it achieved 82.7 percent precision on 1,350 experimental spectra, correctly identifying small molecules de novo and outperforming traditional spectral-matching tools.

VAE-based frameworks are particularly effective for learning compact and interpretable latent representations, capturing data uncertainty, and integrating heterogeneous biological measurements. Their probabilistic formulation provides a principled basis for statistical inference and facilitates downstream analyses such as clustering or trajectory reconstruction. However, because of the latent space regularization, the generated samples often exhibit smoother variations compared to those produced by adversarial models.

15.3.3 Diffusion Models for Biological Data

Diffusion-based generative models (DDPMs) have recently emerged as a robust alternative to adversarial training, demonstrating superior fidelity and denoising capacity in omics and spatial data.

scDiffusion: Luo et al. [20] combined diffusion processes with a pretrained autoencoder to generate high-quality scRNA-seq samples. Conditional classifiers guide the denoising trajectory, allowing cell-type-specific and even novel-type synthesis. Across five single-cell datasets, scDiffusion matched or exceeded GAN-based models in data realism and diversity.

HiCDiff: Wang et al. [31] applied diffusion modeling to denoise single-cell HiC contact matrices. Using a residual convolutional network, HiCDiff achieved higher PSNR, SSIM, and HiCRep scores than DeepHiC, HiCSR, and SCHiCEDRN, effectively restoring 3D genomic contact patterns.

stDiff: Liet al. [16] proposed stDiff, a conditional DDPM that imputes missing genes in spatial-transcriptomics data using scRNA-seq references. A diffusion transformer predicts noise conditioned on shared genes, improving gene-level RMSE and clustering metrics (ARI, NMI) compared with Tangram, stPlus, and spaGE.

scDM: Yuet al. [37] introduced scDM for predicting protein expression from RNA data in CITE-seq. The model couples an unconditional diffusion network with an RNA-guided classifier to generate accurate protein counts. It achieved the highest Pearson correlations and lowest MSE against experimental protein data, surpassing Seurat 4, totalVI, and sciPENN.

Diffusion-based models offer stable training dynamics, high-quality data generation, and effective denoising capabilities, which make them well-suited for handling noise and variability in biological datasets. Their principal challenges lie in the substantial computational cost and slow sampling process, though recent advances in accelerated solvers and latent-space diffusion are beginning to address these limitations.

15.3.4 Literature Synthesis Grid

To consolidate the diverse generative modeling frameworks discussed above, Table 15.1 provides a comparative synthesis of representative methods across model families, highlighting the target biological modalities, primary analytical tasks, and reported quantitative outcomes.

TABLE 15.1
Literature Synthesis of Representative Generative AI Methods for Multimodal Biological Data, Highlighting Model Family, Modality, Analytical Task, and Quantitative Examples Reported in the Cited Studies

Model family	Representative methods	Omics / modality	Primary task	Quantitative example (from case studies)
GAN	DR-A; cscGAN; scDEC; LSH-GAN; scIGANs; omicsGAN; scPreGAN	scRNA-seq; scATAC-seq; multi-omics	augmentation; clustering; imputation; perturbation prediction	DR-A reports ~10–15% higher NMI vs. t-SNE/UMAP/scVI.
VAE	scVAE; mtVAE; CAMPA; totalVI; PaCCMannRL; MassGenie	scRNA-seq; metabolomics; spatial proteomics; CITE-seq	latent representation; denoising; integration; molecule design	Improved reconstruction vs PCA/kPCA; improved RNA–protein modeling.
Diffusion	scDiffusion; HiCDiff; stDiff; scDM	scRNA-seq; Hi-C; spatial transcriptomics; CITE-seq	high-fidelity generation; denoising; imputation; cross-modality prediction	Improved PSNR/SSIM and ARI/NMI vs baselines.
Multimodal fusion	scMVAE; SIMO	RNA–ATAC; RNA–methylation; spatial multi-omics	shared latent space; cross-modality translation; spatial alignment	Improved alignment and tissue architecture recovery.

This synthesis highlights a clear progression from adversarial and variational frameworks toward diffusion-based models, reflecting increasing emphasis on training stability, denoising capability, and fidelity in high-dimensional biological data.

15.4 GENERATIVE AI IN BIOLOGICAL RESEARCH: MODELS IN ACTION

Generative artificial intelligence (GenAI) models are increasingly applied to biological data to overcome challenges such as high dimensionality, small sample size, and pervasive noise. By learning the underlying probability distributions of complex molecular systems, these models can generate realistic biological samples, denoise experimental measurements, integrate multiple omics modalities, and simulate perturbation responses. This section surveys representative applications of GAN-, VAE-, and diffusion-based frameworks, emphasizing how they have been tailored to handle diverse omics data types and biological questions.

15.4.1 Synthetic Data Generation and Data Augmentation

Data scarcity is a central limitation in many biological studies, particularly in single-cell, rare-disease, or privacy-restricted datasets. Generative models enable the creation of realistic synthetic data that augment small samples and improve model generalization.

scGAN and LSH-GAN: Early applications such as scGAN [22] and its faster variant LSH-GAN [14] produce synthetic single-cell RNA-seq profiles that mimic real gene–gene correlations. By combining adversarial learning with locality-sensitive hashing, LSH-GAN accelerates convergence and better preserves cell-type structure, enhancing clustering performance and differentialexpression stability.

scDesign2 and diffusion models: The VAE-based scDesign2 [30] generates high-fidelity scRNA-seq data conditioned on cell type and sequencing depth, offering a statistical foundation for benchmarking pipelines. More recently, diffusion-based models such as scDiffusion [20] and stDiff [16] achieve even greater realism by iteratively denoising latent representations, producing synthetic cells that capture continuous developmental trajectories.

These synthetic-data frameworks allow researchers to evaluate algorithms, train classifiers on rare populations, and explore hypothetical perturbations while safeguarding sensitive patient data.

15.4.2 Data Denoising and Imputation

High-throughput sequencing data often suffer from dropout events, technical noise, and incomplete measurements. Generative models are uniquely suited to reconstruct missing values and restore biologically meaningful variability.

scIGANs and GAIN: GAN-based approaches such as scIGANs [33] generate realistic single-cell profiles to impute missing gene expression, avoiding oversmoothing common in nearest-neighbor methods. For tabular or proteomics data, the generative adversarial imputation network (GAIN) [36] employs a discriminator-guided mask to estimate missing entries, achieving superior recovery accuracy across omics benchmarks.

VAE-based denoising: Variational autoencoders, including scVAE [10] and totalVI [7], explicitly model observation noise through probabilistic likelihoods such as the negative binomial distribution. Their latent variables capture biological variation while filtering technical artifacts, leading to improved clustering, trajectory inference, and feature selection.

Diffusion-based restoration: Diffusion models, originally designed for image denoising, are now applied to omics domains for robust signal recovery. HiCDiff [31] restores chromatin contact maps from noisy single-cell Hi-C data by iteratively removing estimated noise, outperforming CNN- and GAN-based baselines in PSNR and HiCRep scores.

15.4.3 Multimodal and Cross-Omics Integration

Modern biological systems are inherently multimodal: a single cell simultaneously expresses RNA, proteins, chromatin accessibility, and spatial organization. Integrating such heterogeneous omics layers is essential for constructing comprehensive molecular portraits, yet it poses major challenges due to differing data scales, sparsity patterns, and noise distributions. Generative models—especially those based on variational and diffusion principles—offer a flexible probabilistic framework for fusing modalities into a unified latent representation, enabling cross-modality translation and improved downstream biological inference.

scMVAE: Xuet al. [38] introduced scMVAE, one of the first deep generative frameworks for integrating multi-omics single-cell data. scMVAE extends the traditional VAE by coupling multiple encoders and decoders corresponding to distinct omics layers (e.g., transcriptomic and epigenomic data), which jointly learn a shared latent representation of cellular states. Each encoder transforms its input modality into a latent Gaussian distribution, while all decoders reconstruct their respective data from the same latent variable z, enforcing cross-modality consistency through a shared evidence lower bound (ELBO):

$$\mathcal{L}_{\text{scMVAE}} = \sum_{m=1}^{M} \mathbb{E}_{q_{\phi_m}(z|x^{(m)})}\left[\log p_{\theta_m}\left(x^{(m)}|z\right)\right] - D_{\text{KL}}\left(q_{\phi_m}\left(z|x^{(m)}\right) \parallel p(z)\right) \quad (15.7)$$

where m indexes each omics modality. This joint optimization allows scMVAE to capture nonlinear dependencies between molecular layers while accommodating modality-specific noise models. In experiments on paired single-cell RNA–ATAC and RNA–DNA methylation datasets, scMVAE learned coherent low-dimensional embeddings that aligned cell clusters across modalities and improved cell-type annotation compared with canonical correlation analysis (CCA) and non-generative baselines. Importantly, the model could generate one modality from another (e.g., predict ATAC profiles from RNA), illustrating the generative power of the shared latent space.

SIMO: Building on this foundation, Yang et al. [34] proposed SIMO (spatial integration of multi-omics), a 2025 framework that extends multimodal generative modeling to spatially resolved single-cell data. While scMVAE integrates molecular layers measured in the same cells, SIMO additionally aligns these layers in spatial context—an emerging frontier in tissue-scale biology. SIMO employs a conditional variational autoencoder architecture in which spatial coordinates act as conditioning variables that modulate latent embeddings of transcriptomic, proteomic, and epigenomic features. The encoder learns cell-level latent codes ziinformed by both molecular and spatial inputs, while the decoder reconstructs expression patterns constrained by local neighborhood structure:

$$p_\theta\left(x_i \middle| z_i, s_i\right) = \mathcal{N}\left(f_\theta\left(z_i, s_i\right), \sigma^2 I\right) \tag{15.8}$$

where sidenotes spatial coordinates. Training jointly minimizes the reconstruction loss across modalities and a KL-divergence regularizer, ensuring spatial smoothness and biological fidelity. Evaluated on human brain and liver spatialomics datasets, SIMO accurately recovered tissue architectures, captured crosslayer gene–protein correlations, and identified spatially coherent cell-type transitions. Notably, it enabled the generation of missing molecular layers for spatial regions lacking complete measurements, offering a powerful tool for imputation and atlas construction.

Together, scMVAE and SIMO exemplify the evolution of multimodal generative modeling—from early probabilistic fusion of paired omics datasets to sophisticated spatially conditioned architectures capable of integrating molecular, spatial, and phenotypic dimensions. These frameworks highlight how generative AI can transform systems biology by reconstructing a unified, interpretable latent representation that bridges molecular modalities and spatial organization within complex tissues.

15.4.4 Disease Modeling, Perturbation, and Drug Response

Simulation generative models also enable virtual experiments that simulate how cells or tissues respond to genetic or chemical perturbations.

scPreGAN and perturbation models: scPreGAN [32] integrates an autoencoder with adversarial generators to predict single-cell expression responses under stimulation or infection, outperforming scGen and trVAE in reproducing perturbed transcriptional states. Such models can generalize to unseen cell types and conditions, supporting hypothesis generation for drug development.

VEGA and PaCCMann^{RL}: The VEGA framework [26] encodes pathway-level gene modules into a VAE decoder, yielding interpretable latent variables that correspond to biological pathways and transcriptional programs. PaCCMannRL [3] extends this concept to therapeutic design, combining VAEs with reinforcement learning to generate compounds optimized for transcriptomic profiles, linking molecular generation directly to disease phenotype.

Diffusion-based therapeutic modeling: Emerging diffusion architectures are beginning to simulate disease progression and treatment outcomes by modeling temporal transitions between cellular states. Their iterative refinement mechanism naturally supports time-conditioned data generation, offering an interpretable bridge between static snapshots and dynamic processes.

15.4.5 Emerging Trends and Limitations

Although generative models have significantly advanced biological data analysis, several challenges remain. GAN-based methods, while capable of generating realistic biological samples, often encounter training instability and limited interpretability [9,2,22]. VAE frameworks provide structured latent spaces but may yield overly smooth reconstructions that obscure subtle or rare-cell variations [13,19,26]. Diffusion models deliver state-of-the-art generative fidelity and strong denoising capacity yet demand substantial computational resources and long sampling times [12,16,20].

Future research is converging toward hybrid frameworks that integrate the interpretability of VAEs, the realism of GANs, and the stability of diffusion-based methods [17,35]. Another promising direction involves graph-informed and foundation models, which incorporate biological network structure or utilize large pre-trained multimodal embeddings for transfer learning across omics domains [4,6,11,25]. Finally, the establishment of benchmarking standards, biological validation protocols, and transparent evaluation metrics remain essential for translating these generative approaches into reliable tools for biomedical discovery [21,24].

REFERENCES

[1] Khandakar Tanvir Ahmed, Jiao Sun, Sze Cheng, Jeongsik Yong, and Wei Zhang. Multi-omics data integration by generative adversarial network. *Bioinformatics*, 38(1):179–186, 2022.

[2] Martin Arjovsky and L´ eon Bottou. Towards principled methods for training generative adversarial networks. arXiv preprint arXiv:1701.04862, 2017.

[3] Jannis Born, Matteo Manica, Ali Oskooei, Joris Cadow, and Mar´ ıa Rodr´ ıguez Mart´ ınez. Paccmannrl: Designing anticancer drugs from transcriptomic data via reinforcement learning. In International Conference on Research in Computational Molecular Biology, pp. 231–233. Springer, 2020.

[4] Hongyang Chen, Can Xu, Lingyu Zheng, Qiang Zhang, and Xuemin Lin. Diffusion-based graph generative methods. *IEEE Transactions on Knowledge and Data Engineering*, 36(12):7954–7972, 2024.

[5] Yongin Choi, Ruoxin Li, and Gerald Quon. sivae: Interpretable deep generative models for single-cell transcriptomes. *Genome Biology*, 24(1):29, 2023.

[6] Haotian Cui, Chloe Wang, Hassaan Maan, Kuan Pang, Fengning Luo, Nan Duan, and Bo Wang. scgpt: Toward building a foundation model for single-cell multi-omics using generative AI. *Nature Methods*, 21(8):1470–1480, 2024.

[7] Adam Gayoso, Zo¨ e Steier, Romain Lopez, Jeffrey Regier, Kristopher L Nazor, Aaron Streets, and Nir Yosef. Joint probabilistic modeling of single cell multi-omic data with totalvi. *Nature Methods*, 18(3):272–282, 2021.

[8] Daniel P. Gomari, Annalise Schweickart, Leandro Cerchietti, Elisabeth Paietta, Hugo Fernandez, Hassen Al-Amin, Karsten Suhre, and Jan Krumsiek. Variational autoencoders learn transferrable representations of metabolomics data. *Communications Biology*, 5(645), 2022.

[9] Ian Goodfellow, Jean Pouget-Abadie, Mehdi Mirza, Bing Xu, David WardeFarley, Sherjil Ozair, Aaron Courville, and Yoshua Bengio. Generative adversarial nets. In *Advances in Neural Information Processing Systems*, pp. 2672–2680, 2014.

[10] Christopher Heje Grønbech, Maximillian Fornitz Vording, Pascal N Timshel, Casper Kaae Sønderby, Tune H Pers, and Ole Winther. scvae: Variational auto-encoders for single-cell gene expression data. *Bioinformatics*, 36(16):4415–4422, 2020.

[11] Graham Heimberg, Tony Kuo, Daryle DePianto, Tobias Heigl, Nathaniel Diamant, Omar Salem, Gabriele Scalia, Tommaso Biancalani, Shannon Turley, Jason Rock, et al. Scalable querying of human cell atlases via a foundational model reveals commonalities across fibrosis-associated macrophages. *BioRxiv*, pp. 2023–07, 2023.

[12] Jonathan Ho, Ajay Jain, and Pieter Abbeel. Denoising diffusion probabilistic models. *Advances in Neural Information Processing Systems*, 33:6840–6851, 2020.

[13] Diederik P. Kingma and Max Welling. Auto-encoding variational bayes. arXiv preprint arXiv:1312.6114, 2013.

[14] Snehalika Lall, Sumanta Ray, and Sanghamitra Bandyopadhyay. LSH-gan enables in-silico generation of cells for small sample high dimensional scrnaseq data. *Communications Biology*, 5(1):577, 2022.

[15] Kyoungmin Lee, Taehyeong Kim, Mookyung Cheon, and Wookyung Yu. Unveiling oasis family as a key player in hypoxia–ischemia cases induced by cocaine using generative adversarial networks. *Scientific Reports*, 12(1):6734, 2022.

[16] Kongming Li, Jiahao Li, Yuhao Tao, and Fei Wang. stdiff: A diffusion model for imputing spatial transcriptomics through single-cell transcriptomics. *Briefings in Bioinformatics*, 25(3):bbae171, 2024.

[17] Eugene Lin, Sudipto Mukherjee, and Sreeram Kannan. A deep adversarial variational autoencoder model for dimensionality reduction in single-cell RNA sequencing analysis. *BMC Bioinformatics*, 21:1–11, 2020.

[18] Qiao Liu, Shengquan Chen, Rui Jiang, and Wing Hung Wong. Simultaneous deep generative modelling and clustering of single-cell genomic data. *Nature Machine Intelligence*, 3(6):536–544, 2021.

[19] Romain Lopez, Jeffrey Regier, Michael B Cole, Michael I Jordan, and Nir Yosef. Deep generative modeling for single-cell transcriptomics. *Nature Methods*, 15(12):1053–1058, 2018.

[20] Erpai Luo, Minsheng Hao, Lei Wei, and Xuegong Zhang. scdiffusion: conditional generation of high-quality single-cell data using diffusion model. *Bioinformatics*, 40(9):btae518, 2024.

[21] Faisal Mahmood. A benchmarking crisis in biomedical machine learning. *Nature Medicine*, 31(4):1060–1060, 2025.

[22] Mohamed Marouf, Pierre Machart, Vikas Bansal, Christoph Kilian, Daniel S. Magruder, Christian F. Krebs, and Stefan Bonn. Realistic in silico generation and augmentation of single-cell RNA-seq data using generative adversarial networks. *Nature Communications*, 11(1):166, 2020.

[23] Jinhee Park, Hyerin Kim, Jaekwang Kim, and Mookyung Cheon. A practical application of generative adversarial networks for RNA-seq analysis to predict the molecular progress of Alzheimer's disease. *PLoS Computational Biology*, 16(7):e1008099, 2020.

[24] Bhuvnesh Rai, Jyotika Srivastava, and Pragati Saxena. Challenges in translating laboratory findings to drug development. In *Protein Misfolding in Neurodegenerative Diseases*, pp. 319–354. Elsevier, 2025.

[25] Emre Sefer. Drgat: Predicting drug responses via diffusion-based graph attention network. *Journal of Computational Biology*, 32(3):330–350, 2025.

[26] Lucas Seninge, Ioannis Anastopoulos, Hongxu Ding, and Joshua Stuart. Vega is an interpretable generative model for inferring biological network activity in single-cell transcriptomics. *Nature Communications*, 12(1):1–9, 2021.

[27] Aditya Divyakant Shrivastava, Neil Swainston, Soumitra Samanta, Ivayla Roberts, Marina Wright Muelas, and Douglas B Kell. Massgenie: A transformer-based deep learning method for identifying small molecules from their mass spectra. *Biomolecules*, 11(12):1793, 2021.

[28] Jascha Sohl-Dickstein, Eric Weiss, Niru Maheswaranathan, and Surya Ganguli. Deep unsupervised learning using nonequilibrium thermodynamics. In International Conference on Machine Learning, pp. 2256–2265. PMLR, 2015.

[29] Hannah Spitzer, Scott Berry, Mark Donoghoe, Lucas Pelkmans, and Fabian J. Theis. Learning consistent subcellular landmarks to quantify changes in multiplexed protein maps. *Nature Methods*, 20:1058–1069, 2023.

[30] Tianyi Sun, Dongyuan Song, Wei Vivian Li, and Jingyi Jessica Li. scdesign2: A transparent simulator that generates high-fidelity single-cell gene expression count data with gene correlations captured. *Genome Biology*, 22(1):163, 2021.

[31] Yanli Wang and Jianlin Cheng. HiCDiff: single-cell Hi-C data denoising with diffusion models. *Briefings in Bioinformatics*, 25(4):bbae279, 06 2024.

[32] Xiajie Wei, Jiayi Dong, and Fei Wang. scpregan, a deep generative model for predicting the response of single-cell expression to perturbation. *Bioinformatics*, 38(13):3377–3384, 2022.

[33] Yungang Xu, Zhigang Zhang, Lei You, Jiajia Liu, Zhiwei Fan, and Xiaobo Zhou. scigans: Single-cell RNA-seq imputation using generative adversarial networks. *Nucleic Acids Research*, 48(15):e85–e85, 2020.

[34] Penghui Yang, Kaiyu Jin, Yue Yao, Lijun Jin, Xin Shao, Chengyu Li, Xiaoyan Lu, and Xiaohui Fan. Spatial integration of multi-omics single-cell data with simo. *Nature Communications*, 16(1):1265, 2025.

[35] Burak Yelmen and Flora Jay. An overview of deep generative models in functional and evolutionary genomics. *Annual Review of Biomedical Data Science*, 6(1):173–189, 2023.

[36] Jinsung Yoon, James Jordon, and Mihaela Schaar. Gain: Missing data imputation using generative adversarial nets. In International Conference on Machine Learning, pp. 5689–5698. PMLR, 2018.

[37] Hanlei Yu, Yuanjie Zheng, and Xinbo Yang. scdm: A deep generative method for cell surface protein prediction with diffusion model. *Journal of Molecular Biology*, 436(12):168610, 2024.

[38] Chunman Zuo and Luonan Chen. Deep-joint-learning analysis model of single cell transcriptome and open chromatin accessibility data. *Briefings in Bioinformatics*, 22(4):bbaa287, 2021.

16 Multimodal Deepfakes

A Comprehensive Understanding of the Role of Multimodal AI in Recent Deepfake Research and Development

Arnab Kumar Das and Ruchira Naskar

16.1 INTRODUCTION

The rapid growth of deep learning has enabled the creation of synthetic media to achieve remarkable levels of realism. Deepfake technology has become a widely used method for generating fake content, including altered videos, images, and audio recordings. This advancement poses a serious challenge to the credibility of digital media and can have more effects on individuals, organizations, and society as a whole.

The literature highlights multiple instances where deepfakes have been used to spread false information, manipulate public perception, and cause personal harm. The rise of deepfakes can be traced back to around 2017, when AI techniques—particularly generative models like generative adversarial networks (GANs) [1]—enabled the manipulation or creation of highly realistic media content.

We have witnessed examples of deepfake videos and photos featuring well-known public figures that appear authentic but are entirely AI-generated. Such synthetic content can be weaponized in damaging ways, including reputational attacks, fake evidence in legal settings, or even child exploitation crimes.

A harmful example involved a Reddit user sharing fake pornographic videos by superimposing the faces of celebrities like Gal Gadot, Taylor Swift, and Scarlett Johansson [2]. More recently, deepfakes have also been used in the political arena, with claims that manipulated videos were circulated during elections. In response, platforms like Facebook took action by banning such content ahead of the 2020 US elections [3]. Research by Shin et al. [4] also showed how deepfake news content can influence public behavior and perception.

Rather than providing benefits, the misuse of deepfake technology has created serious risks. Given that anyone can become a victim, there is an urgent need for a dependable and effective solution to counter this threat.

In this chapter, we discuss different types of methodologies by which the deepfake video or audio is created. Then, we will discuss the different types of benchmark audio

DOI: 10.1201/9781003678359-16

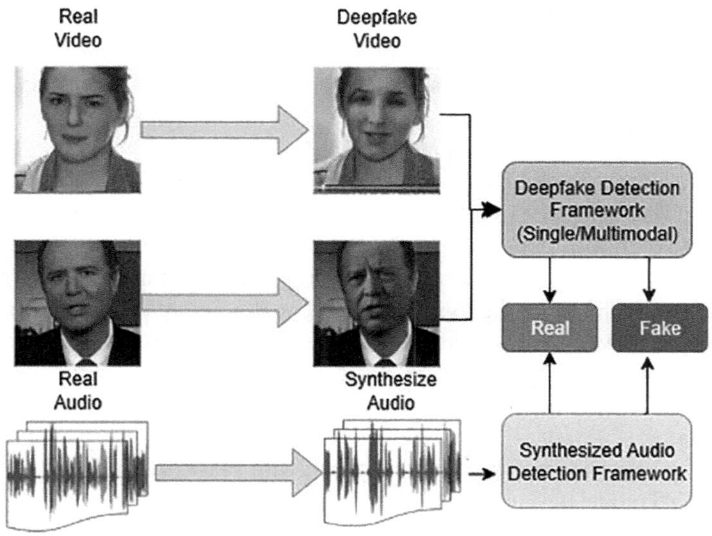

FIGURE 16.1 Workflow diagram of the create and detect the deepfake audio/video.

and video datasets for deepfake identification. Lastly, we have shown the different types of deepfake detectors using single and multi-modality approaches. An example of identifying the deepfake video/audio is illustrated in Figure 16.1, showing how to create the deepfake video or Audio and how to identify it through the detector module.

The following is the structure of the chapter: Section 16.2 explores various deepfake audio/video generation techniques. In Section 16.3, we provide an overview of the widely used and publicly available different types of deepfake videos and the spoofed audio benchmark dataset. A comparative analysis of each dataset discussed is also presented. Section 16.4 explores the deepfake detection frameworks in single and multimodal approaches using machine learning and deep learning. Finally, Section 16.5 offers concluding remarks along with insights into potential directions for future research in this domain.

16.2 MULTIMODAL DEEPFAKE GENERATION TECHNIQUES

Nowadays, deepfake technology, which encompasses a range of advanced digital media creation and manipulation methods using deep learning algorithms and sophisticated artificial intelligence, is now a significant threat because of its potential for abuse to deceive and mislead viewers [5–7]. In this section, we have discussed different types of audio/video generation techniques used to create more realistic video /audio of a target person.

16.2.1 Deepfake Audio Generation

The proliferation of advanced generative models has significantly blurred the line between genuine and fabricated speech, leading to the increasing prevalence of audio deepfakes. To create the spoofed audio, the attacker used different types of models like TTS (text-to-speech), voice cloning, voice conversion, and generative models.

TTS models are designed to transform written content into audible speech, often replicating natural human voice patterns. They process text input and generate corresponding synthetic audio output.

Embarrassingly Easy Text-to-Speech (E2TTS) [8] is a zero-shot one type of a TTS model that operates in a fully non-autoregressive manner, offering natural-sounding speech, high speaker similarity, and clear pronunciation. It processes the text by inserting filler tokens to convert it into a character-level sequence. The core components of the system include a *flow-matching spectrogram generator* and a *vocoder*. The spectrogram generator produces Mel spectrograms by aligning character sequences and is trained using an audio in-filling approach. The vocoder then takes the generated spectrogram and synthesizes it into spoken audio.

Fairytaler Fakes Fluent and Faithful speech with Flow matching (F5TTS) [9] is a zero-shot TTS model designed to produce highly natural and expressive speech, while also supporting effective speed control. The system works by combining a padded sequence of text characters with a reference audio input. F5TTS uses a diffusion transformer [10] paired with ConvNeXt V2 [11], which helps improve the alignment between text and speech during training. The input text is tokenized, and filler tokens are added to maintain the length of the speech signal.

Voice cloning is used to create a synthetic audio of a target person's voice using a machine learning model.

Speech vector to TTS (SV2TTS) [12] is a voice cloning technique that mainly uses a pipeline that includes a TTS system. It is particularly effective for generating moderately high-quality synthetic audio and supports real-time voice cloning (RTVC), even from audio clips just a few seconds long. SV2TTS is composed of three distinct sub-models, each trained independently. The first component is the speaker encoder, which learns to capture the unique vocal characteristics of a speaker. It processes an input audio (typically via its Mel spectrogram) and generates a fixed-length embedding representing the speaker's voice identity. In Figure 16.2, we have shown the screenshot of the SV2TTS model.

Voice conversion [13] changes one person's voice to sound like another without altering the linguistic content. Different types of generative adversarial network (GAN) and variational autoencoder (VAE) will be used in the voice cloning system. GAN has two parts: one is the generator to generate the fake audio, and the other is the discriminator that can check whether the synthesized audio is similar to real audio or not. In VAE, a transformer-based module is used to create synthesized audio. The example of voice conversion is CycleGAN-VC and VAE-based VC.

16.2.2 Deepfake Video Generation

In recent days, many deepfake video generation tools have become available. Some of them are open source, and some of them are used for commercial and academic purposes. In this section, we have discussed the different types of deepfake video generation tools or methodologies.

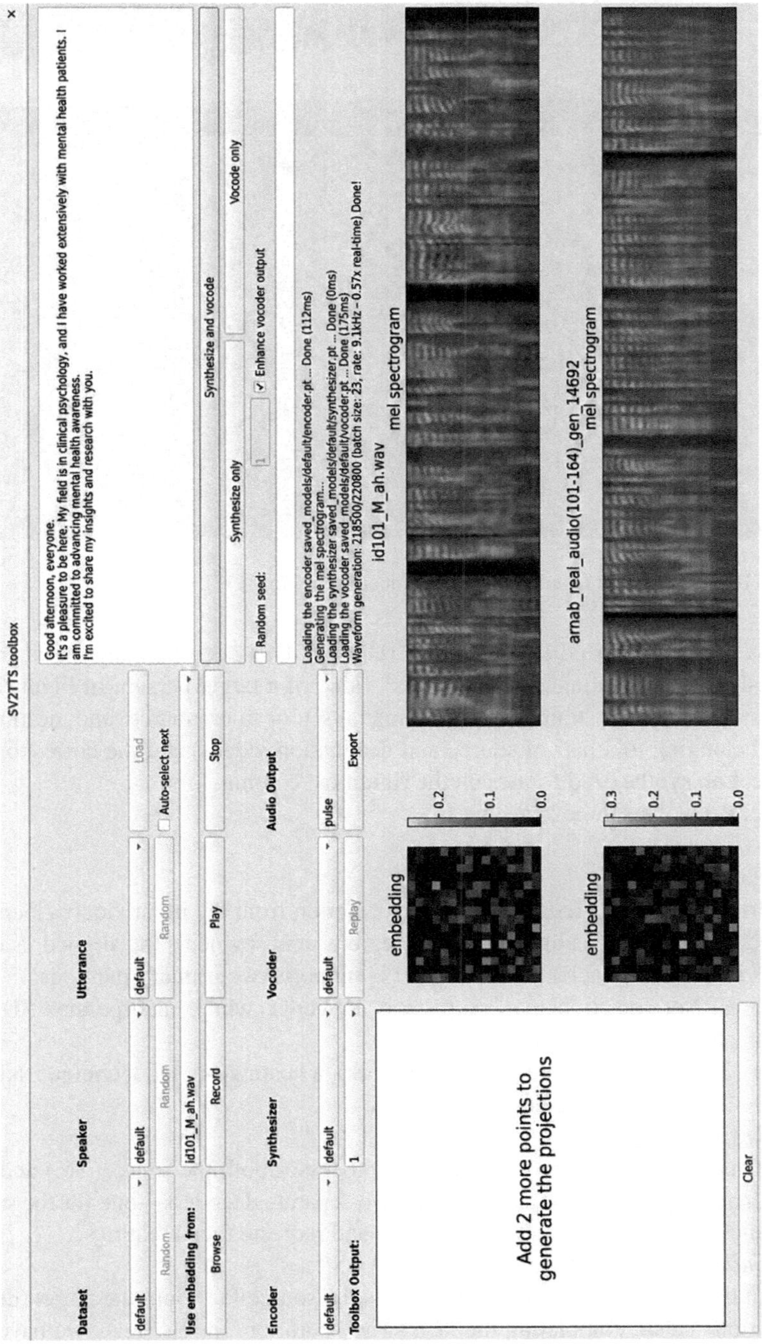

FIGURE 16.2 Text-to-speech conversion using SV2TTS [37].

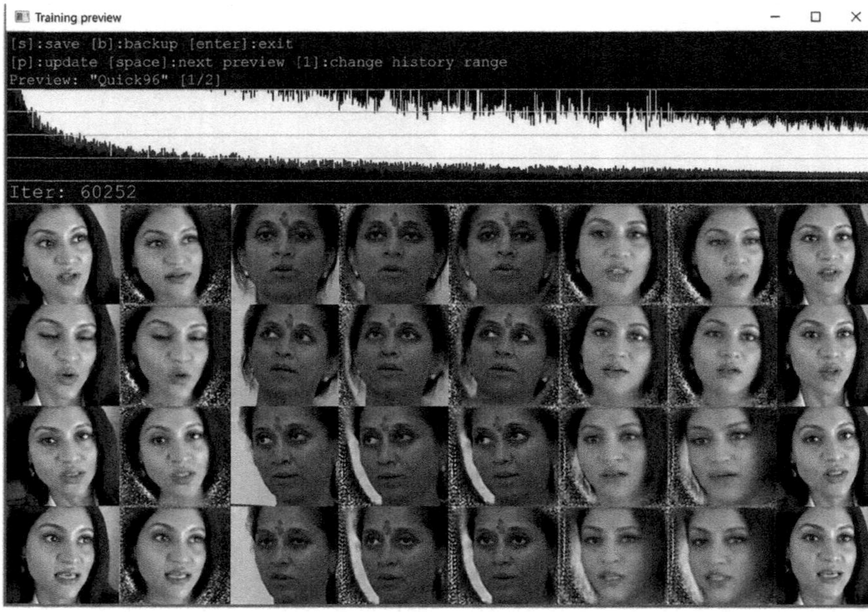

FIGURE 16.3 Snapshot taken during DeepFaceLab training [37].

Petrov et al. introduced DeepFaceLab (DFL) [14], a deep learning-based framework designed for creating the synthesized video of a target person. In Figure 16.3, we show DeeFaceLab training, presenting a set of four source and destination frames, belonging to a pair of source and destination videos, and the corresponding DeepFaceLab synthesized frames on the rightmost column.
The process involves three key stages:

1. *Extraction stage*
   In this phase, DFL carries out face detection from the input videos, identifies facial landmarks, aligns the facial regions, and segments the aligned faces. For extracting facial landmarks, DFL supports two primary methods:
   - PRNet (position map regression network), which incorporates 3D face structure for enhanced accuracy.
   - 2DFAN (2D face alignment network), a heatmap-based technique suitable for frontal or standard facial poses.
2. *Training stage*
   This involves a shared encoder that processes both the source (src) and destination (dst) face data, along with two separate decoders—one for the source and one for the destination—to learn and recreate facial features.
3. *Superimposition stage*
   In the final step, DeepFaceLab blends the source face onto the target (destination) video, completing the face swap operation. In this stage, we have also incorporated or changed the fine-tuning of the parameters to adjust skin tones, brightness, and sharpness.

The FaceSwap [15] model is built using an encoder-decoder structure. It is trained by using a single encoder shared between both videos, and two separate decoders—one for the source and one for the target. The encoder extracts shared feature representations from both the source and target videos, while each decoder learns to reconstruct its respective video from those features. In Figure 16.4, we show a screenshot of the application programming interface (API) version of FaceSwap during the training process. In the FaceSwap method, we can change the number of epochs during the training process or change the batch size and save the checkpoint after completing a fixed number of epochs. Additionally, we can adjust the skin tones, brightness, and other features in the FaceSwap API.

Nirkin et al. introduced FSGAN [16], a well-known tool for facial reenactment. Like FaceSwap and DeepFaceLab, FSGAN also carries out face swapping between a "source" and a "target" video. The process begins by reenacting the target video using selected frames from the source video that closely match the target's facial pose, angle, and expression. It then fills in the missing parts and merges them into the target video.

This method enables fast face-swapping and reduces the training complexity significantly, making it far more efficient than FaceSwap and DeepFaceLab, which rely on heavy computational resources.

Other than the above methodologies, which are open source, some tools are used for commercial and academic purposes.

Synthesia tools are used for commercial purposes. It is an avatar-based commercial AI video creation software frequently used for business presentations.

Wav2Lip [17], launched in 2020, is a widely used and highly accurate model for lip-syncing. It has become a core component in many deepfake generation systems for facial reenactment, as it can accurately align spoken audio with realistic lip movements in videos. The model processes five frames from the input video and their corresponding audio spectrograms to capture and utilize the temporal context of speech effectively. It was used for academic purposes only.

More computational power is needed to create the deepfake video. The iteration in the training process depends on how realistic the video is. With many more iterations, much more realistic videos are made.

16.3 DEEPFAKE AUDIO/VIDEO DATASETS

We must train the model on the respective dataset to identify the audio, video, and audio-video only deepfake content. In this section, we will discuss different types of datasets used to identify deepfake audio/video. In Table 16.1, we have shown comparative analysis of the publicly available audio deepfake datasets.

16.3.1 Spoofed Audio Datasets

The ADD 2023 [18] dataset is a core component of the ADD 2023 challenge, which includes three distinct sub-tasks: fake game (FG), manipulation region localization (RL), and deepfake algorithm recognition (AR). Unlike the previous year's ADD 2022 challenge, which primarily focused on simple binary classification of real versus fake

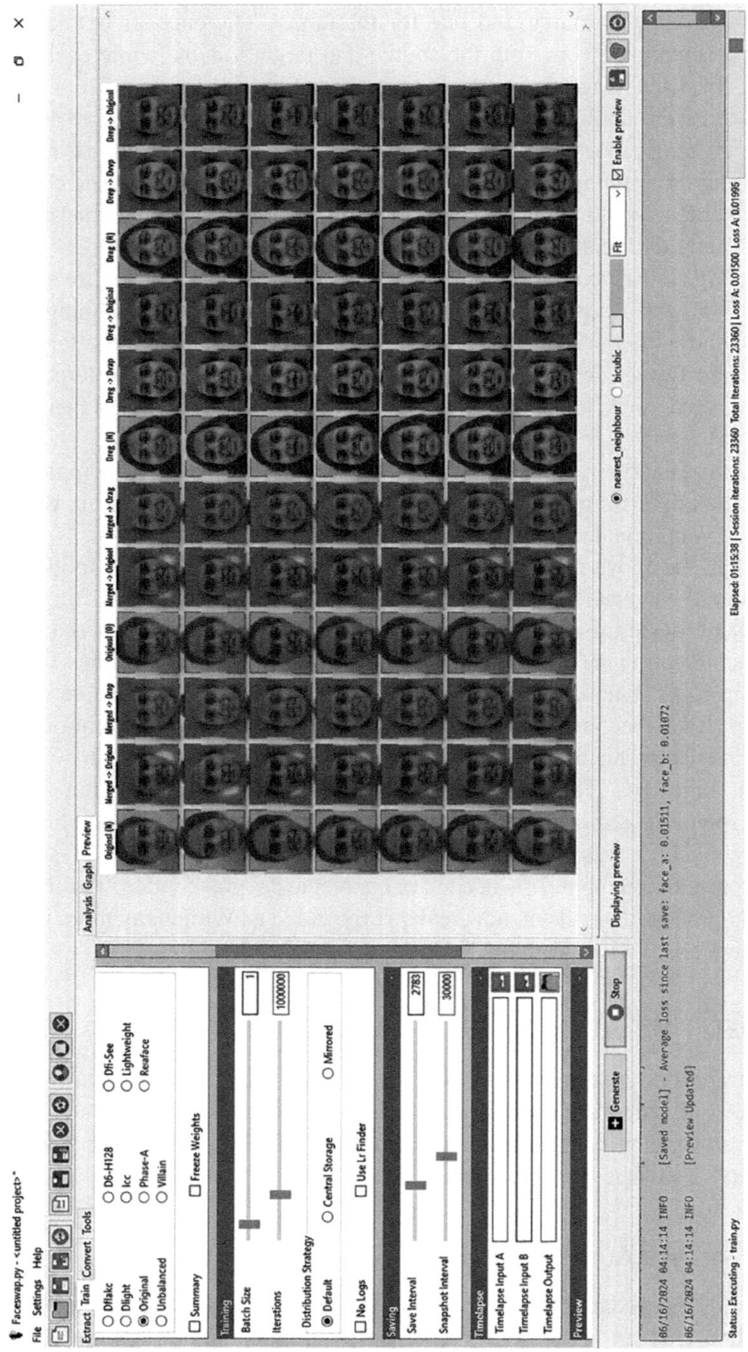

FIGURE 16.4 Snapshot taken during FaceSwap training [37].

TABLE 16.1
Publicly Available Audio Deepfake Datasets: A Comparative Analysis

Dataset	YoP	Methods Used	Real Audios	Fake Audios	Total Audios
ADD 2023 [18]	2023	TTS, VC, Partial fake	LR- 46554 AR- 14907 FG-D- 172819	LR- 65449 AR- 95383 FG-D- 113042	LR- 112003 AR- 110290 FG-D- 285861
In-the-Wild [19]	2022	TTS	19963	11816	31779
ADD 2022 [20]	2022	TTS, VC (various synthesis and conversion models)	LF- 36953 PF- 23897 FG-D- 46871	LF- 123932 PF- 127414 FG-D- 243537	LF- 160885 PF- 151311 FG-D- 290408
CFAD [21]	2022	TTS	115800	Noisy- 115800 Codec- 115800	347400
WaveFake [22]	2021	TTS	-	117985	117985
ASVSpoof 2021 [23]	2021	Speech Synthesis, Voice Conversion	1383820	130032	1513852
FoR (Fake or Real) [24]	2019	Speech Synthesis and TTS	111000+	87000+	198000 +
ASVSpoof 2019 [25]	2019	VC and Speech Synthesis	39000	15600	55200

audio, the 2023 edition shifts attention toward more complex objectives. Specifically, it aims to pinpoint the manipulated segments within partially fake audio and to determine which generation method or algorithm was used to produce the fake content.

In-the-Wild [19] dataset consists of 31,779 audio clips, out of which 11,816 are artificially generated using 19 different TTS algorithms. It features 58 real and 58 synthetic English-speaking voices, primarily of celebrities and political figures. The complete audio content spans approximately 38 hours, with files stored in .WAV format, and total dataset volume is around 7.6 GB.

The ADD 2022 [20] dataset is part of the ADD 2022 challenge, designed for fake audio detection. It includes two main tasks: partially fake (PF) and low-quality fake (LF) audio detection. The LF subset contains 36,953 real audio samples and 123,932 fake recordings, many of which include background noise to mimic real-world conditions. The PF subset consists of 127,414 manipulated or modified audio clips. This publicly available dataset is entirely in Chinese and occupies approximately 3.73 GB of storage. No single model is effective across all types of fake audio in this dataset, highlighting challenges in generalization.

The CFAD [21] dataset comprises 347,400 audio samples, featuring contributions from 1,212 genuine and 1,212 spoofed speakers (as referenced in Table I). The fake and partially manipulated Chinese audio clips were created using 12 different vocoder methods. Audio files are provided in various formats, such as MP3, OGG,FLAC,M4A,WMA, and AAC, with the total dataset size around 29.9 GB.

WaveFake [22] dataset includes a total of 117,985 synthesized audio samples in English and Japanese. The audio clips were generated using various GAN-based TTS models. While the synthetic audio is high in quality and each sample corresponds to only one speaker. All audio files are provided in .WAV format, and the entire dataset occupies about 28.9 GB of storage.

The ASVspoof 2021 [23] dataset offers a rich and extensive collection of 1,513,852 English audio samples. Among them, 130,032 are fake recordings generated using voice conversion (VC) and TTS techniques, incorporating VC attack methods and vocoder models. The dataset maintains a gender-balanced distribution, featuring 41.5 percent male and 58.5 percent female voices. Audio files are available in MP3, OGG, and A4A formats, and the dataset's total size is around 34.5 GB.

The FoR (Fake or Real) [24] dataset, released in 2019, plays a crucial role in the field of audio deepfake detection. It includes over 198,000 audio samples, out of which more than 87,000 are synthetically generated using a variety of TTS systems such as Amazon Polly, Google's traditional TTS, cloud-based TTS, and Microsoft Azure TTS. The dataset, entirely in English, represents about 150.3 hours of audio. It includes 140 genuine speakers and 33 synthetic ones, with all recordings stored in .WAV format. The total dataset size is approximately 16.1 GB.

ASVspoof 2019 [25] dataset is divided into two main subsets: logical access (LA) and physical access (PA). Each subset includes audio recordings from 107 speakers (comprising 61 female and 46 male voices). The LA section features both VC and TTS generated samples, whereas the PA section contains a mix of genuine recordings and replayed (spoofed) audio. Both subsets are further categorized into training, development, and evaluation partitions. The spoofed audio samples were generated using deep learning-based methods. All audio files are stored in FLAC format, and the total dataset size is approximately 23.55 GB.

To summarize, selecting an appropriate audio deepfake dataset requires a careful consideration based on the goals of the research. Researchers must evaluate the available datasets by aligning them with the specific deepfake techniques they intend to study, as well as the desired level of quality and diversity needed to build a reliable and effective model.

Nowadays, it is a good habit to train our model with heterogeneous data, so we need to train our model on multiple datasets from various time periods.

16.3.2 Deepfake Video Datasets

Deepfake video datasets are essential for developing, evaluating, and comparing detection methods. Having a wide variety of these datasets supports the progress of deepfake detection technologies by enabling more robust and comprehensive research. In this section, we show a comprehensive overview of the video and audio-video deepfakes dataset. In Table 16.2, we have included a publicly reported video and audio-video dataset for deepfake identification.

DeepfakeTIMIT [26] dataset, developed in 2018, contains 320 authentic and 640 deepfake videos. The fake videos were generated using the face-swapping GAN (FSGAN) method, focusing on face-swapping techniques. The dataset includes two video resolutions—64×64 pixels (low) and 128×128 pixels (high)—and is provided in AVI format. Most of the deepfakes feature blurred, front-facing subjects, and there is no alteration to the audio. The total dataset size is approximately 226.6 MB.

The UADFV [27] dataset, developed in 2018, consists of 98 video clips—49 real and 49 deepfake samples—primarily centered on face-swapping manipulations. The deepfakes were generated using the FakeApp mobile application. Videos in this dataset have relatively low visual quality, with a resolution of 294 × 500 pixels. The total size of the dataset is around 146 MB.

FaceForensics++ [28] is a widely recognized dataset used for deepfake detection and serves as an enhanced version of the original FaceForensics dataset. It comprises 1,000 authentic and 4,000 manipulated videos. These deepfakes are divided into four categories based on the techniques used: DF (DeepFakes), NT (NeuralTextures), FS (FaceSwap), and F2F (Face2Face). The videos are available in both low and high visual quality, featuring a gender distribution of 60 percent male and 40 percent female. The dataset's original video collection occupies approximately 38.5 GB of storage.

Celeb-DF [29] is a deepfake dataset comprising a total of 6,229 video samples, with 590 real and 5,639 fake instances. It emphasizes face reconstruction using an enhanced deepfake generation algorithm, resulting in high-quality manipulated videos. All videos are in MPEG4.0 format with a resolution of $256_{\times}256$ pixels. The complete dataset size is approximately 9.3 GB.

Released in 2020, the DFDC (deepfake detection challenge) [30] dataset is a publicly available collection of 128,154 videos, 104,500 fake and 23,654 real clips. It features deepfakes generated through both face reenactment and face-swapping techniques. The videos are of high visual quality, stored in .MP4 format, and the total dataset size is almost 471.84 GB.

FFIW (Face Forensics in the Wild) [31], introduced in 2020, includes a total of 20,000 videos, comprising 10,000 real and 10,000 deepfake samples. The dataset focuses on face swapping as the primary manipulation technique. All videos are in .MP4 format with high visual quality and a resolution of 480 pixels or higher.

The FakeAVCeleb [32] dataset consists of 20,000 video samples, including 500 real and 19,500 deepfake videos. These manipulated videos involve techniques such as face swapping, voice cloning, and facial reenactment. All videos are provided in .MP4 format with high-quality visuals and a resolution of $224_{\times}224$ pixels.

Introduced in 2021, the KoDF (Korean DeepFake detection dataset) [33] includes 237,942 video samples, with 62,166 real and 175,776 fake videos. It primarily targets deepfakes generated through face swapping and face reenactment techniques. The deepfakes are of high visual quality, and the entire dataset occupies around 2.6 terabytes of storage.

DeePhy [34], introduced in 2022, is a deepfake dataset containing 5,140 videos—100 genuine and 5,040 manipulated. It includes various deepfake techniques such as face swapping and face reenactment. The videos are formatted in MPEG4.0 with a resolution of 720 pixels, and the total dataset size is approximately 26 GB.

TABLE 16.2
Publicly Available Video and Video-audio Deepfake Datasets: A Comparative Analysis

Dataset	YoP	Nation-Specific	Real Videos	Fake Videos	Total Videos	Modality	Synthesis Methods
DeepfakeTIMIT [26]	2018	N/A	320	640	960	Video Only	2
UADFV [27]	2018	N/A	49	49	98	Video Only	1
FF++ [28]	2019	N/A	1000	4000	5000	Video Only	4
Celeb-DF [29]	2020	N/A	590	5639	6229	Video Only	1
DFDC [30]	2020	N/A	23654	104500	128154	Audio + Video	8
FFIW [31]	2020	N/A	10000	10000	20000	Video Only	1
FakeAVCeleb [32]	2021	N/A	500	19500	20000	Audio + Video	4
KoDF [33]	2021	Korean	62166	175776	237942	Video Only	6
DeePhy [34]	2022	N/A	100	5040	5140	Video Only	3
DF-Platter [35]	2023	N/A	764	132496	133260	Video Only	3
ExDDV [36]	2025	N/A	1000	4369	5369	Audio + Video	6
InDeepFake [37]	2025	Indian	389	4680	5069	Audio + Video	7

DF-Platter [35] is another deepfake dataset comprising 133,260 videos, which includes 764 real and 132,496 fake samples. It mainly targets face swapping and face reenactment techniques. The videos, encoded in MPEG4.0 format, are available in both high (720p) and low (360p) resolutions. With an average blind/referenceless image spatial quality evaluator (BRISQUE) score of 43.25, the dataset reflects high visual quality.

ExDDV (explainable deepfake detection in video) [36] comprises 5,369 real and deepfake videos manually annotated with text descriptions. Out of these videos, 1,000 are real, and the rest are manipulated videos. They have used six synthesis methods to create those fake videos.

The InDeepFake [37] dataset contains 5,069 videos, where 369 videos are real and the rest are synthetic. This is an audio-video and multilingual dataset specially for people of Indian origin. They have used seven different methodologies to create the artificial part. All videos are provided in .MP4 format with high-quality visuals and a 224 × 224 pixels resolution.

In the above, we discuss audio and audio-visual datasets primarily used for deepfake identification purposes. Suppose the forensics researcher wants to experiment with multimodality; in that case, they must choose the audio-visual dataset to extract the features from the audio and video modality.

16.4 DEEPFAKE DETECTION USING AI: UNIMODAL VERSUS MULTIMODAL

In this section, we have discussed different SOTA deepfake detection technologies in single/multi-modality approaches.

16.4.1 Single Modality Approach

In this section, we have discussed the audio-only and video-only deepfake detection techniques. In Figure 16.5, we can only take the input from audio or video for single-modality purposes. After taking the input, most deepfake detectors extract the handcrafted spatial, temporal, and frequency-based features from the audio/video. For audio, the deepfake researcher extracts the Mel-Frequency cepstral coefficients (MFCC), spectrogram, chromogram, pitch difference, fast Fourier transform, discrete Fourier transform, and temporal features. For the video extract, the frequency-based features and the recurrent neural network are used as temporal-based features. After collecting the different types of features, it will go to the classification module for binary (real/fake) classification purposes.

16.4.1.1 Audio-only Deepfake Detection

In general, the audio classification module has two parts: ML (machine learning) based and DL (deep learning) based.

Rodríguez-Ortega et al. [38] proposed two exposers: one is that they proposed an audio deepfake dataset to the deepfake researcher community, and the other is that they proposed an ML model using logistic regression (LR) to detect fake audio. But the audio needs to be preprocessed manually to be fed into that model.

Kumar-Singh et al. [39] proposed a quadratic support vector machine (Q-SVM) model to detect the spoofed audio from real speech audio. This method was compared to other ML methods, like linear SVM, boosted tree ensemble, and K-nearest neighbors (KNN).

Borrelli et al. [40] introduced an SVM model with a random forest (RF) classifier to classify the real and spoofed audio based on the new long-term short-term (LTST) features. This methodology was tested on the automatic speaker verification (ASV) spoof challenge 2019 dataset.

Ballesteros et al. [41] proposed a classification framework called Deep4SNet that visualized the audio dataset from a 2D histogram to classify the audio. This model is a CNN model based.

Modak et al. [42] introduced a vision transformer-based framework that can extract the temporal features from the spectrogram, which is generated from the audio. They have tested their model on four different deepfake audio datasets.

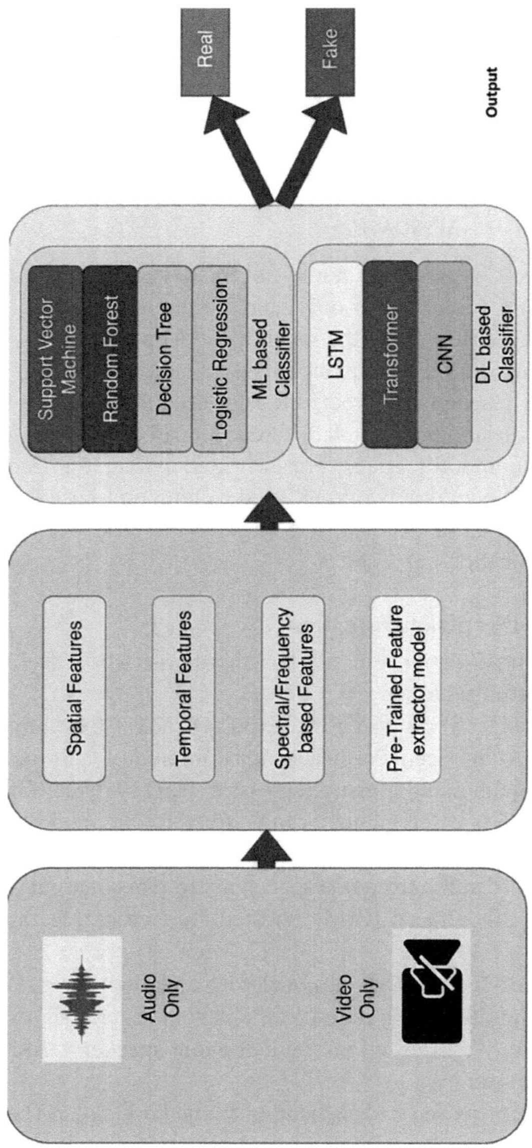

FIGURE 16.5 Audio- and video-only deepfake detection approach.

16.4.1.2 Video-only Deepfake Detection

This section discusses research on deepfake video identification from the spatial, temporal, and frequency domains.

He et al. [43] introduced GAZEFORENSICS, a method that leverages gaze-guided spatial inconsistency detection (such as abnormal eye movement patterns) to enhance deepfake detection performance. The approach employs a 3D gaze estimation network to obtain gaze features, which are then utilized for classification by analyzing the consistency differences between genuine and manipulated gaze behaviors.

Das et al. [44] presented a deepfake detection approach based on a masked autoencoding spatiotemporal transformer within a self-supervised learning framework. The method integrates two vision transformers: a spatial transformer, which captures frame-level visual information from individual red, green, and blue (RGB) frames, and a temporal transformer, which models motion inconsistencies across frames.

Hasanaath et al. [45] proposed frequency-enhanced self-blended images (FSBI), which combines self-blended images with frequency-domain analysis. To transform the images into the frequency domain, the discrete wavelet transform (DWT) was applied to the self-blended images, and the resulting frequency components were processed using a convolutional neural network to extract features, following the standard approach for frequency-based feature extraction.

WaveDIF [46] introduces a deepfake detection framework that leverages wavelet sub-band decomposition to capture subtle frequency-domain artifacts introduced during manipulation. By analyzing multiple sub-bands, the method effectively highlights discrepancies between real and synthetic content. The extracted frequency features are then fed into a deep learning classifier, achieving improved robustness against various deepfake generation techniques.

Spatial domain techniques mainly focus on detecting pixel-level irregularities, whereas frequency domain methods target artifacts within spectral representations (such as DCT, FFT, or DFT). Although spatial approaches have been more widely used, frequency-based methods are becoming increasingly popular due to their efficient feature extraction and robustness against adversarial variations.

16.4.2 Multimodal Approach

Nowadays, identifying deepfake videos using multimodal detection techniques is a hot topic. In this multimodal deepfake detection technique, the researcher finds the inconsistencies between human facial videos, often low synchronization between facial and speech, which affects the ability to identify the deepfake videos. In Figure 16.6, we have shown how the multimodality deepfake video framework worked. After the video and audio preprocessing, the features are collected from the audio as well as the video. The features may be spatial, temporal, or frequency-based. After collecting all the features from different modalities and fusing the features between them, it is fed into the classification module. Several machine learning or deep learning-based classifiers can be used in that module.

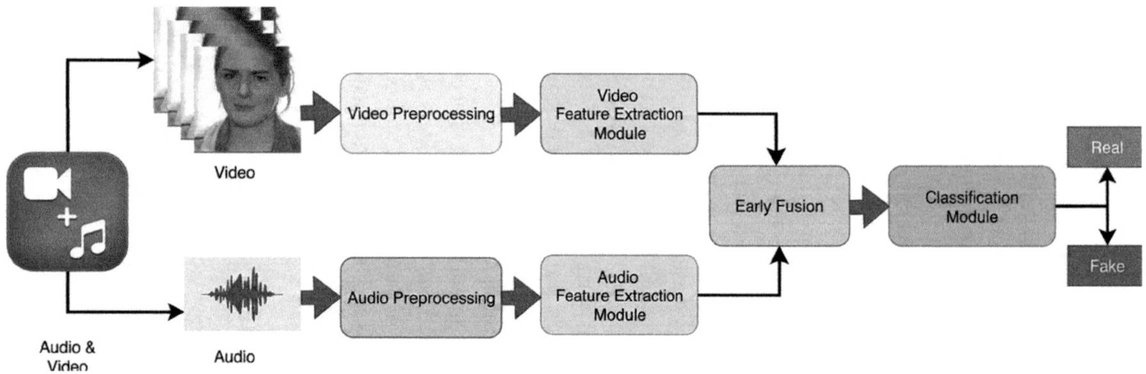

FIGURE 16.6 Deepfake video detection using multi-modal approach.

Yang et al. [47] concentrated on temporal-spatial data and suggested an AVoiD-DF framework that uses a temporal-spatial encoder to identify videos that have been altered.

Ding et al. [48] presented a novel noise aware two-stream network that combines multi-scale RGB data with image noise and utilizes frequency domain features from various human facial areas.

In order to detect deepfakes, Ilyas et al. [49] introduced the AVFakeNet, which is composed of a dense swin transformer (DST) that concentrates on both audio and visual modalities.

Multiple-modality self-distillation (MMSD) [50], audio-visual transformer (AV-transformer) [51], and audiovisual interacting networks (AVIN) [52] are a few supervised techniques that have been investigated recently for multi-modal representation learning using audio-visual cues in deepfake detection.

In a multiple modality framework, detecting deepfake audio and video accuracy is much better than that of a single modality. In this multimodal process, the features come from the dual mode, and after early fusion, these features can easily classify the real/fake videos. However, in this process, we must be careful with the fusion process because we have to fuse one video feature with the corresponding audio feature.

16.4.3 Performance of Multimodal Deepfake Detectors

Some human biometric characteristics, like eye blinking, lip-sync, head and facial motions, head pose, color, texture, and shape signals, can be utilized to identify deepfake media in photos and videos. Current deep learning-based techniques rely heavily on large-scale data, so creating high-quality data sets is crucial. Furthermore, deep learning techniques like GAN, CNN, and RNN are frequently employed in deepfake detection and can automatically extract deepfake features compared to typical machine learning models. Better detection accuracy can also be attained using deep learning-based detection techniques.

In Table 16.3, we have shown the performance of different SOTA deepfake detectors on the deepfake video dataset using a single modality. Table 16.4 also shows the performance of deepfake detectors using multimodality. In Tables 16.3 and 16.4, we have shown the different deepfake detectors along with their methodology and the dataset they have used.

In terms of performance metrics, the SOTA deepfake detectors are used for accuracy, confusion matrix, and F1 score for classification purposes. Precision is used for prediction quality, recall for completeness, accuracy for overall correctness, and F1-score for a fair trade-off between precision and recall. We have summarized the results from different deepfake detectors to compare single-modal and multi-modal approaches. They have calculated the performance using the following formula.

$$\text{Accuracy} = \frac{(\text{TP}+\text{TN})}{(\text{TP}+\text{FP}+\text{TN}+\text{FN})} \tag{16.1}$$

TABLE 16.3
Performance of SOTA Unimodal (Video Only) Deepfake Detectors

Model	Features	Dataset	Modality	Performance Acc (%)
VTD-Net [53]	CNN	Celeb-DF	Video	91.89
NA-VGG [54]	VGG/CNN	Celeb-DF	Video	85.70
Hernandez-Ortega et al. [55]	Convolutional Attention Network	Celeb-DF	Video	91.50
Tarasiou et al. [56]	Lightweight Architecture	Celeb-DF	Video	92.62
OC-FakeDect [57]	VAE	FaceForensics++	Video	97.50
Afchar et al. [58]	MesoInception-4	FaceForensics++	Video	83.10
Rossler et al. [59]	XceptionNet	FaceForensics++	Video	95.73
Das et al. [60]	ResNet101 + CNN	FaceForensics++	Video	99.06

TABLE 16.4
Performance of SOTA Multimodal (Video +Audio) Deepfake Detectors

Model	Features	Dataset	Modality	Performance Acc (%)
AVoiD-DF [61]	Encoder–decoder	FakeAVCeleb	Audio + Video	83.70
VFD [62]	Voice–face matching with cosine similarity	FakeAVCeleb	Audio + Video	81.50
DST-Net [63]	Dense swin transformer	FakeAVCeleb	Audio + Video	78.40
AVAD [64]	Autoregressive transformer	DFDC-p	Audio + Video	93.20
VFD [62]	Encoder	DFDC-p	Audio + Video	80.90

$$\text{Precision} = \frac{\text{TP}}{(\text{TP}+\text{FP})} \tag{16.2}$$

$$\text{Recall} = \frac{\text{TP}}{(\text{TP}+\text{FN})} \tag{16.3}$$

$$\text{F1 Score} = \frac{2 \times \text{Precision} \times \text{Recall}}{\text{Precision} + \text{Recall}} \tag{16.4}$$

where true positive (TP) = fake videos correctly predicted as fake, false positive (FP) = real videos incorrectly predicted as fake, true negative (TN) = real videos correctly predicted as real, and false negative (FN) = fake videos incorrectly predicted as real.

In Table 16.5, we have shown the deepfake detectors' performance against the audio deepfake dataset using a single modality. The equal error rate (EER) is used as

TABLE 16.5
Performance of SOTA Unimodal (Audio Only) Deepfake Detectors

Model	Features	Dataset	Modality	Performance EER (%)
W2V-AASIST [65]	Spectro-Temporal Graph Attention Network	ASVSpoof 2021	Audio	3.6
W2V-LCNN [66]	Self-Supervised Learning, wav2vec	ASVSpoof 2021	Audio	8.1
SLIM [67]	Lightweight projection head classifier	ASVSpoof 2021	Audio	8.3
SpecViT [42]	Vision Transformer, Spectrogram	ASVSpoof 2021	Audio	3.5
DARTS (Best) [68]	Light Differentiable Architecture Search (DARTS)	ADD 2022	Audio	25.57
DCT-DFT spec [69]	Multi-head attention pooling	ADD 2022	Audio	20.11
SpecViT [42]	Vision Transformer	ADD 2022	Audio	17.4

a standard performance metric in the audio deepfake detectors using the following formula.

$$\text{EER} = \frac{(\text{FAR} + \text{FRR})}{2} \tag{16.5}$$

where FAR = false acceptance rate and FRR = false rejection rate.

$$\text{FAR} = \frac{\text{No. of False Acceptance}}{\text{Total No. of Imposter Attempts}}; \text{FRR} = \frac{\text{No. of False Rejection}}{\text{Total No. of Genuine Attempts}} \tag{16.6}$$

Single modality deepfake detectors only analyze one type of data, like audio/visual frames. On the other hand, multimodal deals with audio and video, combining data. So the single modality methods are faster and require fewer resources. Multimodal detectors achieved comparatively higher detection accuracy, however with a higher resource requirement. This performance characteristic is evident from the result presented in Table 16.3–16.5.

In Table 16.3, we have demonstrated the performance of the deepfake detector on the two video datasets, Faceforensics++ and Celeb-DF. In Table 16.4, we also show the performance of the deepfake detector with multimodal data using the FakeAVCeleb and DFDC-p dataset. Similarly, Table 16.5 also shows the detector's performance using single modality on the two audio ASVSpoof 2021 and ADD 2022 datasets.

16.5 CONCLUSIONS AND FUTURE WORK

Deepfakes use advanced artificial intelligence and machine learning techniques to generate or alter audio, visual, or audiovisual content that appears highly realistic. These manipulations can bypass facial and voice recognition systems, spread harmful content in cyber-warfare, and contribute to the spread of fake news and misinformation. As a result, deepfakes pose serious threats to online privacy, security, content authenticity, and the integrity of democratic processes. In this chapter, we have discussed deepfake audio/video generation techniques, deepfake audio/video datasets for identification, and different types of deepfake detection algorithms using single/multi-modality approaches.

In conclusion, this chapter aims to complement existing surveys and motivate a broad audience, including newcomers, researchers, and digital forensic analysts, to prioritize deepfakes as a central topic in their academic and professional endeavors.

In the future, multimodal deepfake detection should aim to develop efficient and compact models that can identify minor mismatches between audio, video, and text in real-time. Combining information from different modalities through self-supervised or few-shot approaches can enhance the system's ability to handle new types of fake content. Moreover, building large, varied multimodal datasets and interpretable detection models will be essential to create reliable and scalable detection systems.

REFERENCES

[1] Ian J. Goodfellow, Jean Pouget-Abadie, Mehdi Mirza, Bing Xu, David Warde-Farley, Sherjil Ozair, Aaron Courville, and Yoshua Bengio. Generative adversarial networks, *Communications of the ACM,* 63(11): 139–144, 2014.

[2] Erkam Temir. Deepfake: New era in the age of disinformation and end of reliable journalism. *Selc¸uk ˙Iletis¸im*, 13(2):1009–1024, 2020.

[3] Andrei O.J. Kwok and Sharon G.M. Koh. Deepfake: A social construction of technology perspective. *Current Issues in Tourism*, 24(13):1798–1802, 2021.

[4] Soo Yun Shin and Jiyoung Lee. The effect of deepfake video on news credibility and corrective influence of cost-based knowledge about deepfakes. *Digital Journalism*, 10(3):412–432, 2022.

[5] P. Swathi and S.K. Saritha. Deepfake creation and detection: A survey. In 2021 Third International Conference on Inventive Research in Computing Applications (ICIRCA), pp. 584–588. IEEE, 2021.

[6] Yisroel Mirsky and Wenke Lee. The creation and detection of deepfakes: A survey. *ACM Computing Surveys* (*CSUR*), 54(1):1–41, 2021.

[7] Thanh Thi Nguyen, Quoc Viet Hung Nguyen, Dung Tien Nguyen, Duc Thanh Nguyen, Thien Huynh-The, Saeid Nahavandi, Thanh Tam Nguyen, Quoc-Viet Pham, and Cuong M. Nguyen. Deep learning for deepfakes creation and detection: A survey. *Computer Vision and Image Understanding*, 223:103525, 2022.

[8] S. E. Eskimez, X. Wang, M. Thakker, C. Li, C.-H. Tsai, Z. Xiao, H. Yang, Z. Zhu, M. Tang, X. Tan, et al., E2 tts: Embarrassingly easy fully non- autoregressive zero-shot tts, in: 2024 IEEE Spoken Language Technology Workshop (SLT), IEEE, 2024, pp. 682–689.

[9] Y. Chen, Z. Niu, Z. Ma, K. Deng, C. Wang, J. Zhao, K. Yu, and X. Chen, F5- tts: A fairytaler that fakes fluent and faithful speech with flow matching, arXiv preprint arXiv:2410.06885 (2024).

[10] W. Peebles and S. Xie, Scalable diffusion models with transformers, in: Proceedings of the IEEE/CVF international conference on computer vision, 2023, pp. 4195–4205.

[11] S. Woo, S. Debnath, R. Hu, X. Chen, Z. Liu, I. S. Kweon, and S. Xie, Convnext v2: Co-designing and scaling convnets with masked autoencoders, in: Proceedings of the IEEE/CVF Conference on Computer Vision and Pattern Recognition, 2023, pp. 16133–16142.

[12] Y. Jia, Y. Zhang, R. Weiss, Q. Wang, J. Shen, F. Ren, P. Nguyen, R. Pang, I. Lopez Moreno, Y. Wu, et al., Transfer learning from speaker verification to multispeaker text-to-speech synthesis, Advances in Neural Information Processing Systems 31 (2018).

[13] O.A. Shaaban, R. Yildirim, and A.A. Alguttar. Audio deepfake approaches. *IEEE Access* 2023, 11, 132652–132682.

[14] I. Perov, D. Gao, N. Chervoniy, K. Liu, S. Marangonda, C. Um´e, M. Dpfks, C. S. Facenheim, L. RP, J. Jiang, et al., Deepfacelab: Integrated, flexible and extensible face-swapping framework, arXiv preprint arXiv:2005.05535 (2020).

[15] I. Korshunova, W. Shi, J. Dambre, and L. Theis, Fast face-swap using convolutional neural networks, in: Proceedings of the IEEE International Conference on Computer Vision, 2017, pp. 3677–3685.

[16] Y. Nirkin, Y. Keller, and T. Hassner, Fsgan: Subject agnostic face swapping and reenactment, in: Proceedings of the IEEE/CVF International Conference on Computer Vision, 2019, pp. 7184–7193.

[17] K. Prajwal, R. Mukhopadhyay, V. P. Namboodiri, and C. Jawahar, A lip sync expert is all you need for speech to lip generation in the wild, in: Proceedings of the 28th ACM International Conference on Multimedia, 2020, pp. 484–492.

[18] J. Yi, J. Tao, R. Fu, X. Yan, C. Wang, T. Wang, C.Y. Zhang, X. Zhang, Y. Zhao, Y. Ren, et al. ADD 2023: The second audio deepfake detection challenge. arXiv 2023, arXiv:2305.13774.

[19] N.M. Müller, P. Czempin, F. Dieckmann, A. Froghyar, and K. Böttinger. Does audio deepfake detection generalize? arXiv 2022, arXiv:2203.16263.

[20] J. Yi, R. Fu, J. Tao, S. Nie, H. Ma, C. Wang, T. Wang, Z. Tian, Y. Bai, C. Fan, et al. Add 2022: The first audio deep synthesis detection challenge. In Proceedings of the ICASSP 2022—2022 IEEE International Conference on Acoustics, Speech and Signal Processing (ICASSP), Singapore, May 22–27, 2022; pp. 9216–9220.

[21] H. Ma, J. Yi, C. Wang, X. Yan, J. Tao, T. Wang, S. Wang, L. Xu, and R. Fu, FAD: A Chinese dataset for fake audio detection. arXiv 2022, arXiv:2207.12308.

[22] J. Frank and L. Schönherr. Wavefake: A data set to facilitate audio deepfake detection. arXiv 2021, arXiv:2111.02813.

[23] X. Liu, X. Wang, M. Sahidullah, J. Patino, H. Delgado, T. Kinnunen, M. Todisco, J. Yamagishi, N. Evans, A. Nautsch, et al. Asvspoof 2021: Towards spoofed and deepfake speech detection in the wild. *IEEE/ACM Transactions on Audio and Speech Language Processing*. 2023, 31, 2507–2522.

[24] R. Reimao and V. Tzerpos, FoR: A dataset for synthetic speech detection. In Proceedings of the 2019 International Conference on Speech Technology and Human-Computer Dialogue (SpeD), Timisoara, Romania, 10 October 2019; pp. 1–10.

[25] J. Yamagishi, M. Todisco, M. Sahidullah, H. Delgado, X. Wang, N. Evans, T. Kinnunen, K.A. Lee, V. Vestman, and A. Nautsch. Asvspoof 2019: Automatic Speaker Verification Spoofing and Countermeasures Challenge Evaluation Plan.

Available online: https://www.asvspoof.org/asvspoof2019/asvspoof2019_evaluation_plan.pdf (accessed on January 15, 2019).

[26] P. Korshunov, S. Marcel, Deepfakes: a new threat to face recognition? assessment and detection, arXiv preprint arXiv:1812.08685 (2018).

[27] X. Yang, Y. Li, S. Lyu, Exposing deep fakes using inconsistent head poses, in: ICASSP 2019–2019 IEEE International Conference on Acoustics, Speech and Signal Processing (ICASSP), IEEE, 2019, pp. 8261–8265.

[28] A. Rossler, D. Cozzolino, L. Verdoliva, C. Riess, J. Thies, M. Nießner, Faceforensics++: Learning to detect manipulated facial images, in: Proceedings of the IEEE/CVF International Conference on Computer Vision, 2019, pp. 1–11.

[29] Y. Li, X. Yang, P. Sun, H. Qi, S. Lyu, Celeb-df: A large-scale challenging dataset for deepfake forensics, in: Proceedings of the IEEE/CVF Conference on Computer Vision and Pattern Recognition, 2020, pp. 3207–3216.

[30] B. Dolhansky, J. Bitton, B. Pflaum, J. Lu, R. Howes, M. Wang, C. C. Ferrer, The deepfake detection challenge (dfdc) dataset, arXiv preprint arXiv:2006.07397 (2020).

[31] T. Zhou, W. Wang, Z. Liang, J. Shen, Face forensics in the wild, in: Proceedings of the IEEE/CVF Conference on Computer Vision and Pattern Recognition, 2021, pp. 5778–5788.

[32] H. Khalid, S. Tariq, M. Kim, S. S. Woo, Fakeavceleb: A novel audio-video multimodal deepfake dataset, arXiv preprint arXiv:2108.05080 (2021).

[33] P. Kwon, J. You, G. Nam, S. Park, G. Chae, Kodf: A large-scale Korean deepfake detection dataset, in: Proceedings of the IEEE/CVF International Conference on Computer Vision, 2021, pp. 10744–10753.

[34] K. Narayan, H. Agarwal, K. Thakral, S. Mittal, M. Vatsa, R. Singh, Deephy: On deepfake phylogeny, in: 2022 IEEE International Joint Conference on Biometrics (IJCB), IEEE, 2022, pp. 1–10.

[35] K. Narayan, H. Agarwal, K. Thakral, S. Mittal, M. Vatsa, R. Singh, Df- platter: Multiface heterogeneous deepfake dataset, in: Proceedings of the IEEE/CVF Conference on Computer Vision and Pattern Recognition, 2023, pp. 9739–9748.

[36] V. Hondru, E. Hogea, D. Onchis, R. T. Ionescu, Exddv: A new dataset for explainable deepfake detection in video, arXiv preprint arXiv:2503.14421 (2025).

[37] A. K. Das, A. Bose, P. Manohar, A. Dutta, R. Naskar, R. S. Chakraborty, Indeepfake: A novel multimodal multilingual indian deepfake video dataset, *Pattern Recognition Letters*, 196: 16-23, 2025.

[38] Y. Rodríguez-Ortega, D.M. Ballesteros, and D. Renza. A machine learning model to detect fake voice. In Florez, H., Misra, S., Eds., *Applied Informatics*, Springer International Publishing, Cham, Switzerland, 2020; pp. 3–13.

[39] A.K. Singh, and P. Singh. Detection of AI-synthesized speech using cepstral and bispectral statistics. In Proceedings of the 2021 IEEE 4th International Conference on Multimedia Information Processing and Retrieval (MIPR), Tokyo, Japan, September 8–10, 2021; pp. 412–417.

[40] C. Borrelli, P. Bestagini, F. Antonacci, A. Sarti, and S. Tubaro. Synthetic speech detection through short-term and long-term prediction traces. *EURASIP Journal of Information Security* 2021, 2021, 2.

[41] D.M. Ballesteros, Y. Rodriguez-Ortega, D. Renza, and G. Arce, Deep4SNet: Deep learning for fake speech classification. *Expert Systems and Applications.* 2021, 184, 115465.

[42] S. Modak, A. K. Das, and R. Naskar, Specvit: A custom vision-transformer based approach for audio deepfake detection, in: ICASSP 2025–2025 IEEE International

Conference on Acoustics, Speech and Signal Processing (ICASSP), IEEE, 2025, pp. 1–5.

[43] Qinlin He, Chunlei Peng, Decheng Liu, Nannan Wang, and Xinbo Gao. Gazeforensics: Deepfake detection via gaze- guided spatial inconsistency learning. *Neural Networks*, 180: 106636, 2024.

[44] Sayantan Das, Mojtaba Kolahdouzi, Levent ¨Ozparlak, Will Hickie, and Ali Etemad. Unmasking deepfakes: masked autoencoding spatiotemporal transformers for enhanced video forgery detection. In 2023 IEEE International Joint Conference on Biometrics (IJCB), pp. 1–11. IEEE, 2023.

[45] Ahmed Abul Hasanaath, Hamzah Luqman, Raed Katib, and Saeed Anwar. Fsbi: Deepfake detection with frequency enhanced self-blended images. *Image and Vision Computing*, 105418, 2025.

[46] Dutta, A., Das, A. K., Naskar, R., and Chakraborty, R. S. (2025). WaveDIF: wavelet sub-band based deepfake identification in frequency domain. In Proceedings of the Computer Vision and Pattern Recognition Conference (pp. 6312–6321).

[47] W. Yang, X. Zhou, Z. Chen, B. Guo, Z. Ba, Z. Xia, X. Cao, K. Ren, Avoid-df: Audio-visual joint learning for detecting deepfake, IEEE Transactions on Information Forensics and Security 18 (2023) 2015–2029.

[48] X. Ding, S. Pang, and W. Guo. Noise-aware progressive multi-scale deepfake detection, Multimedia Tools and Applications (2024) 1–17.

[49] H. Ilyas, A. Javed, and K. M. Malik. Avfakenet: A unified end-to-end dense swin transformer deep learning model for audio–visual deepfakes detection, *Applied Soft Computing* 136 (2023) 110124.

[50] L. Huang, L. Wang, and H. Li. Multi-modality self-distillation for weakly supervised temporal action localization, *IEEE Transactions on Image Processing* 31 (2022) 1504–1519.

[51] Y.-B. Lin and Y.-C. F. Wang. Audiovisual transformer with instance attention for audio-visual event localization, in: Proceedings of the Asian Conference on Computer Vision, 2020.

[52] J. Ramaswamy. What makes the sound?: A dual-modality interacting network for audio-visual event localization, in: ICASSP 2020–2020 IEEE International Conference on Acoustics, Speech and Signal Processing (ICASSP), IEEE, 2020, pp. 4372–4376.

[53] T. Yang et al. (2020) VTD-Net: depth face forgery oriented video tampering detection based on convolutional neural network. In: 2020 39th Chinese Control Conference (CCC).

[54] X. Chang et al. (2020) Deepfake face image detection based on improved VGG convolutional neural network. In: 2020 39th Chinese Control Conference (CCC).

[55] J. Hernandez-Ortega, et al.: DeepFakesON-phys: deepfakes detectionbased on heart rate estimation. arXiv preprint arXiv:2010.00400. (2020).

[56] M. Tarasiou and S. Zafeiriou. Extracting deep local features to detect manipulated images of human faces. In: 2020 IEEE InternationalConference on Image Processing (ICIP), pp. 1821–1825. (2020).

[57] H. Khalid and S.S. Woo (2020) OC-FakeDect: classifying deepfakes using one-class variational autoencoder. In: 2020 IEEE/CVF Conference on Computer Vision and Pattern Recognition Workshops (CVPRW).

[58] D. Afchar et al. Mesonet: A compact facial video forgery detection network. In: 2018 IEEE International Workshop on Information Forensics and Security (WIFS), pp. 1–7. IEEE (2018)

[59] A. Rossler et al.: Faceforensics++: Learning to detect manipulated facial images. In: Proceedings of the IEEE International Conference on Computer Vision, pp. 1–11 (2019).

[60] A. K. Das, S. Mukhopadhyay, A. Dalui, R. Bhattacharya, and R. Naskar. A multi-stage multi-modal classification model for DeepFakes combining deep learned and computer vision oriented features. In *International Conference on Information Systems Security* (pp. 217–226). Cham: Springer Nature Switzerland (2023).

[61] W. Yang, X. Zhou, Z. Chen, B. Guo, Z. Ba, Z. Xia, X. Cao, and K. Ren, Avoid-df: Audio-visual joint learning for detecting deepfake, *IEEE Transactions on Information Forensics and Security* 18 (2023) 2015–2029.

[62] H. Cheng, Y. Guo, T. Wang, Q. Li, X. Chang, and L. Nie, Voice-face homo- geneity tells deepfake, *ACM Transactions on Multimedia Computing, Communications and Applications* 20 (3) (2023) 1–22.

[63] H. Ilyas, A. Javed, and K. M. Malik, Avfakenet: A unified end-to-end dense swin transformer deep learning model for audio -visual deepfakes detection, *Applied Soft Computing* 136 (2023) 110124.

[64] C. Feng, Z. Chen, and A. Owens, Self-supervised video forensics by audio visual anomaly detection, in: Proceedings of the IEEE/CVF Conference on Computer Vision and Pattern Recognition, 2023, pp. 10491–10503.

[65] Hemlata Tak, Massimiliano Todisco, Xin Wang, Jee-weon Jung, Junichi Yamagishi, and Nicholas Evans. Automatic speaker verification spoofing and deepfake detection using wav2vec 2.0 and data augmentation. arXiv preprint arXiv:2202.12233, 2022.

[66] Yuankun Xie, Haonan Cheng, Yutian Wang, and Long Ye. Learning a self-supervised domain-invariant feature representation for generalized audio deepfake detection. In Proc. INTERSPEECH, vol. 2023, pp. 2808–2812, 2023.

[67] Yi Zhu, Surya Koppisetti, Trang Tran, and Gaurav Bharaj. Slim: Style- linguistics mismatch model for generalized audio deepfake detection. arXiv preprint arXiv:2407.18517, 2024.

[68] Chenglong Wang, Jiangyan Yi, Jianhua Tao, Haiyang Sun, Xun Chen, Zhengkun Tian, Haoxin Ma, Cunhang Fan, and Ruibo Fu. Fully automated end-to-end fake audio detection. In Proceedings of the 1st International Workshop on Deepfake Detection for Audio Multimedia, pp. 27–33, 2022.

[69] Rui Yan, Cheng Wen, Shuran Zhou, Tingwei Guo, Wei Zou, and Xiangang Li. Audio deepfake detection system with neural stitching for add 2022. In ICASSP 2022–2022 IEEE International Conference on Acoustics, Speech and Signal Processing (ICASSP), pp. 9226–9230. IEEE, 2022.

Index

D

E

F

G

For Product Safety Concerns and Information please contact our EU representative GPSR@taylorandfrancis.com Taylor & Francis Verlag GmbH, Kaufingerstraße 24, 80331 München, Germany

Batch number: 10725800

Printed by Printforce, the Netherlands